MOBILE COMMUNICATION

BRIJESH VERMA

MOBILE COMMUNICATION

Including GSM, CDMA, EDGE, GPRS

(For B.E. & M.E. Students)

By

Brijesh Verma

Specialist—SCN

Nokia Siemens Networks India Ltd.

Visiting Faculty

IES Academy Jia Sarai, New Delhi

Formerly Sr. Lecturer (ECE)

Lingaya's Institute of Management and Technology

Faridabad (Haryana)

*****

Pragyan Verma

M.Tech (CSE)

S.K. KATARIA & SONS®

Publisher of Engineering Books

4885/109, Prakash Mahal, Dr. Subhash Bhargav Lane,
Opposite Delhi Medical Association, Daryaganj, New Delhi-110002 (INDIA)
Phone: +91-11-43551243 Mobile: +91-9871775858, 7678425209
e-mail: katariabook@gmail.com, katariabooks@yahoo.com
Website: www.skkatariaandsons.com

(e-books, subscriptions and bulk buying also available on our website)

Published By:

S.K. Kataria & Sons®

4885/109, Prakash Mahal, Dr. Subhash Bhargav Lane,
Opposite Delhi Medical Association, Daryaganj, New Delhi-110002 (INDIA)
Phone: +91-11-43551243, Mobile: +91-9871775858, 7678425209
e-mail: katariabook@gmail.com, katariabooks@yahoo.com
Website: www.skkatariaandsons.com

First Edition: 2009-2010, Third Edition: 2014, Reprint: 2026

Price: ₹ 495/-

Laser Typesetting by :
AISA ENTERPRISES

PREFACE

I am very much glad to present Third edition of "Mobile Communication".

As we know there has been phenomenal advancement in the field of telecommunication. But there have been fewer attempts at consolidating the various aspects and application of Mobile Communication.

I hope that the present book is an outgrowth of this need particularly felt at the undergraduate level of ECE, CSE, IT etc. branches.

Having taught the subject for more than 5 years along with valuable experience of 7 years gained in telecom industry, it was felt that a comprehensive look covering all the aspects of the subject as per syllabus of various Engineering Colleges was not available to the students.

This was the inspiration to write the book. This book is intended for use text material and is designed for B.E. in IT, ECE, CSE. Although it does not cover the entire topic of Mobile Communication but a sustained effort has been made to ensure that all the concept introduced have been supported by sufficient number of mathematical proof and solved numerical examples.

Most of the topics included are standard and can be found in various books already published by different authors.

But in this book an attempt has been made to cover in a single volume with all topics in sufficient depth in a suitable form for students. So that students need not search for information in several books.

Any suggestions towards improvement of this book will be highly appreciated.

Brijesh Verma

ACKNOWLEDGEMENTS

I am indebted to a large number of individuals regarding the presentation of this book. First, I would like to thank my publisher M/s S.K. Kataria & Sons, Delhi who gave help during the beginning of this undertaking and were instrumental in the production of the Third edition of the book. My thanks are also extended to Mrs. Poonam Kumari, Manager Network Implementation with NSN Global, Noida, and Mr. Sunil Mishra, Sr. Engineer And also Communication for their help in various areas of wireless communication particularly in GSM and Switching Theory through discussions or comments. I appreciate the suggestions offered by my colleagues. I thank to my students who encouraged me to write the second edition of the book.

Finally, my greatest indebtedness and appreciation is due to my family. The major part of the book was prepared during the time normally devoted to my family; without their understanding, patience, and tolerance, this book would not have been possible.

Brijesh Verma

You can also refer following books by same author and publisher.

1. **Digital and Analog Communication.**
2. **Electronics Communication Question Bank (For GATE, UPSC, PSU etc.) with solution.**
3. **Wireless Communication Systems.**
4. **ATM Network.**
5. **Advance Communication.**

Dedicated to

My Parents

whose sacrifice, love and affection made me capable of writing this book

&

My Wife 'Pragya'

CONTENTS

1. INTRODUCTION TO CELLULAR MOBILE SYSTEMS 1–38

1.0. Introduction 1
1.1. History of Wireless Communication 2
 1.1.1. History of Wireless Technology 2
1.2. Evolution of Mobile Communication 3
 1.2.1. Mobile Wireless Applications 4
 1.2.2. Network Technologies 5
1.3. Examples of Wireless Communication System 5
 1.3.1. Paging Systems 5
 1.3.2. Paging Protocol 7
1.4. Introduction to Basic Cellular System 7
 1.4.1. Basic Cellular System 8
 1.4.2. Network Switching Subsystem (NSS) 11
1.5. Operation of Cellular Systems 12
 1.5.1. Mobile Phone Initialization 15
1.6. Analog Cellular System 16
 1.6.1. Overview of AMPS 17
 1.6.2. Digital Cellular Systems 19
 1.6.3. Digital Cellular System Parameter 19
1.7. Comparison Between Analog and Digital Cellular System 19
 1.7.1. Advantages of the Digital Transmission 20
1.8. Performance Criteria 21
1.9. Overview of Cellular Generations 22
 1.9.1. 1G (First Generation) 23
 1.9.2. Second Generation (2G) Cellular Network 24
 1.9.3. Migrating From 2G to 3G 27
1.10. Third Generation (3G) 28
 1.10.1. Operators and UMTS Networks 28
 1.10.2. 2G to 3G Network Standardization 29
 1.10.3. Layered Network Architecture Advantages 30
 1.10.4. Mobile Technologies 30
 1.10.5. From GPRS to UMTS 30
 1.10.6. Future Network 31
 1.10.7. UMTS, 3G Terminals 31
1.11. 4G (Fourth Generation) 32
 1.11.1. Fourth Generation Standards 32
 1.11.2. Important 4G Components 32
 1.11.3. Access Schemes 32
 1.11.4. 4G True Broadband Wireless System 33
1.12. Planning a Cellular System 33
1.13. Uniqueness of Mobile Radio Environment 35
 ■ **Summary** 37
 ■ **Review Questions** 37

2. ELEMENTS OF CELLULAR RADIO SYSTEMS DESIGN AND INTERFERENCE 39–87

2.0. Introduction 39
2.1. Basic Wireless Concepts 39
2.2. Analog and Digital Transmission 43
2.2.1. Analog Information 43
2.2.2. Digital Information 43
2.3. Access Method: Time Division Multiple Access (TDMA) 43
2.4. General Description of Transmission Problem 44
2.4.1. Solutions to Transmission Problems 47
2.5. Cellular System Design Fundamentals 49
2.5.1. Basic Cellular Concepts 50
2.6. Concept of Frequency Reuse Channels 52
2.6.1. Frequency Reuse and System Capacity 54
2.6.2. Frequency Reuse Distance (D) 55
2.6.3. System Capacity 55
2.7. Interference (Co-Channel Interference) 56
2.7.1. Co-channel Interference (CCI) 56
2.7.2. Worst Case Co-Channel Interference 59
2.7.3. Adjacent Channel Interference (ACI) 60
2.7.4. Cell Capacity—How Many Channels Per Cell 61
2.7.5. Trunking Efficiency and Grade of Service 66
2.7.6. Increasing the Capacity of Cellular Systems 68
2.8. Cell Splitting 69
2.8.1. Sectoring 71
2.8.2. Microcell Zone 75
2.8.3. Antenna and Clusters 76
2.8.4. MicroCells 77
2.9. Antenna 81
2.10. Path Loss Models (Okumara Model, Hata Model) 83
2.10.1. Okumura Model 83
2.10.2. Hata Model 84
2.11. Brewster Angle 85
2.12. Scattering 85
2.13. Directivity 86
2.13.1 Definition 86
2.13.2 Directivity and Gain 87
■ **Review Questions** 87

3. CELL COVERAGE FOR SIGNAL AND ANTENNA STRUCTURE 88–108

3.0. Introduction 88
3.0.1. Predection Model Considers Following Conditions 89
3.1. Mobile Point to Point Model 89
3.1.1. Generate a Standard Condition 90
3.1.2. Area to Area Prediction Model 90
3.2. Propagation Over Water and in Flat Area or in Free Space 92
3.2.1. Two Ray Model 93

3.3. Foliage Loss 95
3.4. Long Distance and Near Distance Propagation of Mobile Signal 95
3.4.1. Propagation in Near-in Distance 95
3.4.2. Long Distance Propagation 96
3.5. Point to Point Prediction Model 97
3.5.1. Obstructive Conditions 97
3.5.2. In Nonobstructive Conditions 98
3.6. Cell Site Antenna Height 99
3.7. Cell Coverage Area 99
3.8. Mobile to Mobile Propagation 101
3.9. Characteristics of Basic Antenna Structures 103
3.9.1. Radiation Patterns 103
3.9.2. Antenna Types 103
3.9.3. Parabolic Antenna 104
3.9.4. Antenna Gain 104
3.9.5. Effective Area 104
3.9.6. Relation Between Gain and Effective Area 105
3.9.7. Critical Frequency 105
3.9.8. Maximum Usable Frequency 105
3.10. Antenna at Cell Site (Omnidirectional antenna) 105
3.10.1. Antenna at Cell Site (Directional Antenna) 105
3.11. Mobile Antenna 107
3.11.1. Room Mounted Antenna 108
■ **Review Questions** 108

4. FREQUENCY MANAGEMENT & CHANNEL ASSIGNMENT, HANDOFF & DROPPED CALLS 109–141

4.0. Introduction 109
4.1. Handoff 109
4.2. Channel Assignment Schemes and Handoff 110
4.2.1. Channel Assignment Strategy 110
4.2.2. Fixed Channel Assignment Strategy 110
4.2.3. Dynamic Channel Assignment Strategy 111
4.2.4. Handoff Strategies 111
4.2.5. Network Initiated Handoff 112
4.2.6. Mobile Assisted Handoff (MAHO) 113
4.2.7. Prioritizing Handoffs 113
4.2.8. Queuing of Calls 113
4.2.9. Practical Handoff Considerations 114
4.2.10. Type of Handoff (Hand Over) 116
4.3. GSM (Global System for Mobile Communication) 117
4.3.1. The SIM Card 118
4.3.2. History 118
4.3.3. Radio Interface 119
4.4. GPRS (General Packet Radio Service) 120
4.4.1. GPRS Architecture 121
4.4.2. GPRS Terminals 123
4.4.3. GPRS BSS 123

4.4.4. GPRS NSS 123
4.4.5. Classes of Gprs Terminals 124
4.4.6. GPRS Multislot Classes 125
4.4.7. GPRS Call Flow 126
4.4.8. Security in GPRS 128
4.4.9. Features of GPRS 129
4.4.10. Applications for GPRS 130
4.5. Edge Technology 135
4.5.1. GPRS vs EDGE/EGPRS 135
4.5.2. Modulation Technique 136
4.5.3. Coding Schemes 137
4.5.4. Packet Handling 137
4.5.5. EDGE Benefits 138
4.6. Dropped Call Rates and Their Evaluations 139
4.6.1. Drop Call Rate for Handoff 139
■ **Summary** 140
■ **Review Questions** 140

5. MODULATION METHODS AND CODING FOR ERROR DETECTION AND CORRECTION 142–221

5.0. Introduction 142
5.1. Analog Modulation 143
5.1.1. Amplitude Modulation (AM) 144
5.1.2. Frequency Modulation (FM) 144
5.1.3. Phase Modulation 144
5.1.4. Pulse Modulation 144
5.2. Digital Modulation 144
5.2.1. Digital Modulation and Demodulation Techniques 145
5.2.2. Amplitude Shift Keying (ASK) 146
5.2.3. Frequency Shift Keying (FSK) 149
5.2.4. Phase Shift Keying (PSK) 151
5.2.5. Differential Phase Shift Keying 154
5.3. Modulation Method in Cellular Wireless System 155
5.3.1. Quadrature Phase Shift Keying (QPSK) 156
5.3.2. QPSK Signalling 156
5.3.3. Minimum Shift Keying 161
5.3.4. GMSK Modulation 165
5.4. Spread Spectrum 166
5.4.1. History of Spread Spectrum 166
5.4.2. How Spread Spectrum Works 167
5.4.3. Advantages of Spread-Spectrum Wireless Systems 170
5.4.4. Modulation 173
5.4.5. Demodulation 174
5.4.6. Spread Spectrum (SS) Communication Concepts 174
5.4.7. Spread Spectrum Block Diagram 176
5.4.8. Frequency Hopping (FH) 177
5.4.9. Applications of Spread Spectrum 178
5.5. Orthogonal Frequency Division Multiplexing (OFDM) 179
5.6. Error Detection and Error Correction 180
5.6.1. Transmission Errors 181

5.6.2. Error Control 183
5.6.3. Error Detection 183
5.6.4. Block Sum Check (BSC) 194
5.7. Error Correction 196
5.7.1. Single Bit Error Correction 196
5.7.2. Hamming Code 197
5.7.3. Burst Error Correction 200
5.8. Data compression 200
5.8.1. Character Suppression 201
5.8.2. Huffman Coding 201
5.8.3. Shannon-Fanno Encoding 204
5.9. Block Codes 205
5.9.1. Binary Code Space 206
5.10. Convolutional Codes 207
5.11. Turbo Coding 211
■ **Review Questions** 221

6. MULTIPLE ACCESS TECHNIQUES ..... 222—257

6.0. Introduction 222
6.1. Multiple Access Techniques 222
6.1.1. Frequency Division Duplexing (FDD) 223
6.1.2. Time Division Duplexing (TDD) 224
6.2. Multiple Access Techniques For Wireless Communication 224
6.2.1. Principles of Operation of an FDMA 224
6.2.2. Principles of Operation of TDMA 227
6.3. Capacity of a Cellular System 231
6.4. Comparison of Spectrum Efficiency for FDMA and TDMA Systems 234
6.4.1. FDMA 234
6.4.2. TDMA 234
6.4.3. Capacity of Digital Cellular CDMA 236
6.5. Summary of FDMA, TDMA and CDMA 239
6.6. Wireless TDMA Design Consideration for mobile phone 240
6.7. Space Division Multiple Access (SDMA) 243
6.8. Code Division Multiple Access (CDMA) 244
6.8.1. CDMA Background 245
6.8.2. Principles of CDMA 245
6.8.3. CDMA Information, Theory and Codes 246
6.8.4. Generation of Code for CDMA : Code Division Multiple Access 247
6.9. Bluetooth 249
6.10. Comparison Between Wireless Communication System 250
■ **Summary** 255
■ **Review Questions** 256
■ **Numerical Based Problems** 256

7. SECOND GENERATION, DIGITAL AND WIRELESS SYSTEMS ..... 258—352

7.0. Introduction 258
7.1. AMPS and ETACS 258
7.1.1. Frequency Bands 259
7.1.2. Introduction of Digital TDMA 259
7.1.3. Introduction of GSM and CDMA 259

7.1.4. The Future of AMPS 260
7.1.5. Analog System in Europe 260
7.1.6. AMPS and ETACS Specifications 260
7.2. United States Digital Cellular (Is-54 and Is-136) 260
7.2.1. IS-54 and IS-136 (D-Amps) 261
7.2.2. IS-136 262
7.2.3. IS-95 262
7.2.4. IS-2000 262
7.3. Introduction to GSM, the Global System for Mobile Communication 263
7.3.1. Introduction: The Evolution of Mobile Telephone Systems 263
7.3.2. GSM (Global System For Mobiles Communication) 264
7.3.3. GSM Specifications 265
7.3.4. GSM Phases 265
7.3.5. GSM Geographical Network Structure 266
7.3.6. GSM Frequency Ranges 267
7.3.7. Various Types of Number for Network Identity 271
7.3.8. GSM System Architecture 273
7.3.9. Traffic Cases 282
7.3.10. GMSK Modulation 293
7.3.11. GSM Network Areas 294
7.3.12. GSM Air Specifications 295
7.3.13. GSM Subscriber Services 297
7.3.14. GSM Security 298
7.4. Setup Channels 299
7.4.1. GSM Traffic Channels (TCHs) 299
7.4.2. GSM Control Channel 299
7.5. Frame Structure of GSM 302
7.6. Signal Processing in GSM 304
7.6.1. Speech Coding 304
7.6.2. Channel Coding 305
7.6.3. Interleaving 305
7.6.4. Ciphering 305
7.6.5. Burst Formatting 306
7.6.6. Modulation 306
7.7. Interim Standard 95 (IS-95) 306
7.7.1. Protocol Revisions 306
7.7.2. Forward Traffic Channels 308
7.7.3. Capacity 309
7.8. Difference between GSM and CDMA 309
7.9. Personal Communication Access System (PACS) 311
7.9.1. PACS System Architecture 311
7.9.2. PACS Radio Interface 312
7.10. Personal Digital Cellular/Pacific Digital Cellular (PDC) 312
7.10.1. Introduction 313
7.10.2. Technology 313
7.11. Personal Handy-phone System 314
7.11.1. Technology 314

7.12. The Industrial, Scientific and Medical (ISM Band) ... 315
7.13. Wireless Personal Area Network (WPAN) ... 315
7.14. Network Database ... 315
7.14.1 History ... 316
7.15. Spectrum Allocation ... 316
7.15.1. Frequency Utilization Measurement Factor ... 317
7.15.2. Spectrum Efficiency ... 317
7.15.3. GSM Spectrum Allocation in India ... 318
7.16. Frequency Hopping ... 320
7.16.1. Frequency Hopping Spread Spectrum ... 320
7.17. Walsh code ... 321
7.18. CDMA (Code Division Multiple Access) ... 321
7.18.1. Usage in Mobile Telephony ... 324
7.18.2. Coverage and Applications ... 325
7.18.3. Technical Details ... 326
7.18.4. Soft Handoff ... 329
7.18.5. CDMA Features ... 329
7.18.6. Comparison of Mobile Phone Standards ... 330
7.18.7. CDMA 2000 ... 331
7.19. Design Concept of CDMA ... 333
7.19.1. Design of a CDMA System ... 334
7.20. Direct Sequence CDMA (DS-CDMA) ... 335
7.21. 3G (WCDMA) Architecture ... 336
7.21.1. 3G Services ... 338
7.21.2. Potential Applications ... 338
7.21.3. Applications Categorisation From the Business Area Point of View ... 339
7.21.4. 3G—Services and Required Data Rates ... 339
7.21.5. Comparison between 2G and 3G Technologies ... 340
7.21.6. Features of WCDMA ... 340
7.22. Overview of Handover in Mobile Networks ... 342
7.22.1. Types of Handover in 3G WCDMA Systems ... 342
7.22.2. Objectives of Handover ... 343
7.22.3. Handover Measurements and Procedures ... 344
7.22.4. Soft Handover (SHO) ... 345
7.22.5. Principles of Soft Handover ... 345
7.23. Vocoders ... 347
7.23.1. Vocoder Theory ... 347
7.23.2. Channel Vocoders ... 348
7.23.3. Cepstrum Vocoders ... 348
7.23.4. Formant Vocoders ... 349
7.23.5. Voice Excited Vocoder ... 349
7.24. Dropped Call Rate ... 349
7.24.1. Dropped Call Rate Type ... 350
■ **Summary** ... 352
■ **Review Questions** ... 352
APPENDICES ... **353–374**
PREVIOUS YEARS EXAMINATION PAPERS ... **375–381**
REFERENCES ... **382**
INDEX ... **383–389**

1

CHAPTER

Introduction to Cellular Mobile Systems

1.0. INTRODUCTION

INSIDE THIS CHAPTER

1.0. Introduction
1.1. History of Wireless Communication
1.2. Evolution of Mobile Communication
1.3. Examples of Wireless Communication System
1.4. Introduction to Basic Cellular System
1.5. Operation of Cellular Systems
1.6. Analog Cellular System
1.7. Comparison Between Analog and Digital Cellular System
1.8. Performance Criteria
1.9. Overview of Cellular Generations
1.10. Third Generation (3G)
1.11. 4G (Fourth Generation)
1.12. Planning a Cellular System
1.13. Uniqueness of Mobile Radio Environment

Just start to think about mid-1970s. Imagine the first time somebody used a remote control to operate his TV set or any other device. It must have been a thrilling experience, well, technology has grown by leaps and bounds since then. Today, you could remotely monitor birds in a sanctuary, wirelessly control your home and share life's unforgettable moments with your loved ones all across the globe-quickly and effortlessly, using your smart mobile phone. Advancement in wireless technologies has made all these possible and is promising even more.

The day is not far off, when your washing machine will send an e-mail to your cellphone informing you about the wash program it's going to use. But since you are driving and cannot read the e-mail, your car audio will connect to your cellphone, using Bluetooth and read out the e-mail.

You can then dictate your e-mail reply, just in case want to modify the program. And when you reach home, you will find your laundry all done, while you were away.

If you have watched 'Black' the movie in Hindi, you could not have been deeply moved by Amitabh Bachchan's portrayal of an old man suffering from Alzheimer's. This combination of social isolation, inactivity and failing nutrition is alarmingly common among today's ageing population (according to the Alzheimer's Association, the U.S.A. alone has 4 million people suffering from it). Technology may just offer to change all that. Let's revisit 'Black'. Imagine that there's a tiny sensor embeded in one 'Debraj Sahai's' walking shoes and another on the Peg where he hangs his coat. Whenever these and other miniature sensing devices throughout his home, 'notice' him moving in the direction of the front door, they use a wireless network to alert Michelle that her 'teacher' is about to go for a morning walk. Michelle gets the signal through a small vibrator (which communicates with the other sensors) fitted on to her wrist band.

The scenario may seem for fetched today but is fairly close to being realised. The key is the wireless sensor networks–a new technology model in which behavioural and biological data is collected and analysed for customised proactive computing applications.

Sensor networks hold promise for meeting these challenges because they have the potential to revolutionise healthcare.

1.1. HISTORY OF WIRELESS COMMUNICATION

Mobile wireless technology describes a wide range of telecommunications services and portable devices that send and receive data, text, voice and video. Cellphones and pagers are the most widely used mobile wireless products today, and many routine daily activities rely on the voice communications and messaging services supported by these devices. New technologies are emerging that will support wireless remote monitoring, wireless financial transactions, wireless Internet access, and wireless control of appliances and devices. Mobile wireless developments are evolving so rapidiy, that it is difficult to keep up with the latest advances and to make purchasing decisions. This section strives to provide a brief background on the history of wireless development, a general overview of mobile wireless technologies that are available today, and some predictions for the future.

1.1.1. History of Wireless Technology

Marconi's Radio: The origins of wireless communications date back to the 1890s with Marconi's work in radio and wireless telegraphy. Since then, advances have come through a **sequence of discoveries and advancements**. In 1898, Marconi produced the first commercial radio service (data only) linking a radio transmitter with a receiver over 7.5 miles away. In December of 1900, the first voice radio link was demonstrated. Radio communications were of natural interest to the shipping industry, and the government established ship to shore communications requirements in 1910. By 1918, 5700 ships had wireless telegraphy.

WIRELESS TELEGRAPH: The "Telimco" Complete Outfit, comprising 1 inch Spark Coil, Strap Key, Sender, Sensitive Relay, Coherer, with Automatic Decoherer and Sounder, 4 Ex. Strong Dry Cells, all necessary wiring, including send and catch wires, with full instructions and diagrams, $8.50. Guaranteed to work up to one mile. Send for Illust. Pamphlet & 64-page catalogue. **ELECTRO IMPORTING CO,** 32 Park Place, New York.

Fig. 1.1. Marconi Radio System.

Another valuable wireless communications service was initiated when the Detroit Police institute issued the first radio dispatch in 1921, operating in 2 MHz band.

On Oct. 17, 1919, the Radio Corporation of America (RCA) was incorporated, and consumer radio broadcasts for news and entertainment became popular in the early 1920s. The advent of Frequency Modulation (FM) radio in 1935 brought reductions in the size and improvements to the quality of radio equipment. It has not been until the past decade that mobile telephone service has become widely available and affordable, but the first public mobile telephone service (MTS) was actually implemented in 1946 in St. Louis, and the FCC approved non-wireline carriers in 1949. These early wireless phone services were expensive and unavailable.

Improvements in the 1960s and 70s generated renewed interest in wireless producing the Advanced Mobile Telephone Service (AMPS). Another development in the 1970s was the widely popular Citizen's Band (CB) radios which offered inexpensive mobile communications in a shared radio channel (800 MHz).

Fig. 1.2. Ericsson's first hand held mobile phone in 1987.

Pages to Cellphones: In the 1980s, the Regional Bell Operating Companies (RBOC) emerged after the breakup of the AT&T, and they began to offer low-data rate wireless text services for pagers. Cellular Digital Packet Data (CDPD) and Mobitex were data services that used a packet data network based upon cellular structure. During this period, a wave of new analog cellphones and service options became widely available and affordable.

In the 1990s, digital signal processing led wireless advances that supported more efficient use of bandwidth, new waveforms, and smaller devices that consumed less power.

1.2. EVOLUTION OF MOBILE COMMUNICATION

The origins of wireless communication followed quickly behind the invention of radio in the late 1800s.

The first applications of mobile radio were related to the navigation and safety of ships at sea. As radio concept used as communication tool. The major milestones in the development of wireless communications are summarized in Table 1.1.

Table 1.1. Major Milestones in Mobile Communication.

Year	Activity
1921	The Detroit police department used a 2 MHz frequency in the departments as first vihicle mobile radio. The system was one way.
1930	Amplitude modulation concept used and formed a two way mobile system, that took advantage of newly developed mobile transmitters and utilized a "Push to talk" or half duplex transmitter.
1935	Invention of frequency modulation (FM), which improved the audio quality. FM eliminated the need for large AM transmitters and resulted in radio equipment which required less power to operate. This made the use of transmiters in vehicles more practical.

Year	Activity
1940	Federal communication commission (FCC) recognized a communication service and named it domestic public land mobile (DPLM) radio service.
1946	The first DPLM system was established in St. Louis in 1946 and it utilized the 150 MHz band. The following year, a "Highway" system was developed along the New York—Boston Corridor using the 35–40 MHz band.
1947	D.H. Ring, working at Bell Laboratories envisions the cellular concept.
1948	Shocklly, Bardeen and Brittain, at Bell Laboratories, invent the transistor which enables electronic equipment, including the radio to be miniaturized.
1949	Radio Common Carriers (RCC) were recognized.
1949-58	Bell systems made broadband proposals.
1964-68	AT&T introduces Improved Mobile Telephone System (IMTS).
1969	The FCC began to address issue of new US spectrum requirements. Nordic Countries of Denmark, Finland, Iceland, Norway and Sweden agree to form a group to study and recommend areas of telecommunication. This led to the standardization of telecommunications for all members of the Nordic Mobile Telephone (NMT) group, the first Comprehensive International Standardization group.
1973	The NMT group specifies a feature allowing mobile telephones to be located within and across networks. This feature would become the basis for roaming.
1979	The FCC authorized the installation and testing of the first developed cellular system in the USA.
1981	Ericsson Launches the world's first cellular system in Saudi Arabia based on the analog NMT 450 standard.
1991	The first digital Cellular Standard (GSM) is launched.
1998	The number of mobile subscribers world-wide has grown to over 200 million.
2005	The number of mobile subscribers world-wide become around 600 million. And different modern technologies for mobile communications has been invented.
2011	More than 5 billion subscribers are using mobile phone.

1.2.1. Mobile Wireless Applications

Communication:

Mobile wireless phones

e-mail, paging, messages

Information (Internet everywhere):

Education: school, health, financial

Access: news, weather, shopping, recreation

Control:

Actions: open doors, elevators, lights, heat, ...

Transactions: ATM's, vending, voting, shopping

Monitoring:

Personal: medical devices, emergency alerts

Environmental: equipment maintenance, security

Location finding:

Global Positioning System (satellite-based) GPS
Enhanced 911 (cellphone-based)

1.2.2. Network Technologies

A variety of different network technologies and service providers support mobile wireless applications today. These include networks for cellular and paging communications, for local and wide area networks, for location finding, and for wireless control. An important change is emerging in the telecommunications world between two fundamentally different approaches for communications: point-to-point versus network. The point-to-point communications approach follows the telephone system's switched model in which two (or more) participants call to establish a communications link. In the earliest days of telecommunications, a human operator at a switchboard established the connection between two parties. The participating sites are only connected during their session, and other users wait until the circuit is clear to make their calls. Complex statistical models are employed to balance the demand for a connection. The network approach for communications follows the Internet model in which many or all sites are connected continuously, and they exchange information by sending packets of data addressed to a specific site. In this network approach, the communications lines are shared. In the wireless world today, many services remain point-to point, but plans for future generations (3G, 4G,...) of wireless are to implement always on packet based networking.

Various access technologies for various generations are as follows:

First Generation (1G), AMPS → FDMA (Frequency Division Multiple Access)
Second Generation (2G), GSM → FDMA and TDMA (Time Division Multiple Access) Both
Third Generation (3G → CDMA (Code Division Multiple Access)
CDMA 2000, WCDMA (Implemented over GSM)

1.3. EXAMPLES OF WIRELESS COMMUNICATION SYSTEM

We are familiar with number of wireless communication systems, as well as with mobile radio communication systems.

For Example: T.V. Remote, Car door Openers, Cordless Telephone, Walkee-Talkee, Pager etc., are the examples of wireless communication system. While cellular telephone are all example of mobile radio communication system, but still it is wireless communication system, only difference is service offered by the systems.

The term mobile means, the device that could move during operation, so all the example, mentioned above are the example, of mobile communication system, but technically the term mobile used for a cellular telephone. Mobile radio transmission system may be half duplex or full duplex or simplex it depends upon technology and application of device.

1.3.1. Paging Systems

It is an elecronic device, which is used to receive the message from PSTN line through paging operator. Message may be numeric, alphanumeric or voice message in form of text message. Therefore, paging system is used to inform a subscriber of having paging device, there is a need to call a particular number or to meet a place to get another instruction. In modern communication system, news headlines or fax may be sent. In India mobile paging service was introduced in 1993 by Modi page link. In which subscriber having a mobile pager can get such information.

Voice transmission or reception was not possible with the pager system and display unit was also very small. If a person want to send some message to a paging subscriber, then first he dial a paging number through a Public switching telephone network (PSTN) line and then he told his message to a paging operator and then paging operator typed his message and transmit to the pager subscriber. A process of sending message shown in Fig. 1.3.

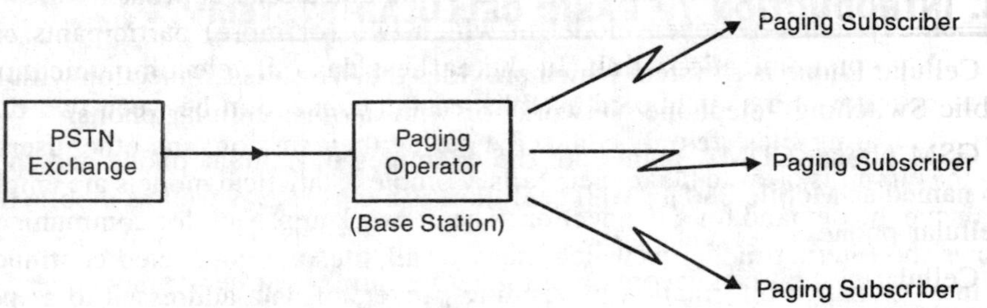

Fig. 1.3. Paging operation.

There was no privacy in the paging communication because you can send your message only through paging operator.

These are the biggest drawbacks of the pager system, so within 3 years pager service became obsolete from the market.

(*a*) Numeric pager.

(*b*) A pager that is in use for emergency services.

(*c*) Motorola pager used in Brazil, in the 90's, operated by Teletrim.

Fig. 1.4. Pager devices.

The issued message is called a page. Paging system covers a limited range of 2 to 5 Km or may be confined within a buildings, while wide area paging system can provide world-wide coverage. Pager systems are simple and inexpensive but transmission system was very complex. Various pager devices are shown in Fig. 1.4.

1.3.2. Paging Protocol

Common paging protocols include: TAP, FLEX, ReFLEX, POCSAG, Golay and NTT. Past paging protocols include: Two-tone and 5/6-tone.

Pager technology is now being used in irrigation control systems and for traffic signals. Pagers usually have very simple ring tones and some include a vibrating alert.

1.4. INTRODUCTION TO BASIC CELLULAR SYSTEM

Cellular Phone is a system which provides the Wireless Connection with PSTN (Public Switching Telephone Network) or with another cellular phone.

GSM phone, CDMA phone are the examples of cellular phone and they are also named as Mobile Phone. Walkee-Talkee is a wireless as well as mobile but not a cellular phone.

Cellular system can accommodate a larger number of subscribers over a large Geographical area (Divided into cells) on a pre-specified frequency range.

The quality of reception or transmission of a Call or Message is very good as compared to a Basic Telephone System. Geographical Area used by a Cellular System is divided into cells. Each cell has its own frequency. But if two cells are far away then same frequency could be allotted.

A cellular system consists of a mobile station (mobile phone or cell phone), base stations and a mobile switching Centre (MSC), GMSC. Mobile station consists of a transmitter, Receiver and an Antenna. A mobile station may be portable or hand held like a simple cellular phone.

Here mobile station (MS) is nothing but a simple mobile phone that you use.

Each Base station consists of several transmitter and receiver which simultaneously handle full duplex communication and generally have towers which support several transmitting and receiving antennas.

Base station works as bridge between a mobile station (MS) and a MSC. A number of Base station is connected with a MSC via microwave link or through a Telephone cable depending upon number of users; distance between Base station and MSC.

Any mobile user, whenever made a call it is routed through Base station via radio link and the call may be handed-off to one base station or more than one base station throughout the call duration.

The MSC coordinates all the activities of Base station with PSTN and connect entire cellular system with PSTN.

A typical MSC may handle around 1000000 subscribers and around 500000 simultaneous conversations at a time, and accommodates all the billing function and maintenance as well. GMSC works as gateway of MSC to connect PSTN and other operator MSC.

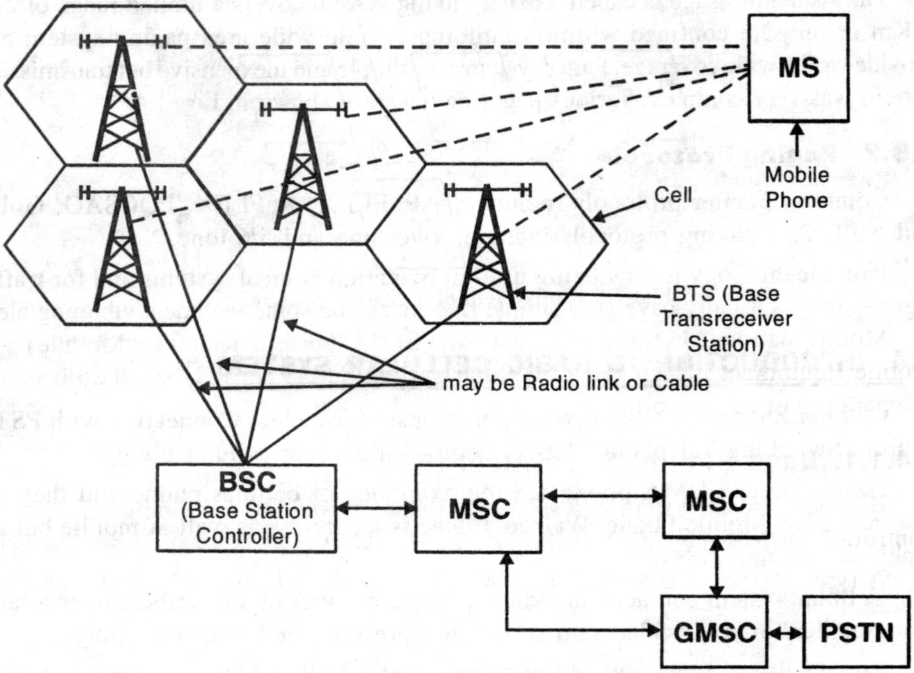

Fig. 1.5. Cellular system.

Communication between a base station and a subscriber is established by CAI (Common Air Interface) that is divided between four channels as follows :

FVC : Forward Voice Channel used for voice transmission between base to mobile.

RVC : Reverse Voice Channel used for voice transmission between Mobile to base.

FCC : Forward Control Channel.

RCC : Reverse Control Channel.

FCC and RCC also named as set-up channels, because they are used for call initiating. A basic wireless cellular system is shown in Fig. 1.5.

All cellular systems provide a service called roaming. This allows subscribers to operate in service areas other than the one from which it is subscribed. When a mobile enters a city or geographic area that is different from its home service area, it is registered as roamer in the new service area.

This is accomplished over FCC.

1.4.1. Basic Cellular System

Basic cellular system of a cellular network is divided in three part:

(*i*) Base Station Subsystem (BSS)

(*ii*) Network Switching Subsystem (NSS)

(*iii*) Network Management Subsystem (NMS).

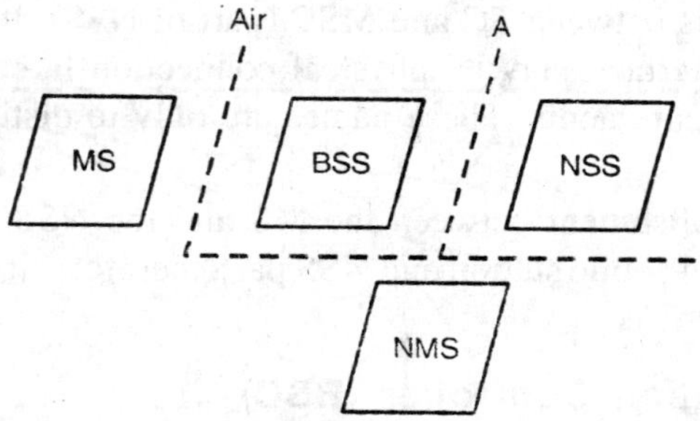

Fig. 1.6. Subsystem of a cellular system.

Mobile station (MS) which consists of a SIM (Subscriber Identity Module) and Mobile Equipment is also part of cellular system but as per ITU standard it is not a separate part of network.

1.4.1.1. Base Station Subsystem (BSS)

Base station subsystem is responsible for managing the radio network and it is controlled by NSS part. One NSS particularly MSC/VLR controlled several BSS.

A typical BSS consists of one BSC and one BTS, which is shown in Fig. 1.7.

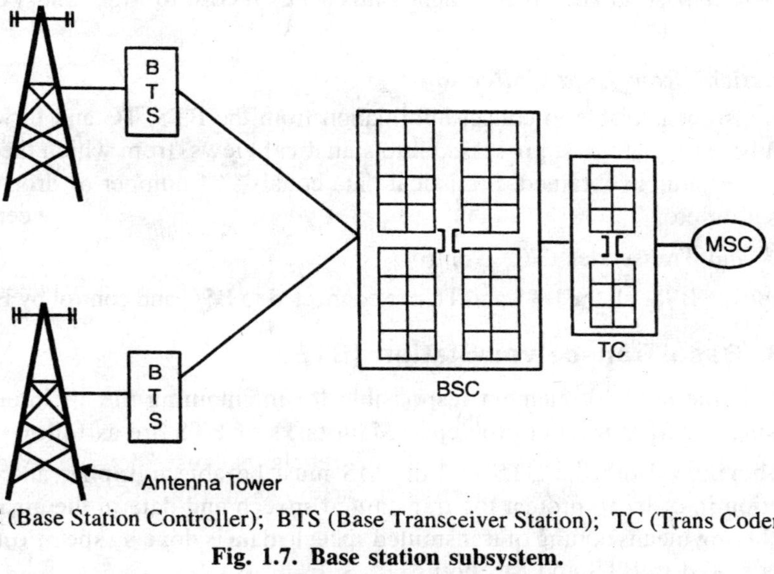

BSC (Base Station Controller); BTS (Base Transceiver Station); TC (Trans Coder)

Fig. 1.7. Base station subsystem.

Some of the most important BSS tasks are listed as follows:

Radio Path Control: In radio cellular network particularly in GSM, BSS is responsible for taking care of radio resources, that is radio channel allocation and quality of radio connection.

Air and A Interface Signalling: In order to establish a call, the MS (mobile station) must have a connection through BSS.

Air interface is between mobile station and BTS which means radio link.

While between BTS and BSC is A_{bis} interface. Bis is a franch word which means second while between BSC and TC is there. Ater interface (ter mean third interface).

And A interface is between TC and MSC (part of NSS). It is important to note that Abis, Ater, A interface may be physical connection like (optical, co-axial or radio link) as per requirements. These names are only to distinguish various parts of network.

Connection establishment between the MS and the NSS Part: BSS works as intermediate between Mobile station and NSS part, therefore, it plays very important role in call establishment.

1.4.1.2. Base Station Controller (BSC)

BSC is a cabinet which consists of various functional units and it is a central network element of the BSS and it controls the radio network. Important tasks of BSC are as follows.

➤ ***Call Establishment***

All calls to and from the MS are connected through the switching functionality of the BSC.

➤ ***Mobility Management***

Mobility management means keeping record of MS movement. BSC does not have any data base to keep record hence this sends record to MSC and VLR keeps record.

➤ ***Statistical Raw Data Collection***

It is also responsible to collect information from the BTS, TC and forward it to NMS. Where they are post processed into statistical views, from which the network quality and status is obtained. Statistical data consists of number of drop call, call condensation etc.

BTS and Transcoder (TC) Control

Inside the BSS, all the BTS and TC are connected to BSC and control by BSC only.

1.4.1.3. Base Transceiver Station (BTS)

BTS is the network element responsible for maintaining the air interface and minimising the transmission problems. Main tasks of BTS are as follows.

Ciphering : Both the BTS and the MS must be able to cipher and decipher information in order to protect the transmitted speech and data in the air interface. This ciphering means coding of transmitted speech, data is done by specific algorithm which is stored in BTS and MS both.

Speech Processing: Speech processing refers to all the functions the BTS performs in order to guarantee an error free connection between the MS and the BTS. This include tasks like speech coding (Digital to analogue in the downlink direction and vice versa), channel coding (for error protection interleading) to enable a secure transmission, and burst information (adding information) to the coded speech/data in order to achieve a well organised and safe transmission.

Modulation and De-Modulation: User data is represented with digital values 0 and 1. These bit values are used to change one of the characteristics of an analogue radio signal in a predetermined way. By altering the characteristics of a radio signal for every bit in the digital signal. We can "translate" an analogue signal into a bit stream in the frequency domain. This technique is called modulation. In GSM, Gaussian Minimum Shift Keying (GMSK) is applied.

Base station can control several TRxs (Transceivers), each supporting one pair of frequencies for transmitting and receiving information. These TRxs are kept inside BTS cabinet. And BTS cabinet is connected to Antenna tower through TRx.

Antenna may be omnidirectional, sectored as per requirements.

Note that one TRx supports only one pair of frequencies. One BTS cabinet may support one TRx to 12 TRx as per requirement.

Transcoder (TC): In the air interface between MS and BTS, the media carrying the traffic is a radio frequency. To enable an efficient transmission of digital speech information over the air interface, digital speech signal is compressed. We must, however, also be able to communicate with and through the fixed network, where the speech compression format is different. Somewhere between the BTS and the fixed network, we, therefore, have to convert from one speech compression format to another, and this is where the transcoder comes into picture.

For transmission over the air interface, the speech signal is compressed by the mobile station to 13 kbps, 5.6 kbps, 12.2 kbps.

It is important to note TC is connected at NSS site with MSC but as per GSM architecture TC is part of BSS. In Fig. 1.7, TC is shown with BSS as per architecture but it will be always with MSC.

1.4.2. Network Switching Subsystem (NSS)

Network switching subsystem is a cabinet which consists MSC, GMSC, VLR, HLR, AC and EIR. The main functions of NSS are:

Call Control: This identifies the subscriber, establishes a call, and clears the connection after the conversation is over.

Charging: This collects the charging information about a call (the numbers of the caller and the called subscriber, the time and type of the transaction etc.) and transfers it to the billing centre.

Mobility management: This maintains information about the subscriber's location. BSC does this operation but MSC keep records.

Signalling: This applies to interfaces with the BSS and PSTN.

Subscriber Data Handling: This is the permanent data storage in the HLR (Home Location Register) and temporary storage of relevant data in the VLR (Visitor Location Resistor).

1.4.2.1. Mobile Services Switching Centre (MSC)

MSC is a cabinet which consists of various functional units and responsible for controlling calls in the mobile network. It identifies the origin and destination of a call (mobile station or fixed telephone), as well as the type of a call.

MSC is responsible for several important tasks, such as follows.

Call Control: MSC identifies the type of call, destination, and the origin of a call. It also set-up, supervises, and clear connections.

Initiation of Paging: Paging is the process of locating a particular mobile station subscriber in case of a mobile terminated call (a call to a mobile station).

Charging Data Collection: MSC generates CDRs (charging Data Records), which contains information about the subscribers usage of the network.

Visitor Location Register (VLR): VLR is integrated with the MSC cabinet. VLR is a data base which contains information about subscribers currently being in the service area of the MSC/VLR, such as :

- ➤ Identification numbers of the subscribers.
- ➤ Security information for authentication of the SIM Card and for ciphering.
- ➤ Services that the subscriber can use.

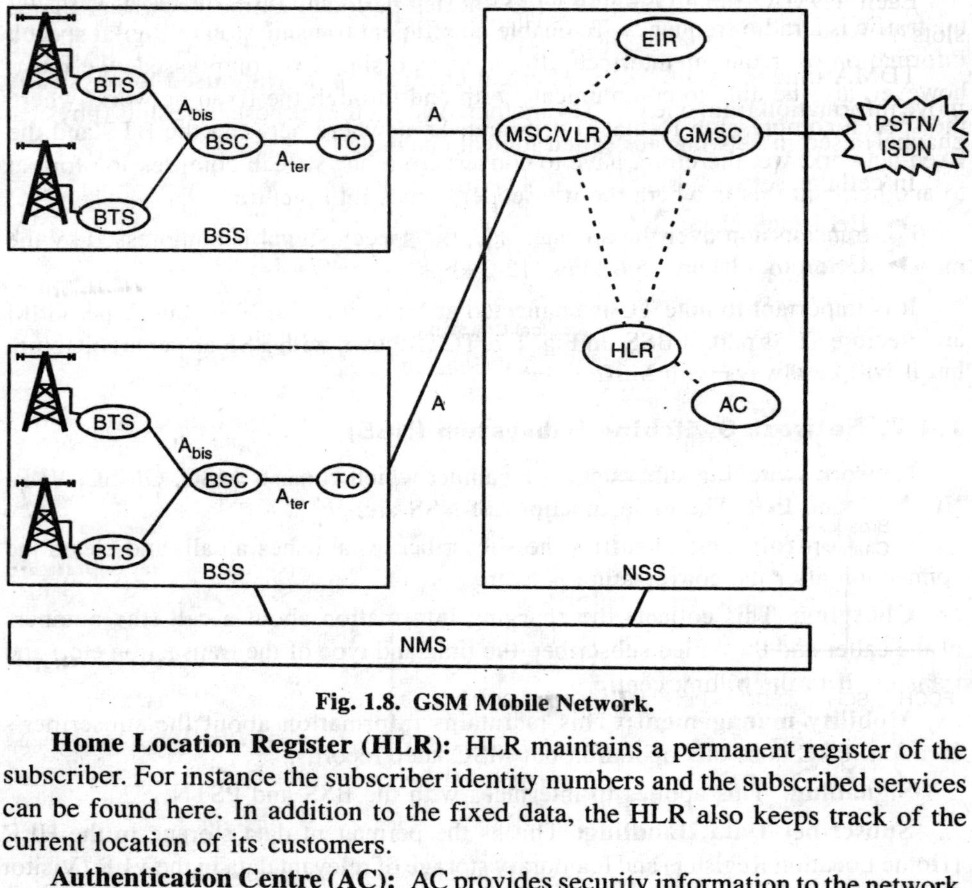

Fig. 1.8. GSM Mobile Network.

Home Location Register (HLR): HLR maintains a permanent register of the subscriber. For instance the subscriber identity numbers and the subscribed services can be found here. In addition to the fixed data, the HLR also keeps track of the current location of its customers.

Authentication Centre (AC): AC provides security information to the network, so that we can verify the SIM cards (authentication between the mobile station and the VLR, and cipher the information transmitted in the air interface (between the MS and the BTS).

Equipment Identity Register (EIR): EIR is responsible for IMEI checking (checking the validity of the mobile equipment) but in India this service is not used by network operators. Some vendor include EIR as integral part of HLR.

1.5. OPERATION OF CELLULAR SYSTEMS

Operation of cellular system is fully automatic and no manual actions require rather than calling and answering. There are various types of channel for various purposes between mobile phone and network which are uses for call conversation or for data transmission. Basically channels are divided in two parts:

➤ Physical channel

➤ Logical channel.

As we know from previous discussion cellular system particularly GSM is based on TDMA and FDMA both. Time Division Multiple Access (TDMA) divides one radio frequency channel into consecutive periods of time, each one called a "TDMA frame".

Each TDMA frame contains eight shorter periods of time known as "Time slots".

TDMA time slots are called physical channel, as they are used to physically move information from one place to another. And for the purpose time slots (physical channel) used in network are called logical channels.

In cellular network logical channels are divided into two types:

➤ Dedicated channels.

➤ Common channels.

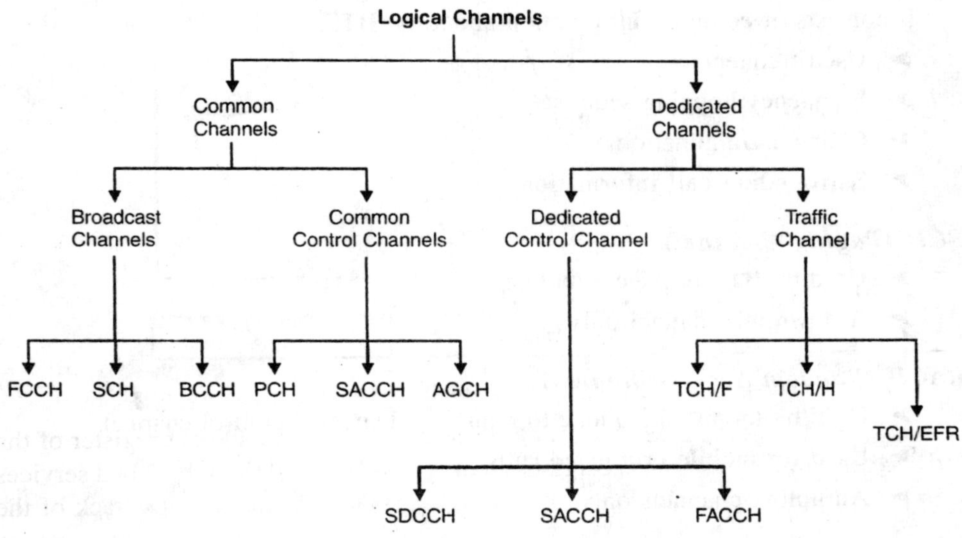

Fig. 1.9. Logical channel classification.

Dedicated channels allocated to mobile phone for call conversation, data transmission and if it is allocated once then it cannot be shared with another mobile phone subscriber untill mobile phone terminates his call.

While common channels use for keeping record of subscriber position, authentication, searching for subscriber. Following are more classifications of common channels and dedicated channels.

These channels are used between mobile phone and BTS and as we know one frequency channel (one TDMA frame) is divided into 8 parts (time slot). So, we can have 8 channels only but we require more. In practical first time slot of TDMA frame is used for common channels and dedicated control channel on time sharing basis and rest 7 channels are used for traffic channel. Traffic may be voice or data. Because common and dedicated control channels are required for searching, authentication of mobile subscriber, so they are required for very less time.

FCCH (Frequency Correction Channel)

It is a pure sine wave string of 142 zero. When mobile phone switches on then firstly mobile phone starts to detect this channel. Every network operator tower transmit this string of 142 zero but of different frequency hence your mobile phone will tune to that frequency of which you have purchased SIM. If more than one tower is in the range of your operator then mobile phone detects strongest signal. It is a Down link channel (BTS to Mobile Phone).

SCH (Synchronisation Channel)

After locking to the frequency the MS synchronises with the SCH.

The SCH contains the Base Station identity code of the BTS and the TDMA frame number (used in encryption).

BCCH (Broadcast Control Channel)

It consists of common information about the BTS.

- Used frequencies
- Frequency hopping sequences
- Channel combination
- Surrounding call information.

PCH (Paging Channel)

- Used by BTS to page a mobile.
- A downlink channel only.

RACH (Random Access Channel)

- Used by the mobile phone to request a dedicated control channel.
- Used for mobile originated calls.
- An uplink channels only.

AGCH (Access Grant Channel)

- Used by the BTS to assign a dedicated control channel.
- A downlink channel only.

Dedicated Channels

Dedicated channels may be in uplink or in downlink but once it is allocated it cannot change or share.

SDCCH (Stand Alone Dedicated Control Channel)

- It is Bidirection channel.
- Used for call set-up procedures for example authentication.
- The traffic channel (TCH) is assigned by using SPCCH.

SACCH (Slow Associated Control Channel)

- Associated with SDCCH and TCH
- Measurement reports
- Mobile Phone Power Control
- Timing alignment (when mobile phone subscriber is on moving).

FACCH (Fast Associated Control Channel)

- Associated with TCH.
- For quick control communication for example handover.
- Physically replaces 20 milli second of speech (It is time which is required to complete handover process).

Traffic Channels

Traffic channels are also dedicated channels and used for voice conversation or for data transfer.

Full Rate

- Bi-direction channel.
- Used for speech or data transmission.
- User data bit rate 13 kbps.

Half Rate

- Bi-direction channel
- Used for speech or data transmission
- User data bit rate 5.6 kbps.

Enhanced Full Rate (EFR)

- Bi-Direction channel
- Used for high quality speech transmission.
- User data bit rate 14.6 kbps.

Note: Whole of the operation can be understood by studying following points.

1.5.1. Mobile Phone Initialization

Whenever mobile phone turns on, it starts searching strongest channel (FCCH). Note that mobile phone automatically selected only that antenna of the cell, within which it will operate. After selecting FCCH a link is established between mobile phone and MSC via BTS and BSC. This link is used to identify the user and register its location. If phone is not static at a point then this link will be continued but if phone subscriber is moving and he enters in a new cell then a new BTS is selected that is belonging from new cell. Following are the simplified steps to complete a call.

Call From Mobile Phone to Fixed Land Line Number (PSTN Number)

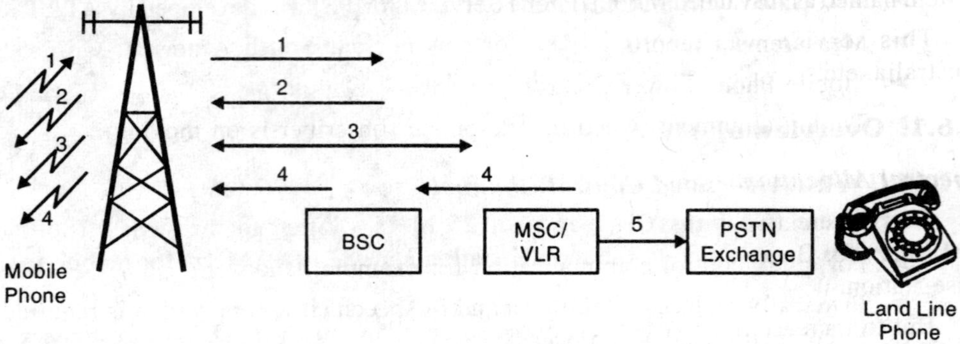

Fig. 1.10. Mobile originated call.

1. The mobile phone tuned to FCCH then SCH and then uses RACH to ask for a signalling channel.
2. The BSC allocates a signalling channel using AGCH.
3. Mobile phone sends a call set-up request via SDCCH to the MSC/VLR. Over SDCCH all signalling preceding a call takes place which includes.
   - ➤ Marking the mobile phone as "active" in the VLR.
   - ➤ Authentication procedure
   - ➤ Start ciphering
   - ➤ Equipment identification
   - ➤ Sending the dialed number (landline number) to the network.
4. MSC/VLR instructs the BSC/BTS to allocate an idle TCH. BTS and Mobile phone are told to Tune to the specified TCH.
5. MSC/VLR forward the Dialed number to the B-number (called subscriber of PSTN) to an exchange in the PSTN, which establishes a connection to the subscriber.
6. If the called subscriber answers, the connection is established.

Call Termination

When one of the two users hangs up or terminates the call or switch off the mobile phone, the MSC is informed and the traffic channels at the two BSC/BTS are released. Generally it is choice of user, when he wants to terminate the call.

Call Drop

During a connection due to interference or weak signal if BTS cannot maintain the minimum required signal. Call will drop and informed to the MSC.

Calls to/from Fixed and Remote Mobile Subscriber

The MSC connects to the PSTN, so MSC can setup a connection between a mobile user in its area and a fixed subscriber via the telephone network.

1.6. ANALOG CELLULAR SYSTEM

Analog Cellular System is a First Generation (1G) Cellular system.

In early 1980, when cellular system was on evolving stage then the first 1G system named as Advanced Mobile Phone Service (AMPS) was developed by AT&T.

This service was employed in North America, South America, China, Australia, etc.

1.6.1. Overview of AMPS

Spectral Allocation

There were two frequency bands of 25 MHz allotted to the AMPS (869-894 MHz) for Base station to the mobile station and (824-849 MHz) for mobile to base station.

But to increase the competition between operator for the benefit of customers the above frequency band splited in two parts.

One operator is allocated only 12.5 MHz in each direction for its system. The channels are spaced 30 kHz apart, which allows a total of 416 channels per operator.

Out of this 416 channels (21 for control and 395 to carry calls). The control channels are data channels operating at 10 kbps. The modulation technique was also Analog frequency modulation. But to support the major customer based market and to increase the facilities, the number of channels was very less. So there is need to change. Either we use less bandwidth per conversation or to reuse frequencies to increase number of channels. For AMPS reuse of frequency is exploited.

Operation

Each AMPS includes a Numeric Assignment Module (NAM) in read only memory.

The NAM contains the telephone number of the phone, which is assigned by the service provider, and serial number of phone, which is assigned by manufacturer. When the phone is turned ON, it transmits its serial number and phone number to the NSS. NSS maintains the database of user for billing purpose or for blocking the call.

How Call is made

Following steps involved in making a call.

1. The subscriber dialed a number and push the send button.
2. NSS verifies that the telephone number is valid and subscriber is authorized to use the service.
3. NSS issues a message to the user's cellphone indicating which traffic channels to use for sending and receiving.
4. The NSS sends out a ringing signal to the called party.
5. When the called party gives the information, the NSS establishes the relation between two parties and starts the billing operation.
6. When one party hangs up, the NSS releases the circuit and free the radio channels.

1.6.1.1. Analog Cellular Phone Parameter

Table 1.2.

Base station transmission band (Downlink)	869-894 MHz
Mobile station transmission band (Uplink)	824-849 MHz
Spacing between uplink and downlink	45 MHz
Channel bandwidth	30 kHz
Number of Full duplex voice channel	790
Number of Full duplex control channel	42
Mobile unit maximum power	3 watt
Cell size, radius	2 to 20 km
Modulation, voice channel	FM, 12 kHz peak deviation
Modulation, control channel	FSK, 8 kHz Peak deviation
Data transmission Rate	10 kbps
Error control coding	BCH (48, 365)

Each AMPS service includes 21 Full-duplex 30 kHz control channels, consisting 21 Reverse Control Channel (RCCs) from subscriber to base station, and 21 forward channels base station to subscriber.

These channels transmit digital data using FSK in both channels, data are transmitted in frames.

Some terms and their function involved in analog cellular system.

Home Mobile Station (or Mobile unit or Mobile phone or Mobile station): A mobile unit, that is subscribed in cellular system.

Land Station: A station other than a mobile unit, which links to the mobile station.

Control Channel: A channel used for the transmission of Information from a land station to a mobile station or vice versa.

Forward Control Channel (FDCC): Used for land station to mobile station.

Reverse Control Channel (RECC): Mobile station to a land station.

Forward Voice Channel (FVC): A voice channel used between land station to a mobile unit.

Reverse Voice Channel (RVC): Voice channel between mobile to land line.

Setup Channels: A number of designated control channel.

Access Channel: Always access from mobile station to cell unit to obtain service.

Paging Channel: When an incoming call from landline has been placed to it.

Signaling Tone: A 10 kHz tone transmitted by the mobile station on a voice channel.

Handoff: The act of transferring a mobile station from one voice channel to another voice channel. There are two kinds of handoffs.

(*a*) Interhandoff; from one cell to another cell.

(*b*) Interhandoff; within a cell.

1.6.2. Digital Cellular Systems

In 1992, the first Digital Cellular System, GSM was deployed in Germany. GSM is a European standard system. In the USA, an NA-TDMA System (IS-54) and a CDMA System (IS-95) have been developed. NA-TDMA was deployed in 1993 and CDMA is planned for deployment in 1995. A Japanese system, PDC (Personal Digital Cellular) was deployed in Osaka in June 1994. Digital Cellular system is a Second Generation Cellular System like GSM, CDMA Cordless phone, etc.

1.6.3. Digital Cellular System Parameter

Table 1.3.

S. No.	Particulars	GSM	IS-136	IS-95
1.	Year introduced	1990	1991	1993
2.	Access method	TDMA	TDMA	CDMA
3.	Base station transmission band	935 to 960 MHz	869 to 894 MHz	869 to 894 MHz
4.	Mobile station transmission band	890 to 915 MHz	824 to 849 MHz	824 to 849 MHz
5.	Spacing between forward and reverse channels	45 MHz	45 MHz	45 MHz
6.	Channel bandwidth	200 kHz	30 kHz	1250 kHz
7.	Number of duplex channels	125	832	20
8.	Mobile unit maximum power	20 W	3 W	0.2 W
9.	Users per channel	8	3	35
10.	Modulation	GMSK	π/4 DQPSK	QPSK
11.	Carrier bit rate	270.8 kbps	48.6 kbps	9.6 kbps
12.	Speech coder	RPE-LTP	VSELP	QCELP
13.	Speech coding bit rate	13 kbps	8 kbps	8, 4, 2, 1 kbps
14.	Frame size	4.6 ms	40 ms	20 ms
15.	Error control coding	Convolutional 1/2 rate	Convolutional 1/2 rate	Convolutional 1/2 rate forward; 1/3 rate reverse

1.7. COMPARISON BETWEEN ANALOG AND DIGITAL CELLULAR SYSTEM

Digital Traffic Channel

The first generation (AMPS) designed to support voice channel using FM.

While 2G system provided digital traffic channels. These readily support digital data, voice is first encoded in digital form by using modem.

Encryption

The first generation system send user traffic without any security while 2G system provides this capability.

Error Detection and Correction

There is no such facilities in 1G system. So voice reception was not clear.

While 2G system provided such capability, so voice was good.

Channel Access: In first generation system, each cell supports a number of channels.

At any given time a channel is allocated to only one user.

2G systems also provide multiple channel per cell but each channel is dynamically shared by a number of users using Time Division Multiple Access (TDMA) or Code Division Multiple Access (CDMA).

1.7.1. Advantages of the Digital Transmission

Mobile communication followed the trend set in fixed networks in the mid-1980's under the term Integrated Services Digital Network (ISDN). Following are several advantages that are correlated with the introduction of 2G digital transmission :

1. Network Capacity: Compression of digitized speech information can considerably increase the capacity of mobile communication networks. Speech compression must be weighed against a reduction in speech quality however. Compression in speech from 64 kbit/s (digital fixed network transmission) to 2.4 – 13 kbit/s is used in the different 2G systems for transmission over the air interfaces.

2. Security Aspects: Unlike analog signals, digital information can be very easily ciphered, preventing unauthorized eavesdropping of user data.

3. Supplementary Services: Digital data transmission greatly simplifies the transfer of signaling information thereby allowing the introduction of a wide range of supplementary services not confined to just pure speech and data transmission.

4. Cost Factor: Digital devices are less expensive to produce than analog devices thanks to better options for the use of large-scale integrated microelectronic components. Purchasing costs, as well as operating and maintenance costs, are lower and opened the way for the 2nd generation to the mass market.

5. Miniaturization: Microelectronics for digital information transmission allows a HW reduction that is relatively simple compared to analog HW elements. In this way, the size and weights of Mobile Stations MS could be reduced very much from 1G to 2G, allowing turning over from car phone to handhelds. The weight of handhelds decreased during the 1990's from more than 500 gm to less than 100 gm.

6. Transmission Quality: During transmission across the air interface the signals experience considerable fading, distortion and corruption. Digital signals can be treated easily with redundancy, can be better regenerated and offer therefore significantly better transmission quality than analog signals. Analog signals can only be amplified (including all disturbances).

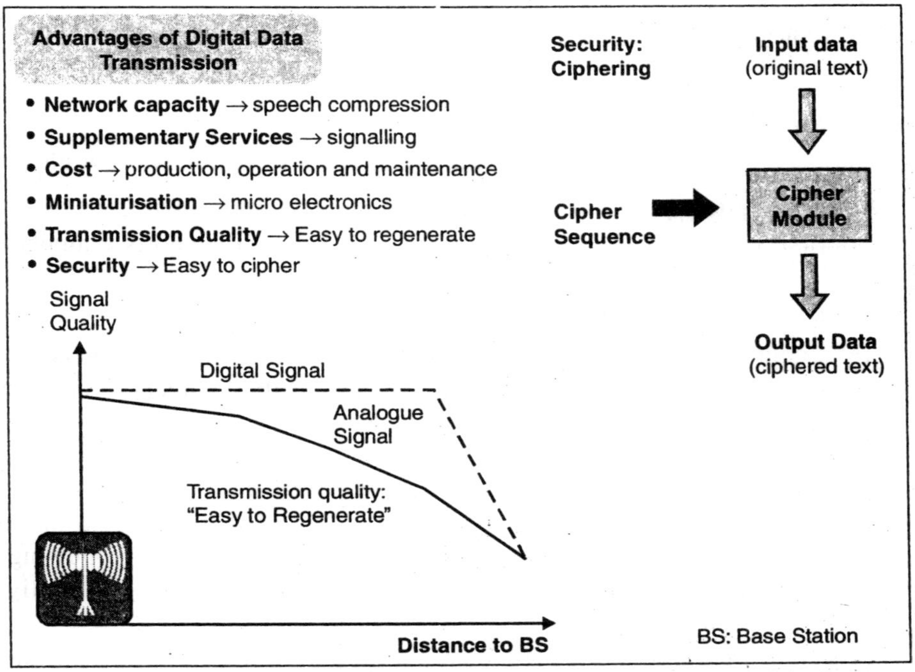

Fig. 1.11. Advantages of Digital data transmission.

1.8. PERFORMANCE CRITERIA

There are so many parameters to decide performance of cellular system but as per International Telecom Union following are main criteria.

Voice Quality

Voice quality is complicated parameter for design engineers because it depends upon person to person and also on mobile equipment. So, in this area designer cannot decide that how to build a system without knowing the voice quality that will satisfy the users.

In Military, Airforce Communication this is not a problem because Armed forces must use the assigned equipment.

But in general the voice quality depends upon the following criterion.

A set value x at which y per cent of customers rate the system voice quality is good or excellent (from transmitter to receiver).

Generally following scale are used for Circuit Merits (CM) in respect of voice quality.

Circuit merit	Score
CM 1	1 (Unsatisfactory) not understandable
2	2 Poor (understandable but repetitions are required).
3	3 Fair (occasional repetitions required)
4	4 Good (understandable but some noise)
5	5 Excellent.

If during survey percentage of customers choosing CM4, CM5 then cost of system increases.

To make balance between voice quality and cost of operation Mean Opinion Score (MOS) is between CM4 and CM5 choosen by network operator.

Service Quality

Following parameters are requires to judge the service quality.

1. Coverage area.
2. Grade of service
3. Number of dropped calls.

Coverage Area

If a system serves as far as possible large area it is good. But it is not possible to serve 100 per cent area due to irregular Geographical structure.

Due to following reasons full coverage of area is not possible.

1. The transmitted power must be very high to illuminate weak spots (where reception is not faithful) which increases the cost.
2. The higher the transmitted power higher the interference. Therefore, a system usually covers 90% area in flat parts while 60 to 75% in a hill part.

Grade of Services [GoS]

The grade of service is very good if number of drop calls is 2 or less than 2 out of 100 calls in peak hour. However, the blocking probability at each cell site is different. To decrease the block calls or blocking probability requires a good system plan and sufficient number of radio channels as well as number of BTS.

Dropped Calls

To measure the dropped calls, there is a parameter named call drop rate.

If during Q calls, Q-1 calls are completed then call drop rate is 1/Q, if Q-2 calls are completed then call drop rate is 2/Q.

As far as possible call drop rate must be low. A high drop rate can be caused either due to coverage problems or handoff problems related to channel availability.

1.9. OVERVIEW OF CELLULAR GENERATIONS

The evolution of cellular communications system is commonly known by 1G, 2G and 3G, 4G designations, the latest being 3G, which is designed for high speed data transmission.

1G introduced in the late 1970 and early 1980s, the first cellular systems were analog. They were used for data like an ordinary telephone line is used with an analog modem. With an adapter, one could plug a laptop modem into some cellphone models and transfer data while travelling.

The second generation (2G) refers to digital voice cellphone systems deployed in the 1990's which are GSM and CDMA. While so called $2G^{+}$ or 2.5 G, technologies have provided e-mail and Internet access. These include GPRS and EDGE for GSM and IS-95 for CDMA. And third generations 3G is designed for high speed multimedia data with speeds ranging from 128 Kbps to several megabits per second. Defined by the ITU under the IMT 2000 global framework.

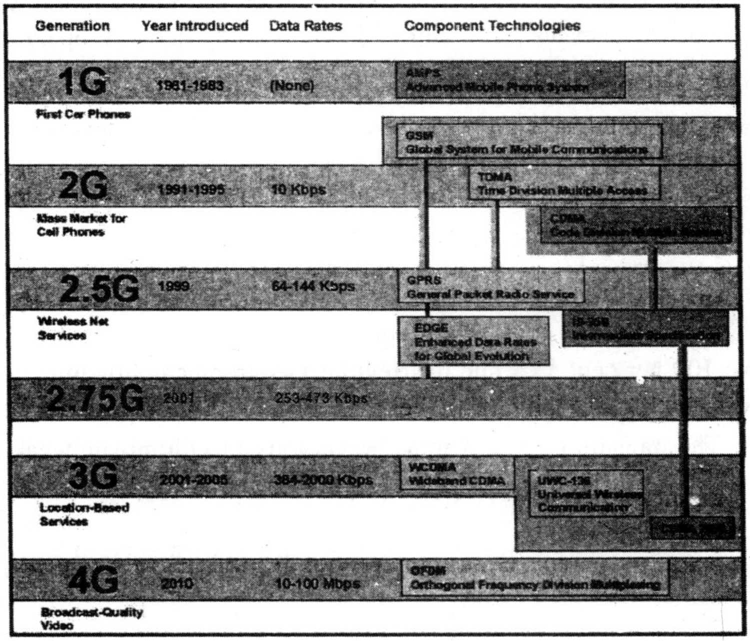

Fig. 1.12. Evolution of various generations.

4G cellphones will support all the service to the 3G with the addition of movies and TV. 4G services based on Wi-max and it offer speed between 1 Gbps to 10 Gbps at air interface. Watching of T.V., Browsing of Internet will have greater experience with 4G. It will get offered by operator in many countries by year 2012.

1.9.1. 1G (First Generation)

1G (or 1-G) is short for first generation wireless telephone technology, cellphones. These are the **analog** cellphone standards that were introduced in the 1980s and continued until being replaced by 2G **digital** cellphones. The main difference between two succeeding mobile telephone systems, 1G and 2G, is that the radio signals that 1G networks use are analog, while used by 2G networks are digital. Note that both systems use digital signaling to connect the radio towers (which listen to the handsets) to the rest of the telephone system. But the call itself is encoded to digital signals in 2G whereas 1G is only modulated to higher frequency (typically 150 MHz and up). Following are the 1G Mobile Phone Standards.

- NMT
- AMPS
- CDPD (Cellular Digital Packet Data)

One such standard, NMT (Nordic Mobile Telephone), used in Nordic countries, Switzerland, Netherlands, Eastern Europe and Russia. Others include AMPS (Advanced Mobile Phone System) used in the United States, TACS (Total Access Communications System) in the United Kingdom, C-450 in West Germany, Portugal and South Africa.

1.9.1.1. Chronology of First Generation Wireless Systems

1970 Developments of radio and computer technologies for 800/900 MHz mobile communications

1976 WARC (World Administrative Radio Conference) allocates spectrum for cellular radio

1979 NTT (Nippon Telephone & Telegraph) introduces the first cellular system in Japan

1981 NMT (Nordic Mobile Telephone) 900 system introduced by Ericsson Radio System AB and deployed in Scandinavia

1984 AMPS (Advanced Mobile Phone Service) introduced by AT&T in North America

1.9.2. Second Generation (2G) Cellular Network

Second generation system was first introduced in the early 1990s and evolved from the first generation of Analog mobile phone systems (like AMPS, ETACS, ITACS). Nowadays many wireless service providers use both first generation and 2G equipment and often provide to customers with subscriber units that can support multiple frequency bands and multiple Air interface standards. But in India, we are not using 1G equipment, neither this facility is possible in 2G equipments.

The first generation cellular system works with FDMA/FDD and Analog FM, while 2G standards use digital modulation formats and TDMA/FDD (Frequency Division Duplexing) and CDMA/FDD multiple Access techniques. 2G standard includes three TDMA standards and one CDMA standards which are as follows.

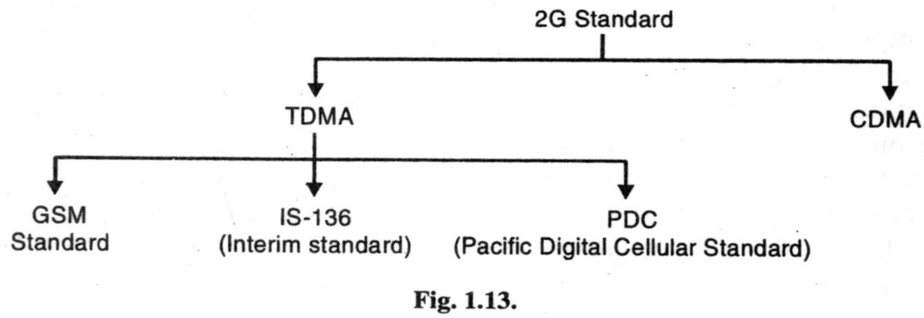

Fig. 1.13.

GSM Standard: It supports eight time slotted users for each 200 kHz radio channel and employed in whole world.

IS-136: It is also known as NADC (North American Digital Cellular) or USDC (US Digital Cellular). It supports three time slotted users for each 30 kHz radio channel. Mostly used in North America, South America, Australia.

PDC: It is a similar to IS-136 but it is Japanese TDMA standard.

CDMA: Also known as (IS-95) or CDMA-one. Which supports upto 64 users that are orthogonally coded and simultaneously transmitted on each 1.25 MHz channel. It is mainly used in North America, Korea, Japan, China, India.

2G network is only technology that is used by most of countries in the world. Because 2G standard are capable to handle the high capacity subscriber. For employing the 2G Network it is required to increase at least 3 times spectrum efficiency (thus overall total capacity increased three times) as compared to 1G network.

Evolution of 2.5G: If we see the table of key specifications, we can conclude that all 2G networks, as originally developed, only support single user data rates on the order of 9.6 kilobits per sec, which is too slow for fast Internet Applications.

This problem arises because 2G standards have been implemented by wireless carriers for cellular and PCS (Personal Communication System), before widespread use of the Internet. So there is need to restandardise the 2G Network. For using internet application mean to increase the data rates.

2G network was able to support the limited internet browsing and Short Messaging Service (SMS). But in many countries it was not possible due to different types of technologies and Network owners, and SMS was presently possible between same network.

In an effort to restandardise the 2G for compatibility with increased data rates that are required to support Modern Internet applications, new data standards have been developed that can be overlaid upon existing 2G technologies.

These new standards represent 2.5G technology and allow existing 2G equipment to be modified and supplemented with new base station.

So by implementing 2.5G standards we became able to use web browsing, e-mail traffic, mobile commerce. The 2.5G technologies also support a popular new web browsing format language, called Wireless Application Protocol (WAP), that allows standard web pages to be viewed in a compressed format, specifically designed for small, portable hand-held wireless devices.

The appropriate 2.5G upgrade path for a particular wireless carrier must match the original 2G technology choice made earlier by the same carrier. For example, a 2.5G upgrade solution designed for GSM must be compatible with original GSM air interface standard (Gar99), since it would otherwise be incompatible and require wholesale equipment changes at each base station. For this reason, a wide range of 2.5G standards have been developed to allow each of the major 2G technologies (GSM, CDMA) to be upgraded incrementally for faster Internet Data rates.

In 2.5G standard that is commonly called GPRS (General Packet Radio Service), we are getting speed up to 53 kbps only that is very slow for Internet Browsing. While Internet application and services is going to popular very much day by day.

So there is need to increase access speed for mobile phone but it was not possible with existing technology. To achieve higher speed some new coding scheme was introduced and some hardware and software changes also took place in mobile phone and network side. With this new change, we got speed for data access as 253 to 473 kbps. And this new feature was put in 2.75 Generation and technology named EDGE (Enhanced data for GSM evolution).

In India almost every operator is providing this service but EDGE enabled mobile phone is very costly, therefore, EDGE services are not so popular.

Note that SMS (Short Message Service) is also data service but now this service is inbuilt with voice service so if you take a connection then SMS service you will get automatically with voice feature. There is no need to take GPRS or EDGE services from network operator.

But if you want to use your mobile phone for MMS (Multimedia Message Services, Internet browsing) then you will have to take these services from operator but then you will have to pay extra money.

It is important to note that GPRS, EDGE are mobile phone and network operator enabled services not mobile phone alone itself.

Table 1.4. Second Generation Cellular Telephone Systems.

S. No.	Particulars	GSM	IS-136	IS-95
1.	Year introduced	1990	1991	1993
2.	Access method	TDMA	TDMA	CDMA
3.	Base station transmission band	935 to 960 MHz	869 to 894 MHz	869 to 894 MHz
4.	Mobile station transmission band	890 to 915 MHz	824 to 849 MHz	824 to 849 MHz
5.	Spacing between forward and reverse channels	45 MHz	45 MHz	45 MHz
6.	Channel bandwidth	200 kHz	30 kHz	1250 kHz
7.	Number of duplex channels	125	832	20
8.	Mobile unit maximum power	20 W	3 W	0.2 W
9.	Users per channel	8	3	35
10.	Modulation	GMSK	π/4 DQPSK	QPSK
11.	Speech coding bit rate	13 kbps	8 kbps	8, 4, 2, 1 kbps
12.	Frame size	4.6 ms	40 ms	20 ms
13.	Error control coding	Convolutional 1/2 rate	Convolutional 1/2 rate	Convolutional 1/2 rate forward; 1/3 rate reverse

Chronology of Second Generation Wireless Systems

1982 CEPT established GSM to define future Pan-European Cellular Radio

1990 IS-54 (USDC) adopted by TIA

1990 IS-19B (NAMPS) adopted by TIA

1991 Japanese PDC system standardized by the MPT

1992 Phase I GSM system is operational

1993 IS-95 (CDMA) adopted by TIA

1994 IS-136 adopted by TIA

1995 PCS Licenses issued in North America

1996	Phase II GSM operational
1997	North American PCS deploys GSM, IS-54, IS-95
1999	IS-54: North America
	IS-95: North America, Hong Kong, Israel, Japan, China, etc.
	GSM: 110 countries

1.9.3. Migrating From 2G to 3G

Upgrading on 2G GSM systems include high-speed circuit-switched data (HSCSD), general packet radio service (GPRS) and an Enhanced data rate for GSM evolution. (EDGE)—all of which allow for higher data transmission rates.

The goal of GSM migration is to reach the universal mobile telecommunication system (UMTS), which is part of the ITU's IMT-2000 vision of a global family of 3G mobile communication systems. All these generation systems are now well on their way to development and deployment and the question now is which one will be the most relevant, versatile, cost-effective and able to cope with the demands of a complex telecommunications service landscap ?

Which system will succeed in effectively delivering new services ? And will these 'half-steps' toward elusive 3G roll-outs. The need for 3G itself. Most GSM operators feel that the jump from 2G to 2.5G will be more dramertic than from 2.5G to 3G. The big job for the operator is not going from GPRS to UMTS, but actually going from GSM to GPRS.

For one, it has to completely change the business model from time based to volume-based charging and also from traditional services to more internet based services. Although 2.5G technologies were expected to smooth the transition to 3G, it proved to be less than satisfactory.

A sudden jump from today's GSM networks to the 3G networks of tomorrow is unlikely.

GSM-based services depend on digital transmission between base stations and handsets with high-speed to and from the centres equipped with circuit switches. At 9.6 kbps, transmission is slow, and the architecture itself is unsuitable for data traffic or streaming, as it is circuit-switched rather than packet-switched.

While GPRS (General Packet Radio Service) seems to be an obvious migration step for GSM operators.

The next steps require further evoluation. It is also important to note that through the course of the transition, it is not necessarily the case that the early 3G networks when they appear will be packet switched.

The evolution to packet based networks is likely to occur over sometime as systems are tested and proven.

The transition to HSCSD (High Speed Circuit Switch Data) is not difficult for existing 2G operators, and typically necessitates only a software upgrade of the Base Station Subsystems (BSS) and Network switching subsystems (NSS).

1.10. THIRD GENERATION (3G)

3G describes the updating of cellular mobile telecommunications networks around the world to use new 3G technologies. This process is taking place over the period 1999 to 2010. Japan is the first country having introduced 3G nationally, and in Japan the transition to 3G is being largely completed during 2005/2006. 3G technologies enable network operators to offer users a wider range of more advanced services, while achieving greater network capacity through improved spectral efficiency. If we define 3G then 3G is based on CDMA but implemented over GSM architecture. Because 2G, GSM became very much popular and operator has been invested lot of money so practically it was not possible to implement 3G by replacing 2G completely.

1.10.1. Operators and UMTS Networks

As of 2005, the evolution of the 3G networks was on its way for a couple of years. The main reason for these changes are basically the limited capacity of the existing 2G networks. The second generation of networks were built mainly for telephone calls and slow data transmission. Due to the rapid changes in technology, these factors do not meet the requirements of today's wireless revolution. The developments of so-called "2.5G" technologies such as i-mode data services, camera phones, HS CSD and GPRS have been ways of bridging the oncoming change to 3G networks, but are not permanent solutions. They are merely stepping stones towards the new technology. These stepping stones were built to introduce the possibilities with the future wireless application technology to the end consumers. These procedures are necessary to ensure that the operators and the infrastructure itself have a healthy ground to operate on.

The evolution of networks from the second generation of technologies to the third generation technologies could not be done without the help of network operators. In 2005 there were about 23 networks worldwide that operated on 3G technologies, the most advanced being KDDI in Japan. Some of these networks were only for test use but some were already in consumer based use.

Basically network operators need to find the many answers to three questions before they can start operating in the new field. What are the things to be taken into consideration during the change ? How to manage the change? What is the future role for the network operators ?

Note: Detailed description is available in Chapter 7.

Network operators have invested huge amounts of money into existing 2G networks. These networks have been around only for 10-15 years, and the investments made have not all paid off. Network operators need to find out ways of reusing their investments to build the 3G network. Because of the financial situation of the world, network operators do not necessarily have new resources to invest in the future. They must recycle the old ones first.

Another thing network operators need to understand is that their roles are changing dramatically. They are becoming not only network providers, but also service providers. Network operators need to differentiate themselves in the markets, and one way is to concentrate on the content of the service and products. It is widely believed that the markets will consist of content oriented service providers, since 3G technology allows anyone willing to build software and sell it directly to end consumers. Thus network operators need to adapt to this change too.

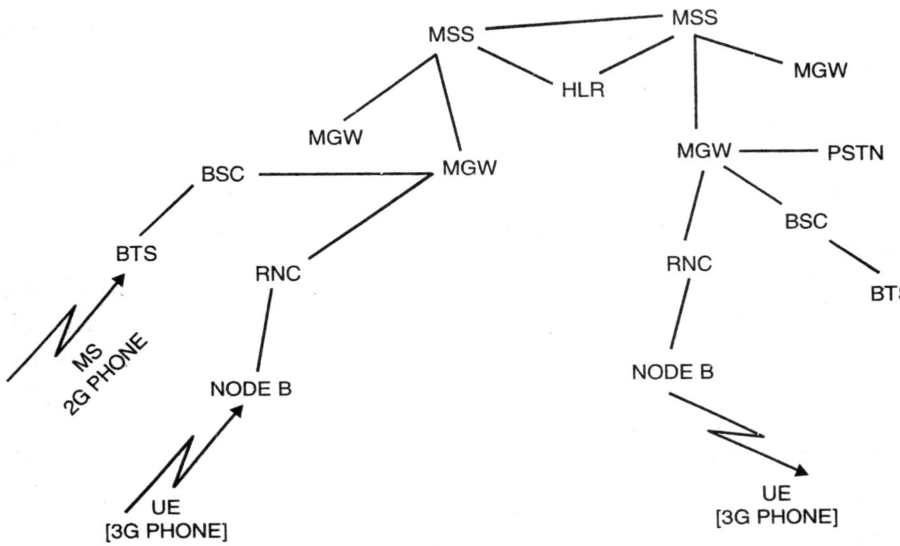

Fig. 1.14. WCDMA Architecture.

1.10.2. 2G to 3G Network Standardization

The **International Telecommunication Union (ITU)** has defined the demands for third generation mobile networks with the IMT-2000 standard. An organisation called 3GPP has continued that work by defining a mobile system that fulfils the IMT-2000 standard. This system is called **Universal Mobile Telecommunications System (UMTS)**. The evolution of the system will move forward with so called releases. In each release new features will be introduced. The following features are just examples of many others in these new releases.

Release '99 (Commonly known as Release 3)

- Bearer services
- 64 kbit/s circuit switched
- 384 kbit/s packet switched
- Location services
- Call services: GSM-compatible. USIM-based

Release 4

- EDGE radio
- Multimedia messaging
- Improved location services

Release 5

- IP Multimedia Subsystem (IMS)
- IPv6 IP transport in UTRAN (UMTS Radio Access Network)

- Improvements in GERAN (GSM, EDGE Radio Access Network), etc.
- HSDPA

Release 6

- WLAN integration
- Multimedia broadcast and multicast
- Improvements in IMS
- HSUPA

There are several different paths from 2G to 3G. In Europe the main path starts from GSM when GPRS is added to a system. From this point it is possible to go to the UMTS system. In North America the system evolution will start from TDMA going to EDGE and from there to UMTS.

In Japan, there are two 3G standards used: W-CDMA (which is compatible with UMTS) by NTT DoCoMo. Vodafone KK, and by new entrants, and CDMA2000 which is very successfully used by KDDI. Transition to 3G is being largely completed in Japan during 2005/2006. In India operators are implementing 3G Release 4 directly. It was expected that by June 2008 all operators will implement 3G Architecture.

1.10.3. Layered Network Architecture Advantages

The UMTS system is based on layered services, unlike GSM. On the top there is the services layer, which will give advantages like fast deployment of services and centralized location. In the middle there is the control layer, which will help upgrading procedures and allow the capacity of the network to be dynamically allocated. On the bottom is the connectivity layer where any transmission technology can be used and the voice traffic will transfer over ATM/AAL2 (ATM Adaption Layer 2) or IP/RTP.

1.10.4. Mobile Technologies

The first new technology when going from GSM towards UMTS is General Packet Radio Service (GPRS). It is the trigger to 3G services. The main point is that the network connection is always on, so the subscriber is online all the time. From the operator's point of view, it is important that GPRS investments are reused when going to UMTS. Also capitalizing on GPRS business experience is very important.

From GPRS, operators could go directly to UMTS, but they could also invest in an EDGE system. One advantage of EDGE is that there is no new licence needed as in UMTS The frequencies will also be reused and no new antennas are needed. The main issue is that subscribers will have to buy new EDGE terminals.

1.10.5. From GPRS to UMTS

The key point when going to UMTS is the use of the existing mobile network. From GSM core network side, the following network elements are reused:

- MSC (Mobile switching centre) (vendor dependent)
- AUC (Authentication centre)

- ➤ HLR (Home location register)
- ➤ VLR (Visitor location register)
- ➤ EIR (Equipment identity register)

From GPRS network, the following network elements will be reused:

- ➤ SGSN (Serving GPRS Support Node) (vendor dependent)
- ➤ GGSN (Gateway GPRS Support Node)

From GSM radio network, the following network elements can NOT be reused. Note, however, they can remain in the network and be used in dual network operation where 2G and 3G networks co-exist while network migration and new 3G terminals become available for use in the network.

- ➤ BSC (base station controller)
- ➤ BTS (base transceiver station)

The UMTS network introduces new network elements that give functionality as given in the 3GPP specifications:

- ➤ Node-B (base station)
- ➤ RNC (Radio Network Controller)
- ➤ MGW (Media Gateway)

The functionality of MSC and SGSN changes when going to UMTS. In a GSM system the MSC handles all the circuit switched operations like connecting A- and B-subscriber through the network. SGSN handles all the packet switched operations and transfers all the data in the network. In UMTS the MGW (Media gateway) will take care of all data transfer in both, circuit and packet switched networks. MSC and SGSN will act as "brains" of the system and they will control MGW operations. The name of the nodes will change into MSC-server and GSN-server.

1.10.6. Future Network

When UMTS networks are in commercial use and the services have found users, we need to check whether the capacity given by UMTS is enough or do we need more capacity. To increase capacity WLAN could be one solution that would be cost-efficient. Integration with UMTS would not be so difficult to do. When so called "hot services" will be found in UMTS, we can see the demands for the network after UMTS. Before that it is hard to do any presumption how the network should evolve.

1.10.7. UMTS, 3G Terminals

3G handsets are far more advanced and have much more functionality than conventional 2G handsets. 3G handsets usually include cameras, music players, video players, contactless smartcards for payment functions (wallet phones), web browsers, e-mail clients and more. This shows that UMTS system is based on layered services and future applications can be supported without too much impact to the underlying radio access network.

The future of UMTS terminals sees a change coming. So far the UMTS technology in Europe has primarily been used in very similar terminals as the GSM technology. The UMTS terminals have been very closely linked with GSM phones in Europe. This factor is now slowly starting to change— and has not been the case in Japan and South Korea, where 3G introduction is several years ahead of Europe. 3G handsets are far more advanced and have much more functionality than conventional 2G handsets. 3G handsets usually include cameras, music players, video players, contactless smart cards for payment functions (wallet phones), web browsers, e-mail clients and more options.

1.11. 4G (FOURTH GENERATION)

4G (or **4-G**) is short for **fourth-generation**, the successor wireless access technology to 3G. It is not used consistently, but generally describes several different, but overlapping ideas. The IEEE (Institute of Electrical and Electronics Engineers) official name for 4G is "3G and beyond".

1.11.1. Fourth Generation Standards

WiMax and WiBro.

1.11.2. Important 4G Components

(*a*) **Multi-Antenna Systems:** To faster the growing data rate needs of 4G, deploying mulitple antennas at the transmitter and at the receiver will increase the data rate.

(*b*) **Software Defined Radio (SDR):** SDR is one form of open wireless architecture (OWA). Since 4G is the collection of wireless standards, the final form of the 4G device will constitute all standards. This can be realized using SDR technology.

(*c*) **Adaptive Modulation and Coding:** The modulation and coding techniques change according to the network resource, user requirement and physical channel conditions. Cross layer techniques are being proposed for 4G.

1.11.3. Access Schemes

The scarced resource frequency and network infrastructure is accessed using the channel accessing schemes. The existing wireless standards use TDMA, FDMA, CDMA and combinations of these. Recently, new access schemes like OFDMA (Orthogonal Frequency Division Multiplexing) and MC-CDMA gain more importance in 802.16e and 802.20 standards.

IPv6

It is generally believed that 4th generation wireless networks would support a great number of wireless devices that are addressable and routable. Therefore, in the context of 4G, IPv6 is an important network layer technology and standard that can support a great number of wireless enabled devices. In addition to increasing the number of IP addresses, IPv6 also removes the need for Network Address Translation (NAT), a technique used in 3G and other networks to make private IP addresses work with Internet applications.

In the context of 4G, IPv6 also enables a number of applications with better multi-cast, security and route optimization capabilities. With the available address space and number of addressing bits in IPv6, many innovative coding schemes can be developed for 4G devices and applications that could aid deployment of 4G networks and services.

1.11.4. 4G True Broadband Wireless System

The hope for 3G to become a true broadband service has all but it vanished the hope.

It is apparent that 3G systems, while maintaining the possible 2-Mbps data rate in the standard, will realistically achieve 384 Kbps rates.

To achieve the goals of true broadband cellular service, the systems have to make the leap to a fourth generation (4G) network.

The goal is to have data rates up to 20 Mbps, even when used in a vehicle, aeroplane etc.

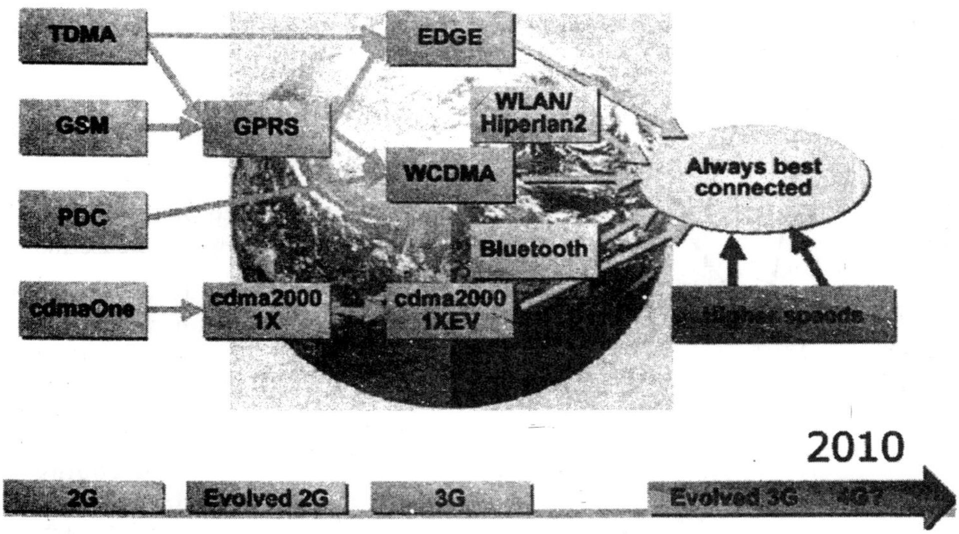

Fig. 1.15. Evolution of 4G.

1.12. PLANNING A CELLULAR SYSTEM

If a network operator is planning to implement a cellular system then following specifications should be followed.

(1) General Description of a GSM PLMN

Network operator should collect data about geographical area and number of subscribers because then only it is possible to decide that how many MSC, BSC, BTS are required and how many frequencies to provide full coverage.

(2) Services

It should be decided that what are services to be offered. There are two types of services:

(*a*) Voice service.

(*b*) Data service {GPRS, EDGE, FAX}

(3) Network Functions

It should be decided that whole geographical area will be covered by own equipment or it will be shared by other operator.

If it is required to share then at what points it will be shared.

(4) MS-BSS Interface

It indicates that what type of interface will be used between mobile station and BTS.

Interface will be always air but at what frequency and what type access like TDMA, FDMA, CDMA will be used. It will decided by services and technology.

(5) Radio Path

Radio path is decided by frequency of uplink and down link and bandwidth of channel.

(6) Speech Processing Function

It indicates that what type of coding technique will be used for speech processing function.

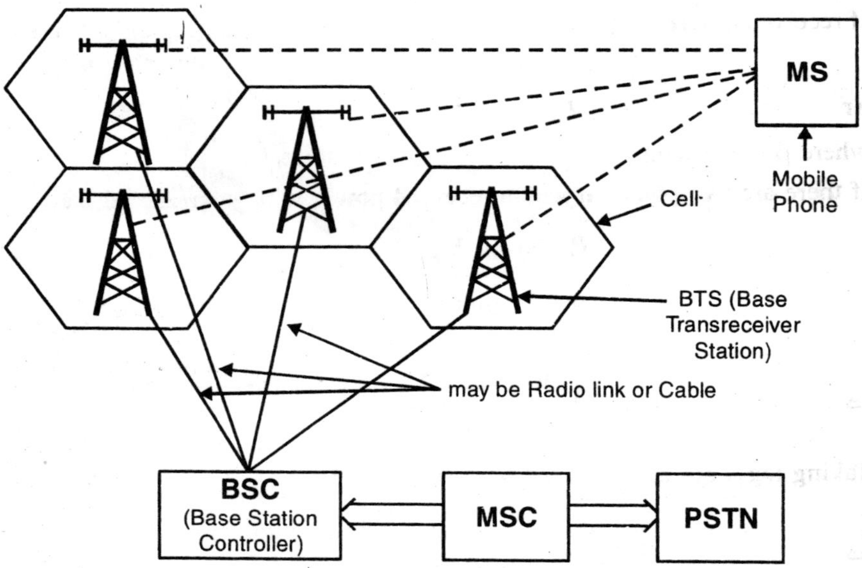

Fig. 1.16. Cellular system.

(7) BSS-MSC Interface

It indicates about BSS-NSS interface that it will be air or physical media.

(8) Network-Interworking

It indicates interworking of network compatibility.

(9) Type Approval Procedure

It is mandatory to get approval from government authority so that services can be offered to subscriber without any problem.

(10) Operation and Maintenance

There should be operation and maintenance node in network so that services can be continued even in faulty situation.

1.13. UNIQUENESS OF MOBILE RADIO ENVIRONMENT

As we know mobile communication uses air as communication channel and maximum error cause in this communication channel.

If communication channel will not be free of noise then signal quality will not be error free and due to that quality and reception of signal will not be proper.

It is impossible to make error free communication because air noise is not possible to control but we can optimize our network in such a way so that effect of noise can minimise.

Propagation Attenuation

Propagation attenuation is very important parameter which attenuates signal strength.

Propagation attenuation is directly proportional to frequency and distance.

If received power is P_r and R is distance between transmitter and receiver then

$$P_r \propto R^{-4} \qquad ...(1.1)$$

or

$$P_r = \beta R^{-4} \qquad ...(1.2)$$

where β is constant.

If there are two points at which received power will be measured then

$$P_{r_1} = \beta R_1^{-4} \qquad ...(1.3)$$

$$P_{r_2} = \beta R_2^{-4} \qquad ...(1.4)$$

$$\Rightarrow \quad \frac{P_{r_1}}{P_{r_2}} = \left(\frac{R_1}{R_2}\right)^{-4} \qquad ...(1.5)$$

Taking $\log_{10}$ of both sides

$$\Rightarrow \quad \log \frac{P_{r_1}}{P_{r_2}} = \log_{10} \left(\frac{R_1}{R_2}\right)^{-4} \qquad ...(1.6)$$

or

$$\log_{10} \frac{P_{r_2}}{P_{r_1}} = \log_{10} \left(\frac{R_2}{R_1}\right)^{-4} \qquad ...(1.7)$$

$$\Rightarrow \quad (P_{r_2} - P_{r_1})\text{dB} = 40 \log \frac{R_1}{R_2} \qquad ...(1.8)$$

or

$$\Delta P_r = 40 \log \frac{R_1}{R_2} \qquad ...(1.9)$$

if $R_2 = 2R_1$ then $\Delta P_r = -12$ dB ...(1.10)

which indicates that as distance increases power decreases.

But this is suitable for mobile radio communication and in general for air propagation

$$P_r \propto (R)^{-r} \quad ...(1.11)$$

and r lies between 2 and 5 for mobile communication depending upon physical situation of antenna height and frequency of operation.

Model of Transmission Medium

A cellular mobile signal is characterized by long term fading and short term fading or Rayleigh fading.

$$s(t) = m(t)\ r(t)$$

where $m(t)$ is long term fading and $r(t)$ is short term fading.

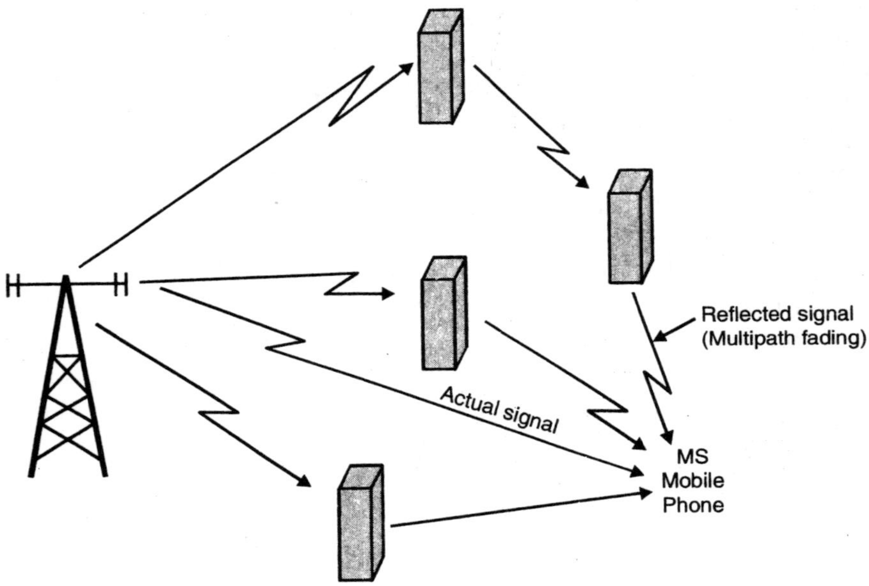

Fig. 1.17. Mode of transmission.

Long term fading $m(t)$ can be obtained from

$$m(t) = \frac{1}{\tau}\int_0^{\tau} s(t)\, dt$$

where t is time interval for averaging $s(t)$ and τ can be obtained on the fading rate of $s(t)$ and fading may cause by signal reflected by surroundings. Generally fading value is 50 to 80 fades.

Rayleigh fading is also called multipath fading in the cellular environment. When wave reflected from surrounding medium then it form standing waves and resultant of standing wave formed irregular wave which becomes noise for actual travelling signal.

SUMMARY

This chapter has provided a brief introduction about the basic concepts of wireless communication, including the limitations of wireless communication. In history of wireless communication, we study that what are the various stages of wireless communication.

After studying the chapter, we can conclude that the wireless communication becomes a ubiquitous means of transport for information in the 21st century. It is only one technology which is increasing by the rate of 100% growth.

In 1990s, the cellphone in India was a symbol status while now-a-days it became necessary part of our life. Cellular communication is only one technology, due to which whole of the world became a small village. No mean of time, place, whenever we want at our simplicity, we can talk to our friends, relatives, due to only wireless technology it become possible to conduct on-line test, interview, video conferencing, etc.

If this growth will continue, then in coming days fixed telephony becomes history, and everyone will have cellphone. With the help of cellphone technology, it became possible to realize a modern electronic home that will be fully automatic and operated by a cellular phone.

In 1G, 2G, 3G, 4G network we have studied that what were the various stages, to develop such a revolutionary technology.

In this century, we cannot think without the cellphone, computer.

These small electronic devices become essential for growth of a nation.

Some abbreviations are as follows.

IrDa (Infrared Data Access)

Bluetooth (Name of Denmark King)

GSM (Global System for Mobile Communications)

CDMA (Code Division Multiple Access

WCDMA (Wide Band Code Division Multiple Access)

Wi-Fi (Wireless Fidelity)

Wi-Bro (Wireless Broadband)

AMPS (Analog Mobile Phone System)

MS (Mobile Station)

BTS (Base Transceiver)

TS (Time Slot)

TDMA (Time Division Multiple Access)

FDMA (Frequency Division Multiple Access)

UMTS (Universal Mobile Telecommunication System)

REVIEW QUESTIONS

1. What are the various stages in Wireless Communication history and when ?
2. What do you mean by term Wireless Communication and how will you describe using Block diagram ?
3. What are the comparison factors between first generation wireless communication network and modern wireless communication ?
4. Describe the various generations of wireless communication system.
5. Give the brief introduction of Bluetooth technology, with appropriate block diagram.

6. Compare the different factors between IrDA and Bluetooth technology.
7. What are the limitations of IrDA over Bluetooth technology ?
8. How IrDA and Bluetooth technology is useful in our practical life, can we implement this technology in Home Automation system ?
9. Between a cellular Mobile Phone, Pager, Cordless Phone, Landline Phone, which battery life is strongest and why, explain ?
10. Why do paging system need to provide low data rate and how does a low data rate provide a larger coverage area ?
11. What do you mean by modulation and why it is required in wireless communication system ?
12. What is the effect of antenna height on cell area coverage ? Explain in detail.
13. What are the propagation effects in following areas :
    (*i*) Over flat terrain (*ii*) Over water
    (*iii*) In an urban area.
14. Explain three propagation mechanisms:
    (*i*) Reflection (*ii*) Diffraction
    (*iii*) Refraction.
15. What are type of a cell site antenna ? Compare them according to geographical area of a cell.

❑❑❑

2

CHAPTER

Elements of Cellular Radio Systems Design and Interference

2.0. INTRODUCTION

There are so many elements in design of cellular system. It may be classified in various category like network element, transmission media, network capacity, subscriber services. But it's very important that we can design system with less of interference. There are various method to design a system to provide subscriber services. Sometime, it is based on location, type of antenna and coverage area. Coverage of antenna and cell area can be improved with cell splitting concept. Cell area is very important because if it will very big then coverage will not be good but if it will be small then it will not be a good optimization of resources.

INSIDE THIS CHAPTER

2.0. Introduction
2.1. Basic Wireless Concepts
2.2. Analog and Digital Transmission
2.3. Access Method: Time Division Multiple Access (TDMA)
2.4. General Description of Transmission Problem
2.5. Cellular System Design Fundamentals
2.6. Concept of Frequency Reuse Channels
2.7. Interference (Co-Channel Interference)
2.8. Cell Splitting
2.9. Antenna
2.10. Path Loss Models (Okumara Model, Hata Model)
2.11. Brewster Angle
2.12. Scattering
2.13. Directivity

2.1. BASIC WIRELESS CONCEPTS

Frequency Concepts

MS communicates with a BTS by transmitting or receiving radio waves, which consists of electromagnetic energy. The frequency of a radio wave is the number of times that the wave oscillates per second. Radio frequencies are used for many applications in the world today. As follows:

Television	300 MHz approx
FM radio	100 MHz approx
Police radios	Country dependent
Mobile network	300-2000 MHz approx

The frequencies used by mobile varies according to the standard being used. In India, an operator applies for the available frequencies to "TRAI", then TRAI allotes particular band to operators at an auction. The following frequencies are used for GSM.

Table 2.1. GSM Standards.7

System	P-GSM 900	E-GSM 900	GSM 1800	GSM 1900
Frequencies Up link	890-915 MHz	880-915 MHz	1710-1785 MHz	1850-1910 MHz
Downlink	935-960 MHz	925-960 MHz	1805-1880 MHz	1930-1990 MHz
Bandwidth	25 MHz	35 MHz	75 MHz	60 MHz
Carrier separation	200 kHz	200 kHz	200 kHz	200 kHz
Radio channels	125	175	375	300
Duplex distance	45 MHz	45 MHz	95 MHz	80 MHz
Transmission rate*	270 kbits/sec	270 kbits/sec	270 kbits/sec	270 kbits/sec

*270 kbps is maximum theoretical speed while practical speed is 9.6 kbps.

In India network operators are using P-GSM 900 and GSM 1800 while GSM 1900 is used by military.

Bandwidth

Bandwidth is term used to describe the amount of frequency range allocated to one application, the bandwidth given to an application depends on the amount of available frequency spectrum. The amount of bandwidth available is an important factor in determining the capacity of a mobile system. The number of calls which can be handled. In 2G, GSM 900, GSM 1800 is having bandwidth of 25 MHz and 75 MHz respectively.

Channels

Another important factor in determining the capacity of a mobile system is the channel. A channel is a frequency or set of frequencies which can be allocated for the transmission, and possibly the receipt of information.

In 2G GSM 900 is having 125 channels and GSM 1800 is divided in 375 channels.

Communication channels may be of following type.

Table 2.2.

Type	Description	Examples
Simplex	One way only	FM radio, T.V.
Half duplex	Two way, only one at a time	Police Radio, Walkee-Talkee
Full Duplex	Two way, both way at same time	Mobile system

Duplex Distance

The use of full duplex requires that the uplink and downlink transmissions must be separated in frequency by a minimum distance. This is the duplex distance. Without it, uplink and downlink frequencies would interfere with each other as shown in Fig. 2.1.

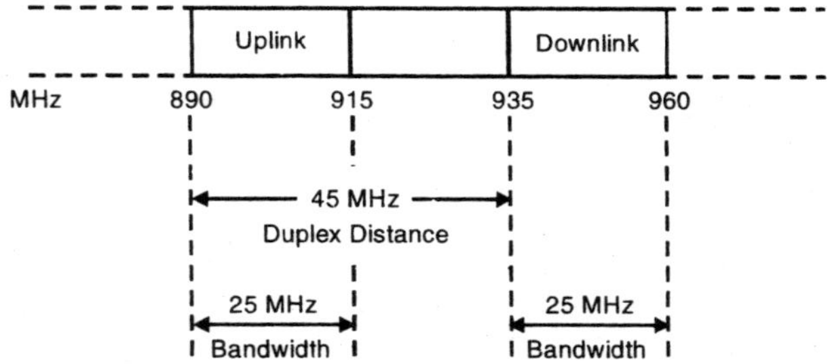

Fig. 2.1. Duplex distance for P-GSM 900.

Carrier Separation

As in Duplex, every mobile system includes a carrier separation (sometimes carrier bandwidth). This is the distance on the frequency band between channels being transmitted in the same direction.

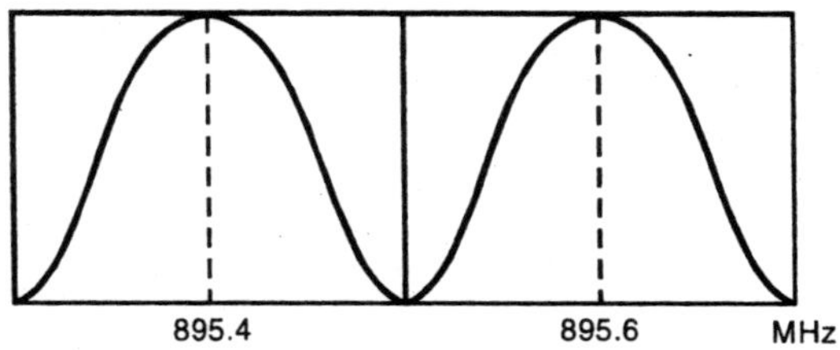

Fig. 2.2. 200 kHz separation.

This is required in order to avoid the overlapping of information in one channel into an adjacent channel.

The length of separation between two channels is dependent on the amount of information which is to be transmitted within the channel.

The greater the amount of information to transmit, the greater the amount of separation required as shown in Fig. 2.2.

From Fig. 2.2, it is clear that the information to be sent is modulated around the carrier frequency of 895.4 MHz for information to be sent is at 895.6 MHz. If less than 200 kHz is used, then interference may take place, and caller on 895.4 MHz may experience crosstalk or noise from the caller on 895.6 MHz. Actually it is centre frequency of GSM 900 which is divided as follows.

GSM 900 is divided in uplink and downlink bandwidth as follows:

Uplink bandwidth (890 MHz–915 MHz)

Downlink bandwidth (935 MHz–960 MHz)

Carrier Pairs are (MHz)

890.0	935.0	124 carriers
890.2	935.2	
890.4	935.4	
890.6	935.6	
890.8	935.8	
⋮	⋮	
915.0	960.0	

890.2 is centre frequency which is used as carrier for transmission of signal in uplink while 935.2 MHz is carrier which is for transmission of signal (voice or data) in downlink direction. It is important to note that duplex distance of 45 MHz must remain same in pair. If uplink is 890.6 MHz then downlink must be 935.6 MHz.

In uplink (mobile to network), we use lower frequency than in downlink (network to mobile) because mobile station has limited power while at network side, we can increase power.

In GSM one carrier frequency is used by 8 persons simultaneously on time sharing basis which is called time slot. And technique is called Time Division Multiple Access.

Capacity and Frequency Reuse

It is the number of frequencies in a cell which determines the cell's capacity. Each company with a license to operate a mobile network is allocated a limited number of frequencies. These are distributed throughout the cells in their network.

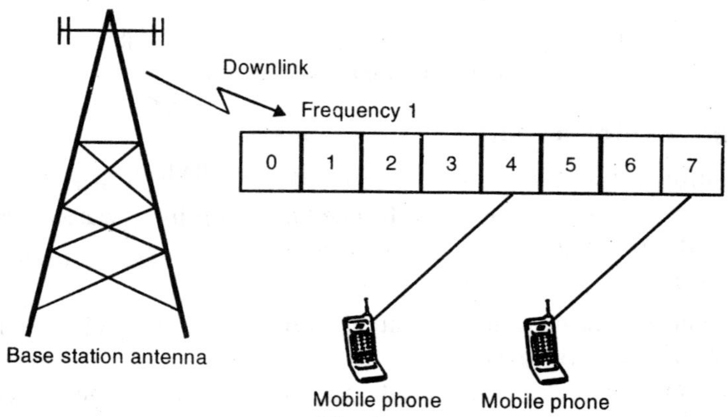

Fig. 2.3.

Depending on the traffic load and the availability of frequencies, a cell may have one or more frequencies allocated to it.

It is important to allocate frequencies such that interference is avoided. Interference can be caused by a variety of factors. A common factor is the use of similar frequencies close to each other. Higher interference lower quality.

To cover an entire country, for example, frequencies must be reused many times at different geographical locations in order to provide a network with sufficient capacity.

The same frequencies cannot be reused in neighbouring cells as they would interfere with each other so special patterns of frequencies usage are determined during the planning of networks.

These frequency reuse patterns ensure that any frequency being re-used is located at a sufficient distance apart.

In GSM, a TDMA frame consists of 8 time slots. It means a radio carrier can carry 8 calls simultaneously.

Here, only the downlink direction is shown. There is also a corresponding frame in the uplink direction.

2.2. ANALOG AND DIGITAL TRANSMISSION

2.2.1. Analog Information

Analog information is continuous and does not stop at discrete values like time. It is continuous and does not stop at specific points.

An Analog signal is continuous waveform which changes in accordance with the properties of the information being represented.

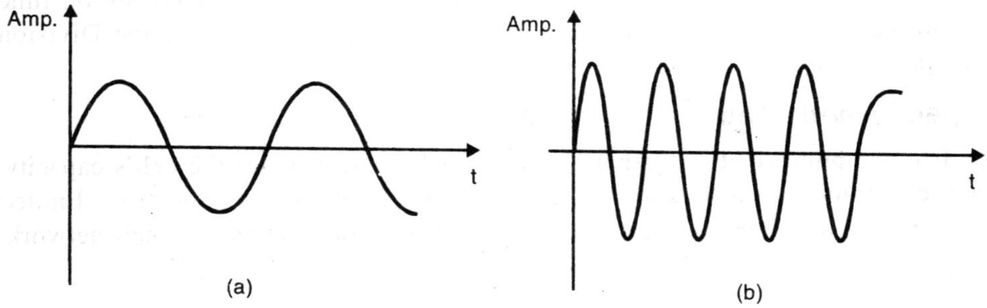

Fig. 2.4. Analog signals.

2.2.2. Digital Information

Digital information is a set of discrete values, time can also be represented digitally. However, digital time would be represented by a watch which jumps from one minute to the next without stopping at the seconds.

Transmission Rate

The amount of information transmitted over a radio channel over a period of time is known as the transmission rate.

Transmission rate is expressed in bits per second. In GSM the bit rate over the interface is 270 kbits/sec.

Modulation Method

In GSM 900, the frequency that is used to transfer the information over the interface is around 900 MHz. Since this is not the frequency at which the information is generated.

Modulation techniques are used to translate the information into the usable frequency band.

The modulation technique used generally in GSM is GMSK (Gaussian Minimum Shift Key). It enables the transmission of 270 kbits/sec within a 200 kHz channel.

2.3. ACCESS METHOD: TIME DIVISION MULTIPLE ACCESS (TDMA)

Most Digital Cellular systems use the technique of TDMA, to transmit and receive speech signals. With TDMA, one carrier is used to carry a number of calls. Each call using that carrier at designated periods in time. These time periods are referred as Time slots.

Each mobile phone on a call is assigned one time slot on the uplink frequency and one on the downlink frequency. The information sent during one time slot is called a **burst**. In GSM, a TDMA frame consists of 8 time slots.

Advantage of Using Digital Technology

Human speech is a form of analog information. It is continuous and changes in both frequency (higher and lower pitches) and amplitude (whispering and shouting).

At first, analog signals may appear to be a better medium for carrying analog information such as speech. Analog information is continuous and if it were to be represented by discrete samples of the information (Digital Signal), then some information would be missing (like the seconds on the digital watch). An analog signal would not miss any values as it is continuous.

All signals, analog and digital, become distorted over distances.

In analog, the only solution to this is to amplify the signal. However, in doing so, the distortion is also amplified. In digital, the signal can be completely regenerated as new without the distortion.

When we convert an Analog signal into Digital, S/N ratio also improves, which is a parameter of better quality.

Fig. 2.5. Transmission and reception of a signal.

In sampling process of a analog signal, if samples are taken often enough, digital signals provide a better quality for transmission of Analog information than Analog signals.

2.4. GENERAL DESCRIPTION OF TRANSMISSION PROBLEM

If a network operator is planning to start wireless service then he must take care of following problems although some problems are out of control like environmental problem, geographical condition.

But if network operator will consider these problems and design network by considering such problems then quality of service can be optimized.

There are many problems in transmissions, some of the most common problems are described as follows:

Path Loss

Path loss occurs when the received signal becomes weaker and weaker due to increasing distance between MS (Mobile Station, Simply Mobile Phone) and BTS (Base Transceiver Station), even if there are no obstacles between the transmitting Tx and receiving Rx antenna. The path loss problem seldom leads to a dropped call because the problem becomes extreme, a new transmission path is established via another BTS.

Shadowing

Shadowing occurs when there are physical obstacles including hills and buildings between the BTS and the MS. The obstacles create a shadowing effect which can decrease the received signal strength. When the MS moves, the signal strength fluctuates depending on the obstacles between the MS and BTS. A signal influenced by fading varies in signal strength. Drops in strength are called fading dips as shown in Fig. 2.6.

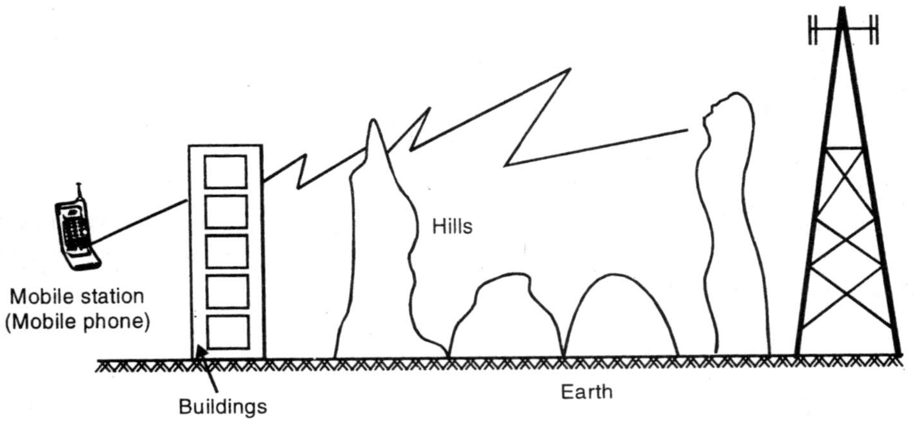

Fig. 2.6. Shadowing.

Multipath Fading

Multipath fading occurs when there is more than one transmission path to the MS or BTS, and therefore, more than one signal arriving at the receiver. This may be due to buildings or mountains, either close to or far from the receiving device.

Time Alignment

Each MS on a call is allocated a time slot on a TDMA frame. This is an amount of time during which the MS transmits information to the BTS. The information must also arrive at the BTS within time slot. The time alignment problem occurs when Part of the information transmitted by an MS does not arrive within the allocated time slot. Instead, that part may arrive during the next time slot, and may interfere with information from another MS using that other time slot. Time alignment is caused by a large distance between the MS and BTS. Effectively, the signal cannot travel over the large distance within the given time.

Example: Let an MS is close to BTS and has been allocated time slot 3 (TS3). During the call, the MS moves away from the BTS causing the information sent from the BTS to arrive at the MS later and later.

The answer from the MS also arrives later at the BTS. If nothing is done, the delay becomes so long that the transmission from the MS in time slot 3 overlaps.

Rayleigh Fading

This occurs when a signal takes more than one path between the MS and BTS Antennas. In this case, the signal is not received on a line of sight path directly from the Transmitter Antenna. Rather it is reflected from buildings, and is received from several different indirect paths. Rayleigh fading occurs when the obstacles are close to the receiving antenna as shown in Fig. 2.7. It is a type of multipath fading.

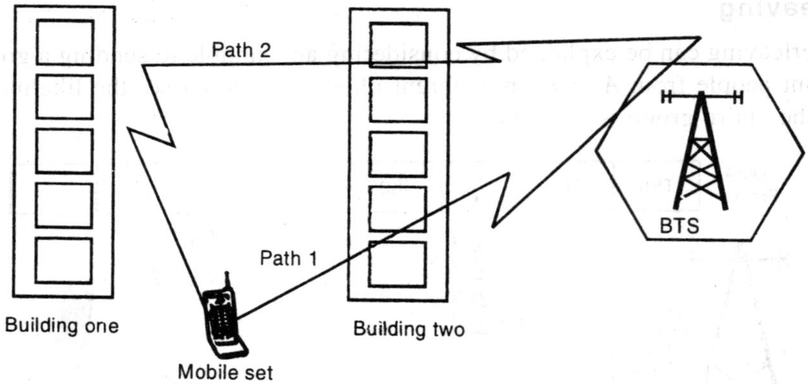

Fig. 2.7. Rayleigh fading.

Time Dispersion

It is another problem that is created due to multipath to Rx antenna of either a MS or BTS. However, in contrast to Rayleigh fading, the reflected signal comes from an object for away from the Rx antenna.

Time dispersion causes consecutive bits to interfere with each other making it difficult for the receiver to determine which symbol is the correct one.

An example of this is shown in Fig. 2.8, where the sequence 1, 0 is sent from the BS.

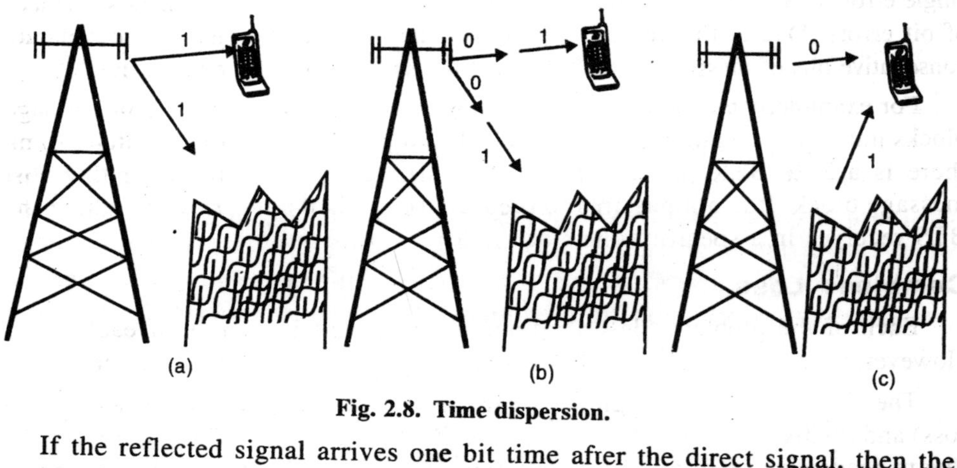

Fig. 2.8. Time dispersion.

If the reflected signal arrives one bit time after the direct signal, then the receiver detects a 1 from the reflected wave at the same time it detects a 0 from the direct wave. The symbol interferes with symbol 0 and the MS does not know which one is correct.

One percentage should be as low as possible. It is not possible to reduce the percentage to zero because the transmission path is constantly changing.

This means that there must be an allowance for a certain amount of errors and at the same time an ability to restore the informations bits are not interpreted as correct. This is especially important during transmission of data, as opposed to speech, for which a higher BER is acceptable. Channel coding is used to detect and correct errors in a received bit stream.

It adds bits to a message. These bits enable a channel decoder to determine whether the message has faulty bits, and to potentially correct the faulty bits.

Interleaving

Interleaving can be explained by considering an example to sending a group of important people from *A* to *B* on different planes. By doing so, the likelihood of losing the entire group is minimised.

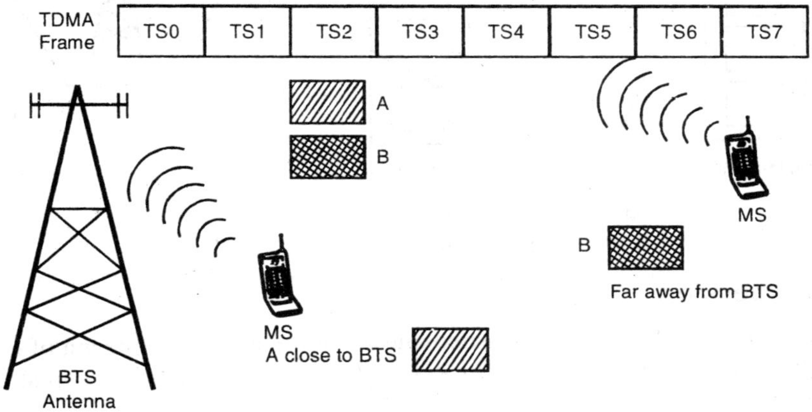

Fig. 2.9.

In reality, bits errors often occur in sequence, as caused by long fading dips affecting several consecutive bits. Channel coding is most effective in detecting and correcting single errors and short error sequences. It is not suitable for handling longer sequences of bit errors. Due to this reason, a process called interleaving is used to separate consecutive bits of message so that these are transmitted in a non-consecutive way.

For example, a message block may consist of four bits (1234). If four message blocks must be transmitted, and one is lost in transmission, without interleaving there is a 25% BER (Bit error rate) overall, but a 100% BER for that lost message block. It is not possible to recover from this the information which the BTS receives in a specific time slot.

Combined Loss

Each of the problems described above occurs independently of each other. However, in most calls some of these problems may occur at the same time.

The signal strength as a global mean value decreases with the distance (path loss) and finally results in a lost connection. Practically, slow variations are present due to shadowing effects and fast variations are present due to rayleigh fading.

2.4.1. Solutions to Transmission Problems

Channel Coding

In digital transmission, the quality of the transmitted signal is often expressed in terms of how many of the received bits are incorrect. This is called Bit Error Rate (BER). BER defines the percentage of the total number of received bits which are incorrectly detected.

Transmitted bits	1	1	0	1	0	0	0	1	1	0	
Received bits	1	0	0	1	0	0	1	0	1	1	
Errors		↑						↑	↑	↑	4/10
											= 40% BER

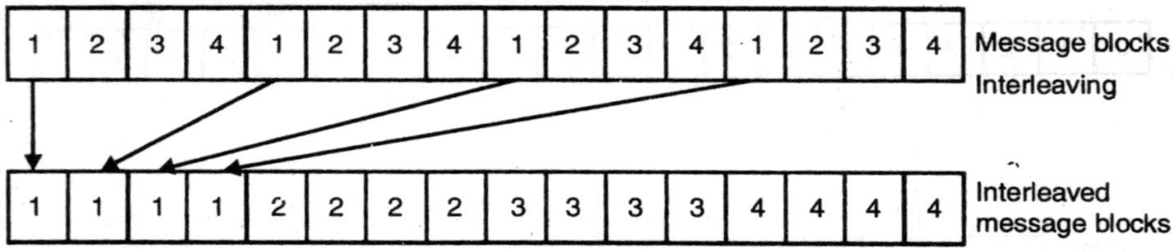

Fig. 2.10.

If interleaving is used, as shown in Fig. 2.10, the bits of each block may be sent in a non-consecutive manner. If one block is lost in transmission, again there is a 25% BER overall.

However, this time the 25% is spread over the entire set of message blocks, giving a 25% BER for each.

This is more manageable and there is a greater possibility that the errors can be corrected by a channel decoder.

1	X	3	4	1	X	3	4	1	X	3	4	1	X	3	4

Fig. 2.11. Received interleaved message blocks.

Time Advance

Timing advance is a solution specifically designed to solve the problem of time alignment. It works by instructing the misaligned Mobile phone to transmit its burst earlier than it normally would.

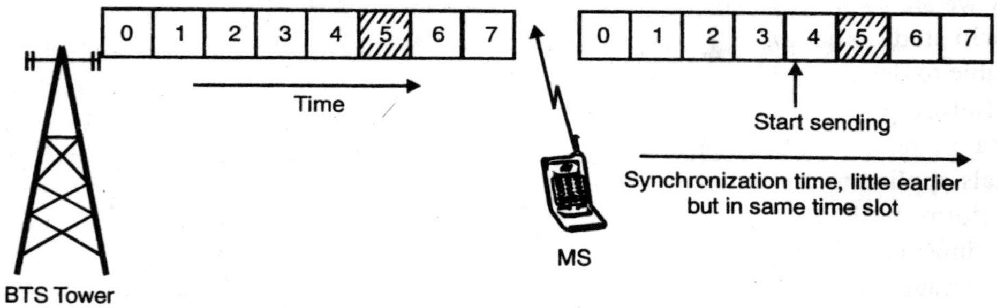

Fig. 2.12. Time advance.

Frequency Hopping

It is possible for the BTS and MS (mobile station) to hop from frequency to frequency during a call. This may avoid some Rayleigh fading.

If you are communicating with your friend and moving also same time but due to any reason your signal strength is going down then network decide to apply frequency hopping so that your conversation can be continue with good quality of signal strength.

Let us take a example, if network allocated to you carrier frequency 890.2 MHz in uplink and 935.2 MHz in downlink and time slot 2 but suddenly your signal strength is become very weak then network switch over your carrier frequency whichever is free with same time slot.

It is important to note in frequency hopping only frequency in uplink and downlink will change not time slot. If same time slot is not free with any carrier frequency then frequency hopping is not possible and finally your call will drop.

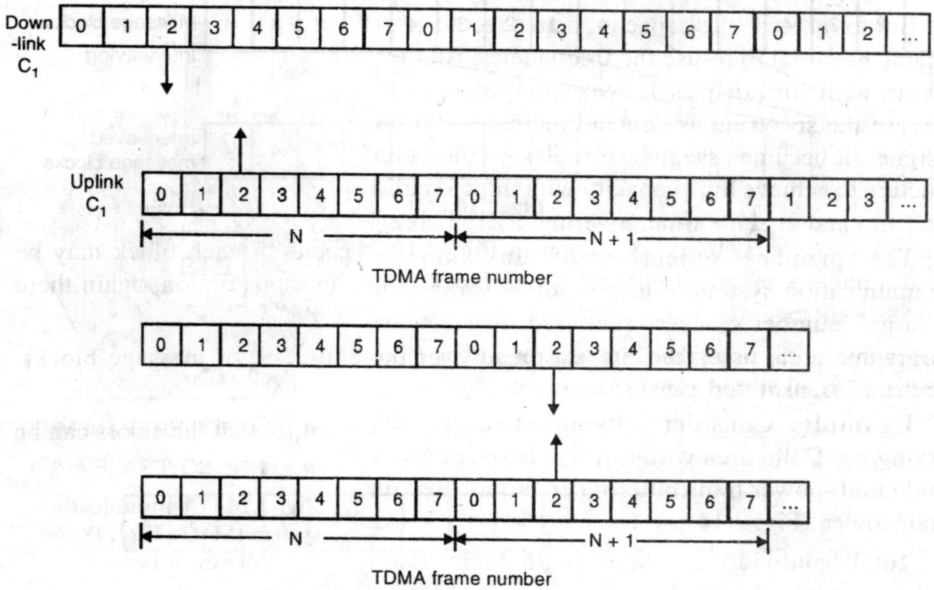

Fig. 2.13. Frequency hoping.

2.5. CELLULAR SYSTEM DESIGN FUNDAMENTALS

In previous section, we have studied the analog and digital cellular systems. Now we are aware with different parameters of both types of system. In this section we will study, what are designing parameters and how we can improve them so we get able to design a system which fullfil all the modern requirements in all respect.

Before going to discuss in detail, one question may arise why we go for a digital system. What are the advantages ? As we know that in an analog system the signals applied to the transmission media are continuous functions of the message waveform. In the analog system, either the amplitude, the phase, or the frequency of a sinusoidal carrier can be continuously varied in accordance with the voice or the message. In digital system, the transmitted signals are discrete in time, Amplitude, phase, or frequency, or in a combination of any two of these parameters. To convert from analog form to digital form, the quantizing noise due to discrete levels should be controlled by assigning a sufficient number of digits for each sample, and a sufficient number of samples is needed to apply the Nyquist rate for sampling an analog waveform.

One advantage of converting message signals into digital form is the ruggedness of the digital signal. The impairments introduced in the medium in spite of noise and interference can always be corrected. This process, called regeneration, provides the primary advantage of digital transmission.

After converting an analog system to digital system, the 2nd aim of designer is to cover a large area by using a single, high power transmitter with an antenna mounted on a tall tower.

This approach was for providing a good coverage area but reception was not good and in this method frequency reuse was not possible and any attempts to reuse the frequency result in interference.

But to increase the capacity, better reception it became essential to re-use the frequencies. And for government authorities it was not possible to increase the spectrum as demand increases. So far designer, it became essential to re-design the radio structure to achieve high capacity with limited radio spectrum and at same time covering a large area.

The primary objective of any mobile communication system is to provide coverage for a "large" number of users scattered over a wide geographic area, using the limited resources (like spectrum, transmitted power) available.

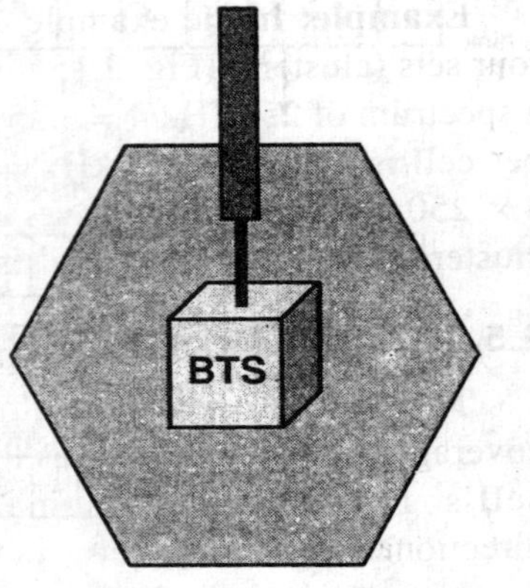

Fig. 2.14. Communication system (Single High Power Transmitter).

Example: Consider a hypothetical system serving the Delhi area with a single base station, a single high-power transmitter/receiver mounted on a tall tower (Fig. 2.14).

Total bandwidth available is 25 MHz. Each active user in the system requires 25 kHz of transmission bandwidth for voice communication. Since one antenna is used to cover the entire town, 25 MHz/25 kHz = 1000 simultaneous users can be supported. System capacity can only be increased by increasing the total bandwidth of the system. Any attempt to reuse the same frequency channels throughout the system would result in interference.

2.5.1. Basic Cellular Concepts

A single, high-power transmitter (Base Station) with many low-power transmitters, each providing coverage to only a small portion of the service area or cell. To each cell (Base Transceiver Station), *i*, a subset of frequencies f_i is assigned from the total bandwidth assigned to the respective mobile network. Cells are grouped into clusters. A cluster contains all the frequencies of the mobile cellular system. Size of a cluster is characterized by the number of cells per cluster *N*. Cells in different clusters that use the same frequencies are known as co-channel cells. Transmissions in co-channel cells will interfere with each other. This is known as co-channel interference (CCI). To reduce co-channel interference, co-channel cells must be physically separated by a minimum distance. Frequency reuse distance is the minimum distance between co-channel cells required to keep co-channel interference below a certain threshold.

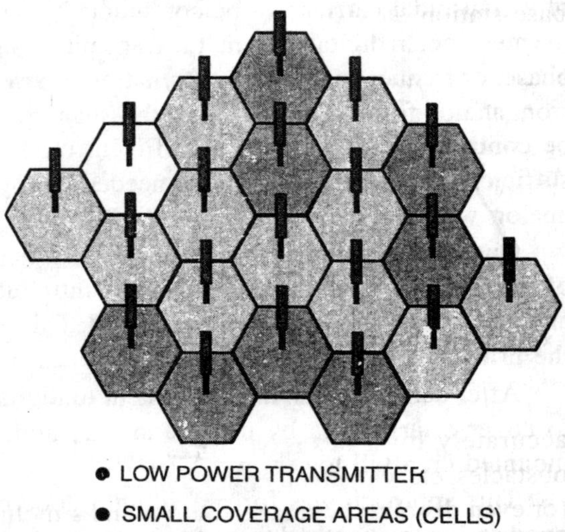

Fig. 2.15. Cellular system with many low power transmitters.

Example: In the example above, now divide total bandwidth of 25 MHz into four sets (clusters) (Fig. 2.15) and assign one set to each base station. Each cell has a spectrum of 25 MHz/4 = 6.25 MHz allocated to it. Number of simultaneous users per cell is 6.25 MHz/25 kHz = 250. Number of simultaneous users per cluster is 4 × 250 = 1000. If the coverage area of the entire system is divided into four clusters, the total number of simultaneous users is 1000 × 4 = 4000.

2.5.1.1. Cell Geometry

The actual cellular radio coverage of a cell is known as the cell's footprint. With omni-directional antennas, the coverage contour is "circular". We can thus divide the coverage area into circular cells, each of radius *R*. For simplicity, consider the radius *R* as the maximum separation that is allowed between a mobile and its base (say from the SNR point of view). In actual system design and planning, *R* depends on other issues such as infrastructure cost per unit area, frequency of handoff, capacity per unit area etc. Unfortunately, circular cells create void areas without coverage. Refer to Fig. 2.16.

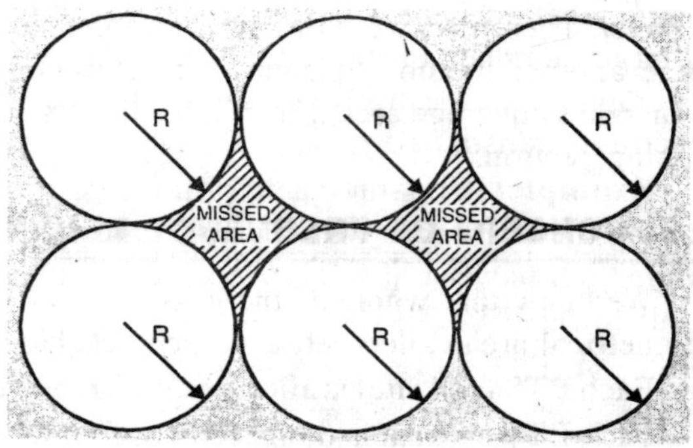

Fig. 2.16. Circular cell.

Uniform area polygons can be used to "tesselate" the plane without overlap or missed areas. The hexagon is usually chosen as the cell shape because it maximizes area coverage, as shown in Fig. 2.17.

If the received signal strength is the same monotonically decreasing function of distance irrespective of location, then a mobile will always be paired up with the base-station closest to it. In this case, the cell shape can indeed be represented

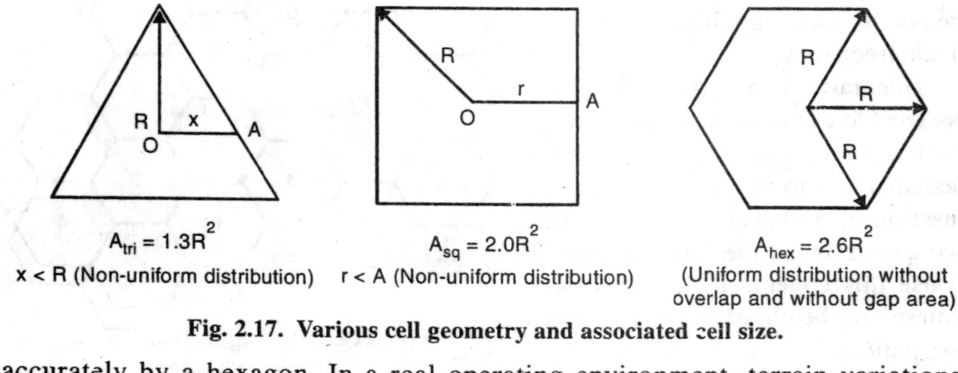

Fig. 2.17. Various cell geometry and associated cell size.

accurately by a hexagon. In a real operating environment, terrain variations, obstacles, etc. can greatly alter the received signal strength. This leads to an irregular (or even disjoint) coverage contour and a mobile will not always be paired with the closest base station. In this case, the concept of cells becomes a bit fuzzy. In other words, a hexagonal cell is just a deterministic model for a probabilistic world. It does, however, help the system engineers to know roughly where the radio resources should be deployed. Note that hexogonal geometary of a cell is hypothetical, practically it is impossible to construct hence practical area is of irregular shape as shown in Fig. 2.18.

It is understood from Fig. 2.18 that some area is overlaping and some area is not possible to cover. Therefore, we will have to implement some another method to control interference, to increase coverage area that will be discuss in coming section.

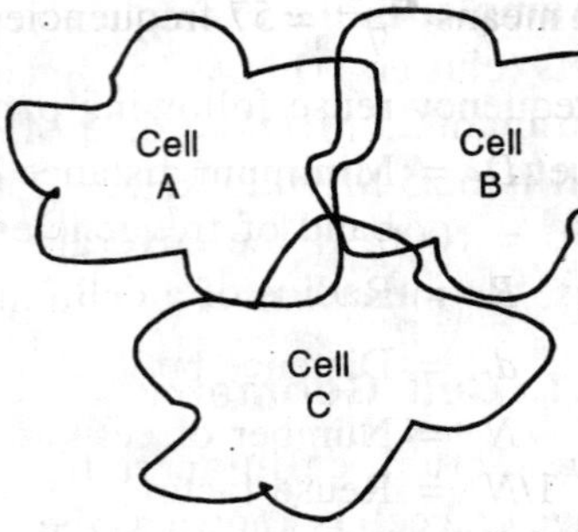

Fig. 2.18. Cell geometry in practical.

2.6. CONCEPT OF FREQUENCY REUSE CHANNELS

We know that whole of the geographical area is divided into hexagonal shape geometrical area called cell and each cell having its own transceiver.

Each BTS (cell site) is allocated different band of frequency or different channel.

Each BTS antenna is designed in such a way that it covers cell area in which it is placed with frequency allotted without interferring other cell signals. By limiting the coverage area to within the boundaries of a cell, the same group of channel or same frequency may be asigned to other cell that are separated from one another by large distances enough to keep interference levels within tolerable limits.

The design process of selecting and allocating channel groups for all of the cellular base station within a system is called frequency reuse.

In Fig. 2.19 the cell having same number shows the same group of channel or same frequency band. When we use the concept of frequency reuse, the transmission power must be carefully controlled so that it allows the communication within cell using a given frequency.

Generally 3 to 4 frequencies are assigned to each cell depending on the traffic expected. When engineers design or try to use this concept one question arises that after how many cell two cells asigned the same frequency, to use this concept different type of pattern may be possible. Let us discuss one pattern.

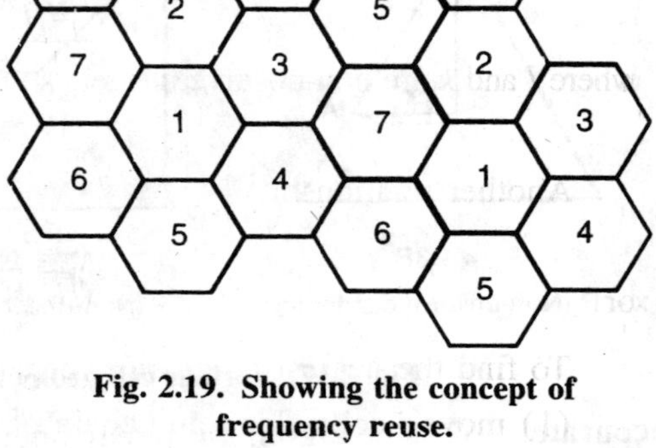

Fig. 2.19. Showing the concept of frequency reuse.

Let N cell asigned having the same frequencies. Each cell can have M/N frequencies, where M is the total number of frequencies alloted to the system.

For AMPS, $M = 395$ and $N = 7$ is the smallest pattern that can provide sufficient isolation between two users of the same frequencies.

It means $\frac{395}{7} \simeq 57$ frequencies is possible, an average per cell to characterizing the frequency reuse following parameter is used.

Let D = Minimum distance between centres of cells that use the same band of frequencies [co-channels].

R = Radius of a cell.

d = Distance between centres of a cell $(d = \sqrt{3}\, R)$.

N = Number of cells in a repetition pattern.

$1/N$ = Reuse factor.

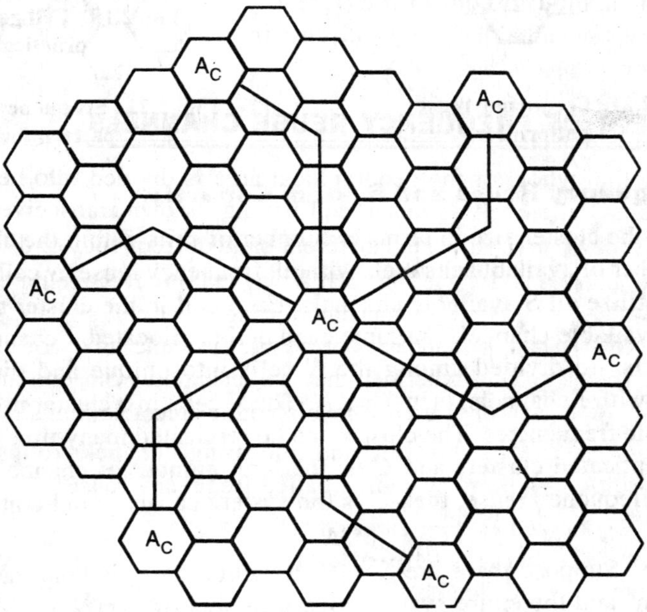

Fig. 2.20. Method of locating co-channel cells in a cellular system. In this example, $N = 19$ (*i.e.* $i = 3, j = 2$).

In a hexagonal pattern only following values are possible for N.

$$\boxed{N = i^2 + j^2 + ij}$$

where l and k are non-positive integers so possible values of N for different i and j.

$$N = 1, 3, 4, 7, 9, 12, 13, 16, 19 \text{ etc.}$$

Another relationship $\left[\frac{D}{R} = \sqrt{3N}\right]$

or $\boxed{\frac{D}{d} = \sqrt{N}}$

To find the nearest co-channel of a particular cell, do the following steps.

(1) move i cells along any chain of hexagons and then turn.

(2) 60 degree anticlockwise and move j cells.

Following Figs. 2.21 and 2.22 illustrate for $N = 4$, $N = 7$.

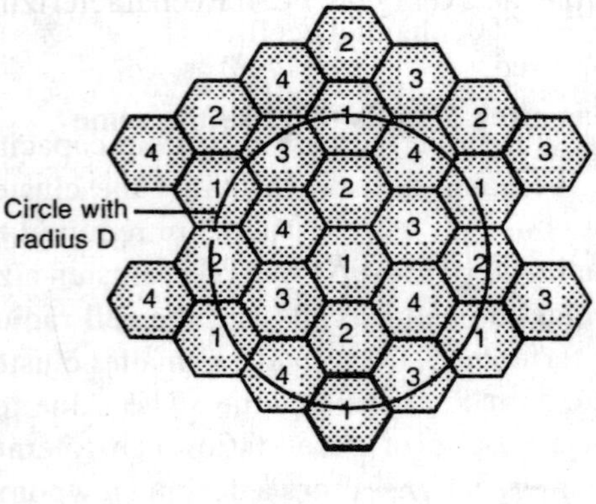

Fig. 2.21. Frequency reuse pattern for $N = 4$.

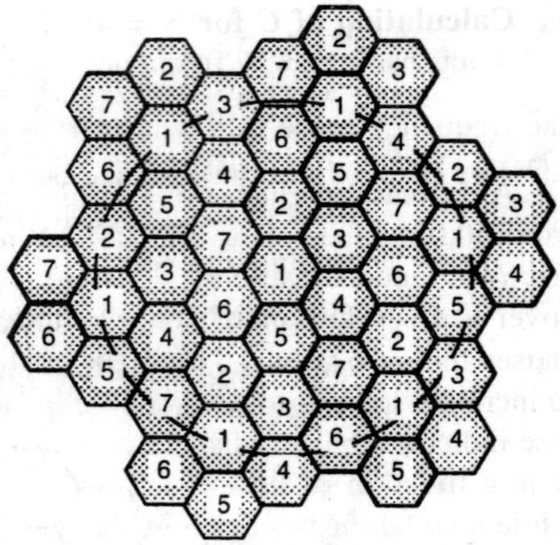

Fig. 2.22. Frequency reuse pattern for $N = 7$.

2.6.1. Frequency Reuse and System Capacity

Let N be the cluster size in terms of number of cells within the cluster and S be the total number of available channels without frequency reuse. N cells in the cluster would then utilize all S available channels. Each cell in the cluster then uses $1/N^{th}$ of the total available channels. Assume each cell is allocated J channels ($J < S$). If the S channels are divided among the N cells into unique and disjoint channel groups, each with J channels, then $S = J N$. The N cells in a cluster use the complete set of available frequencies. The cluster can be replicated many times. Let M be the number of replicated clusters and C be the total number of channels in the entire system with frequency reuse, then C is the system capacity and computed by

$$C = MJN \qquad \text{...(2.1)}$$

Example: Suppose there are $S = 1001$ radio channels, and each cell has an area of 12 km^2 and the entire system covers an area of 4200 km^2.

1. Calculate the system capacity if the cluster size is 7.
2. How many times would the cluster of size 4 have to be replicated in order to approximately cover the entire cellular area ?
3. Calculate the system capacity if the cluster size is 4.
4. Does decreasing the cluster size increase the system capacity ?

Calculation of C for $N = 7$:

Coverage area of a cluster = $N * 7 = 84$ km^2

$$J = \frac{S}{N} = 143 \text{ channels/cell.}$$

The number of times that the cluster must be replicated to cover the entire cellular system is

$$M = 4200/84 = 50;$$

$$C = MJN = 50{,}050 \text{ channels.}$$

Calculation of M for $N = 4$:

$N = 4$, Coverage area of cluster = $N * 4 = 48$ km^2,

$$M = \frac{4200}{48} = 87.5 = 87$$

Calculation of C for $N = 4$:

$$J = \frac{S}{N} = \frac{1001}{4} = 250 \text{ channels/cell.}$$

$$C = 87 * 250 * 4 = 87{,}000 \text{ channels.}$$

A decrease in cluster size from 7 to 4 results in an increase in system capacity from 50,050 channels to 87,000 channels. We have just shown that if the cluster size N is reduced while the cell size is kept constant, more clusters are required to cover a given area, and hence more capacity is achieved. A larger cluster size causes the ratio between the distance between co-channel cells and the cell radius to increase, leading to weaker co-channel interference. Conversely a smaller cluster size indicates that co-channel cells are located much closer together. The value for N is a function of how much interference a mobile or base station can tolerate while maintaining a sufficient quality of communications. From a design viewpoint, the smallest possible value of N is desirable in order to maximize capacity over a coverage area (*i.e.* to maximize C). The frequency reuse factor of a cellular system is given by $1/N$, since each cell within a cluster is only assigned $1/N$ of the total available channels in the system.

2.6.2. Frequency Reuse Distance (*D*)

Distance D between the nearest co-channel cells of adjacent cluster is given by

$$D^2 = 3R^2 (i^2 + ij + j^2) \qquad ...(2.2)$$

where i and j are non-negative integers and are the displacements in number of cells from the reference cell. R, cell radius, is the largest possible distance between the center of a cell (base station) and its edge. The number of cells per cluster, N, is

$$N = (i^2 + ij + j^2) \qquad ...(2.3)$$

For hexagonal cells reuse distance is given by,

$$D^2 = 3R^2 (N)$$

$$D = \sqrt{3N} \,.\, R \qquad ...(2.4)$$

2.6.3. System Capacity

System capacity is a parameter in wireless communication that is directly related to interference. If interference caused by any source is less, system capacity is high.

As we have seen, if the size of each cell is approximately the same and the BTS transmit the same power, the co-channel interference ratio is independent of the transmitted power and become function of the cell radius R and distance between centers of the nearest co-channel cells (D).

$$D = \sqrt{3N}\, R \quad \Rightarrow \quad Q = \frac{D}{R} = \sqrt{3N} \qquad ...(2.5)$$

A small value of Q provide, larger capacity, since cluster size N is small large value of Q improves transmission quality.

Note: To find the nearest co-channels neighbours of a particular cell, do the following steps. (1) Move i cells along any chain of hexagons and then (2) turn 60 degrees counterclockwise and move j cells. [See Fig. 2.19]

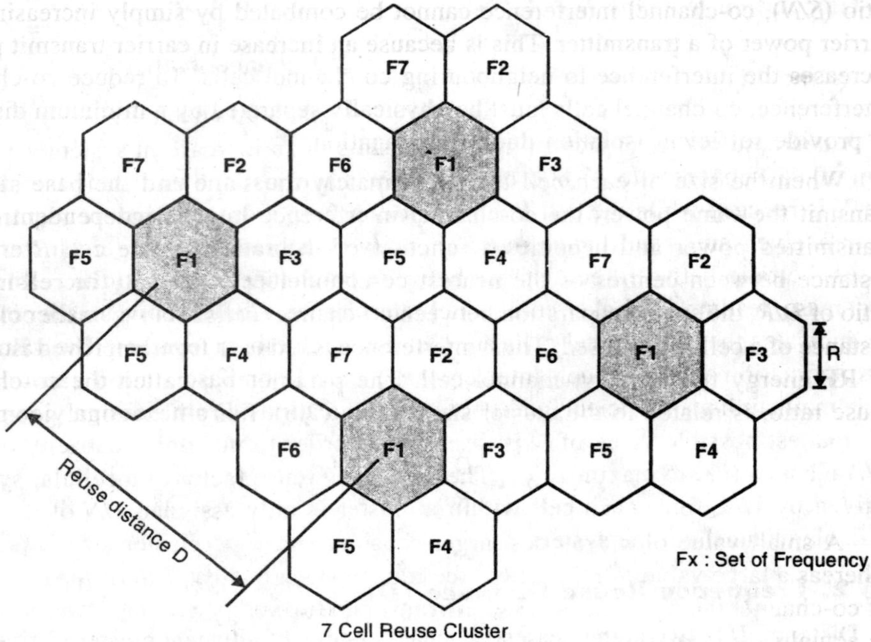

Fig. 2.23. Reuse frequency.

2.7. INTERFERENCE (CO-CHANNEL INTERFERENCE)

Major limiting factor for cellular system performance is the interference. Sources of interference are: another mobile in the same cell, a call in progress in neighbouring cell, and other base stations operating in the same frequency band or any non-cellular systems leaking energy into cellular frequency band Implications of interference are:

(*a*) Interference on voice channels causes cross talk where the subscriber hears interference in the background due to an undesired transmission.

(*b*) On control channels, interference leads to missed and block calls due to error in digital signalling.

Interference is more severe in urban areas, due to greater RF noise floor and the large number of base stations and mobiles. Interference has been recognized as a major bottleneck in increasing capacity and is often responsible for dropped calls.

Two major types of system generated cellular interference are:

(*i*) Co-channel interference.

(*ii*) Adjacent channel interference.

2.7.1. Co-channel Interference (CCI)

Frequency reuse implies that in a given coverage area, there are several cells that use the same set of frequencies. These cells are called co-channel cells, and the interference between signals from these cells is called co-channel interference. Unlike thermal noise which can be overcome by increasing the signal-to-noise ratio (*S/N*), co-channel interference cannot be combated by simply increasing the carrier power of a transmitter. This is because an increase in carrier transmit power increases the interference to neighbouring co-channel cells. To reduce co-channel interference, co-channel cells must be physically separated by a minimum distance to provide sufficient isolation due to propagation.

When the size of each cell is approximately the same and the base stations transmit the same power, the co-channel interference ratio is independent of the transmitted power and becomes a function of the radius of the cell *R* and the distance between centres of the nearest co-channel cells (*D*). By increasing the ratio of *D/R*, the spatial separation between co-channel cells relative to the coverage distance of a cell is increased. Thus, interference is reduced from improved isolation of RF energy from the co-channel cell. The parameter *q* called the co-channel reuse ratio, is related to the cluster size by Eq. (2.6). For a hexagonal geometry

$$q = \frac{D}{R} = \sqrt{3N} \qquad \text{...(2.6)}$$

A small value of *q* provides larger capacity since the cluster size *N* is small, whereas a large value of *q* improves the transmission quality, due to a smaller level of co-channel interference. A trade-off must be made between these two objectives in actual cellular design.

Let i_0 be the number of co-channel interfering cells. Then, the signal-to-interference ratio (*S/I*) for a mobile receiver, which monitors a forward channel, can be expressed by Eq. (2.7)

$$\frac{S}{I} = \frac{S}{\sum_{i=1}^{i_0} I_i} \qquad \text{...(2.7)}$$

where *S* is the desired signal power from the desired base station and I_i is the interference power caused by the *i*th interfering co-channel cell base station. If the signal levels of co-channel cells are known, then the *S/I* ratio for the forward link can be found using Eq. (2.7).

Propagation measurements in a mobile radio channel show that the average received signal strength at any point decays as a power law of the distance of separation between a transmitter and receiver. The average received power P_r at a distance *d* from the transmitting antenna is approximated by

$$P_r = P_0 \left(\frac{d}{d_0}\right)^{-\alpha} \qquad \text{...(2.8)}$$

or

$$P_r(dBm) = P_0(dBm) - 10\,\alpha \log 10\left(\frac{d}{d_0}\right) \qquad \text{...(2.9)}$$

where P_0 is the power received at a close-in reference point in the far field region of the antenna at a small distance d_0 from the transmitting antenna and α is the propagation path loss. The path loss factor depends on the terrain (Table 2.3).

Table 2.3.

Environment	Propagation loss exponent α
Free space	2
Urban cellular radio	2.7 to 3.5
Shadowed urban cellular radio	3 to 5
In building with line of sight	1.6 to 1.8
Obstructed in building	4 to 6

Now consider the forward link where the desired signal is the serving base station and where the interference is due to co-channel base stations. If D_i is the distance of the ith interference from the mobile, the received power at a given mobile due to the ith interfering cell will be proportional to $(D_i)^{-\alpha}$. The path loss exponent (α) typically ranges between two and four in urban cellular systems.

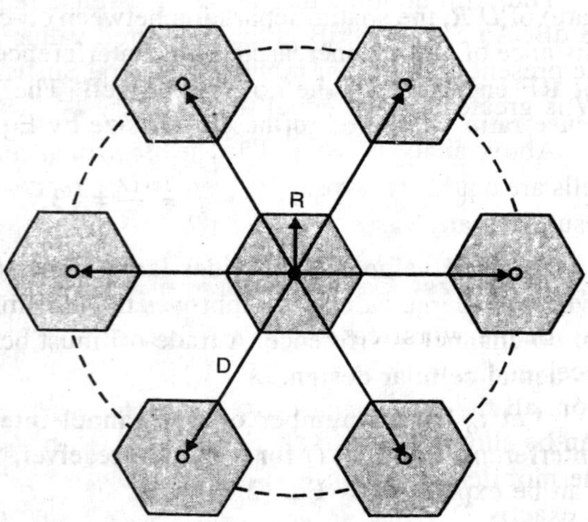

Fig. 2.24. Ideal case geometry.

When the transmit power of each base station is equal and the path loss exponent is the same throughout the coverage area, S/I (Eq. 2.7) for a mobile can then be approximated as

$$\frac{S}{I} = \frac{R^{-\alpha}}{\sum_{i=1}^{i_0} D_i^{-\alpha}} \quad \text{...(2.10)}$$

If all the interfering base stations are equidistant from the desired base station if this distance is equal to the distance D between cell centers, the Eq. (2.10) reduces to

$$\frac{S}{I} = \frac{R^{-\alpha}}{\sum_{i=1}^{i_0} D_i^{-\alpha}} = \frac{R^{-\alpha}}{i_0\, D^{-\alpha}} = \frac{(D/R)^{\alpha}}{i_0} = \frac{q^{\alpha}}{i_0} = \frac{\left(\sqrt{3N}\right)^{\alpha}}{i_0} \quad \text{...(2.11)}$$

Equation (2.11) relates S/I to the cluster size N, which in turn determines the overall capacity of the system from equation ($C = MJN$).

For example, assume that the six closest cells ($N = 6$) are close enough to create significant interference and that they are all approximately equidistant from the desired base station. For an AMPS cellular system which uses FM (Frequency modulation) and 30 kHz channels, subjective tests indicate that sufficient voice quality is provided when S/I is greater than or equal to 18 dB. But for GSM, we can get good voice quality if S/I is greater than –73 dB.

$$q = \left[i_0 \left(\frac{S}{I} \right) \right]^{1/\alpha} \qquad ...(2.12)$$

For path loss $\alpha = 4$,

$$q = (6 \times 63.1)^{1/4} \approx 4.41.$$

Thus, the cluster size N should be

$$N = q^{2/3} = 6.49 \approx 7.$$

Thus a minimum cluster size of seven is required to meet as S/I requirement of 18 dB. An S/I of 18 dB is the measured value for the accepted voice quality from the present day cellular mobile receivers. Sufficient voice quality is provided when S/I is greater than or equal to 18 dB.

Above analysis is based on the hexagonal cell geometry where all the interfering cells are equidistant from the base station receiver, and hence provides an optimistic result in many cases.

2.7.2. Worst Case Co-Channel Interference

In the worst case mobile terminal is located at the cell boundary where it receives the weakest signal from its own cell but is subjected to strong interference from all the interfering cells. Using an exact cell geometry layout, Fig. 2.24, it can be shown for a seven-cell cluster, with the mobile unit at the cell boundary, the mobile is a distance D-R from the two nearest co-channel interfering cells and is exactly $D + R/2$, D, $D - R/2$, and $D + R$ from the other interfering cells in the first tier. The S/I for the worst case can be approximated as

$$\frac{S}{I} = \frac{R^{-\alpha}}{\sum_{i=1}^{i_0} D_i^{-\alpha}} = \frac{R^{-\alpha}}{2(D-R)^{-\alpha} + 2D^{-\alpha} + 2(D+R)^{-\alpha}} \qquad ...(2.13)$$

In terms of co-channel reuse ratio q above equation can be written as

$$\frac{S}{I} = \frac{1}{2(q-1)^{-\alpha} + 2q^{-\alpha} + 2(q+1)^{-\alpha}} \qquad ...(2.14)$$

For $N = 7$, for $\alpha = 4$, the co-channel reuse ratio q (Eq. (2.6)) is 1.6 and the worst case S/I using Eq. (2.14) becomes 49.56 (10 log (49.56) = 17 dB) whereas the exact solution using Eq. (2.10) yields 17.8 dB. Hence for a seven-cell cluster, the S/I ratio is slightly less than 18 dB for the worst case. To design the cellular system for proper performance in the worst case, it would be necessary to increase N to the next largest size, which from Equation (2.3) is found to be 9 (corresponding to $i = 0$, $j = 3$). This obviously entails a significant decrease in capacity, since 9-cell reuse offers a spectrum utilization of 1/9 within each cell, whereas seven-cell reuse offers a spectrum utilization of 1/7. In practice, a capacity reduction of 7/9 would not be tolerable to accommodate for the worst case situation which rarely occurs. From the above discussion, it is clear that co-channel interference determines link performance, which in turn dictates the frequency reuse plan and the overall capacity of cellular systems.

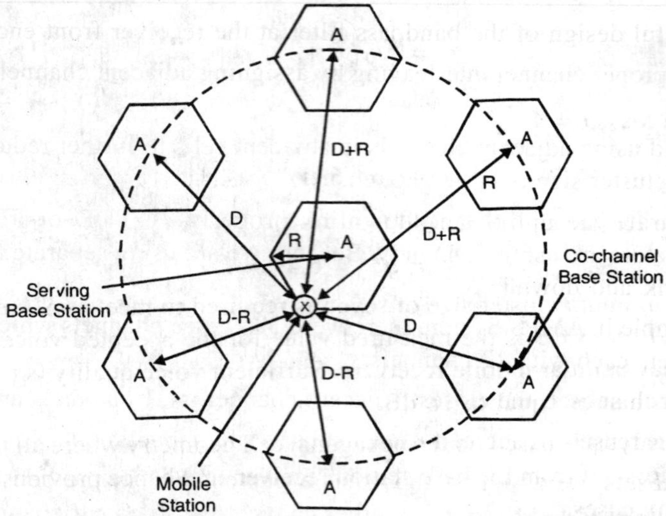

Fig. 2.25. Worst Case Scenarios (where mobile is at the cell boundary).

Adjacent channel interference: Energy from other channels splits over into the carrier's channel.

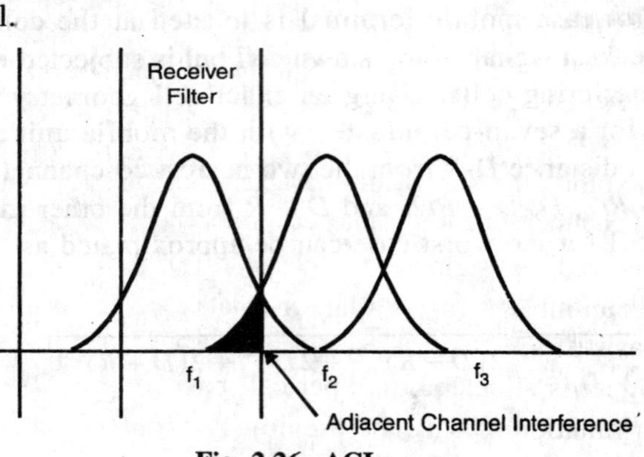

Fig. 2.26. ACI.

2.7.3. Adjacent Channel Interference (ACI)

ACI is caused by the inability of a mobile phone to filter out the signals (frequencies) of adjacent channels assigned to side-by-side cells sites. ACI occurs more frequently in small cell clusters and heavily used cells.

Near Far Effect

Consider the uplink transmissions from two mobile users using adjacent channels, one very close to the base station and the other very close to the cell boundary. Without proper transmission power control, the received power from the mobile close to the base station is much larger than that from the other mobile far away.

This near far effect can increase the ACI from the strong received signal to the weak received signal.

How to Reduce ACI ?

➤ Use modulation schemes which have low out-of-band radiation (*e.g.*, MSK is better than QPSK and OMSK is better than MSK).

- Careful design of the bandpass filter at the receiver front end.
- Use proper channel interleaving by assigning adjacent channels to different cells.
- Avoid using adjacent channels in adjacent cells to further reduce ACI if the cell cluster size is large enough.
- Separate the uplink and downlink properly by TDD or FDD. In India operators are using FDD only. Because we are using separate frequency for uplink and downlink.

For example in AMPS System there were 395 voice channels which are divided into 21 subsets each with 19 channels.

- In each subset, the closest adjacent channel is 21 channels away.
- 7-cell reuse -> each cell uses 3 subsets of channels.
- 3 subsets are assigned such that every channel in the cell is assured of being separated from every other channel by at least 7 channel spacing.

2.7.4. Cell Capacity—How Many Channels Per Cell

Cellular radio systems rely on *trunking* to accommodate a large number of users with a limited number of channels. In a trunked radio system, each user is allocated a channel on a per call basis, and upon termination of the call, the previously occupied channel is immediately returned to the pool of available channels.

Trunking exploits the statistical characteristics and behaviour of users so that a fixed number of channels or circuits may accommodate a large, random user community.

In a trunked radio system

- A large number of users share a small pool of channels in a cell on a per call basis.
- A channel is allocated on a per call basis.
- On termination of call, the previously occupied channel is returned to the pool of available channels.
- When all channels are in use, access by a new user is blocked.

The fundamentals of trunking theory were developed by Erlang, a Danish mathematician who, in the late 19th century, embarked on the study of how a large population could be accommodated by a limited number of servers. Today, the measure of traffic intensity bears his name. Traffic intensity (named after a Danish mathematician) is the average number of calls simultaneously in progress over a certain time. It is a dimensionless unit.

- one hour of continuous use of one channel = 1 Erlang
- 1 Erlang = 1 hour (60 minutes) of traffic

For example, if a group of user made 30 calls in one hour, and each call had an average call duration of 5 minutes, then the number of Erlangs this represents is worked out as follows:

Minutes of traffic in the hour = number of calls × duration

Minutes of traffic in the hour = 30 × 5 = 150

Hours of traffic in the hour $= \frac{150}{60}$

Hours of traffic in the hour = 2.5

Traffic Intensity = 2.5 Erlangs

How to compare the quality of services provided by different service providers ? What is the probability of not being able to make a call ? What is the probability of waiting before a call is connected ? All these can be explained by the Grade (Quality) of Service (GOS). GOS is a measure of the call blocking or the ability to make call during the busiest time. It is typically given as the likelihood that a call is blocked or the likelihood of a call experiencing a delay greater than a certain queuing time. GOS is determined by the available number of channels and used to estimate the total number of users that a wireless network can support. For example, if GOS = 0.05, one call in 20 will be blocked during the busiest hour because of insufficient capacity. It is the wireless designer's job to estimate the maximum required capacity and to allocate the proper number of channels in order to meet the GOS.

In general GOS is measured by looking at traffic carried and traffic offered and calculating the traffic blocked and lost. The proportion of lost calls is the measure of GOS. For cellular circuit groups an acceptable GOS = 0.02. This means that two users of the circuit group out of a hundred will encounter a call refusal during the busy hour at the end of the planning period. GOS is calculated from the Erlang-B formula, as a function of the number of channels required for the offered traffic intensity.

A number of definitions listed in Table below are used in trunking theory to make capacity estimates in trunked systems.

Definitions of Common Terms Used in Trunking Theory

Setup Time: The time required to allocate a trunked radio channel to a requesting user.

Blocked Call: Call which cannot be completed at time of request due to congestion. Also referred to as a lost call.

Holding Time: Average duration of a typical call. Denoted by H (in seconds).

Traffic Intensity: Measure of channel time utilisation, which is the average channel occupancy measured in Erlangs. This is a dimensionless quantity and may be used to measure the time utilization of single or multiple channels. Denoted by A.

Load: Traffic intensity across the entire trunked radio system, measured in Erlangs.

Grade of Service (GOS): A measure of congestion which is specified as the probability of a cell being blocked (for Erlang B), or the probability of a call being delayed beyond a certain amount of time (for Erlang C).

Request Rate: The average number of call requests per unit time. Denoted by λ seconds^{-1}.

The traffic intensity offered by each user is:

$$A = \lambda H \text{ Erlangs} \quad ...(2.15)$$

where H is the average holding time of a call, λ is the average number of call requested/hour. If there are U users and an unspecified number of channels, The total offered traffic intensity is:

$$A_T = UA \text{ Erlangs} \quad ...(2.16)$$

In a trunks system of C channels and equally distributed traffic among the channels, the traffic intensity per channel is:

$$A_c = \frac{UA}{C} \text{ Erlangs/channel}$$

The offered traffic is the volume of traffic offered to the trunked system to be processed and is defined as:

Offered traffic = Carried traffic + Overflow

The carried traffic is the actual traffic carried by a switch and **Overflow** (blocked) traffic is portion of the traffic not processed.

Example 2.1. A call established at 1 a.m. between a mobile and MSC. Assuming a continuous connection and data transfer rate at 30 kbit/s, determine the traffic intensity if the call is terminated at 1.50 a.m.

Solution. Traffic intensity = (1 call) * (50 mins) * (1 hour/60 min) = 0.833

Note that traffic intensity has nothing to do with the data rate, only the holding time is taken into account.

Example 2.2. Consider a PSTN which receives 240 calls/hr. Each call losts an average of 5 minutes. What is the outgoing traffic intensity to the public network ?

Solution.

$$A = \lambda * H \quad \text{...(2.17)}$$

λ = 240 calls/hr and H = 5 minutes

A = (240 calls /hr) × (5 min/call) = 1200 min/hr

Erlang cannot have any unit so:

A = 1200 min/hr * (1 hour/60 minutes)

= 20 Erlangs

So, 20 hours of circuit talk time is required for every hour of elapsed time. (There is 20 hours of continuous use of 20 channels.)

Quality of service (QoS) is expressed in terms of blocking probability as:

$$P = B(A \,.\, C) \quad \text{...(2.18)}$$

where B = Erlang B Formula, A = The traffic intensity, C = No. of channel (lines)

Erlang B formula provides the probability of blockage due to congestion and is derived under the following assumptions:

(1) No waiting is allowed (lost calls are cleared) (*i.e.* they disappear from the system. This assumption is valid for systems that can overflow blocked calls onto another trunk (*e.g.* a high usage trunk).

(2) Traffic originated from an infinite numbers of sources.

(3) Limited Number of trunk (or serving channels) .

(4) Memory-less, channel requests at any time.

(5) The probability of a user occupying a channel is based on exponential distribution.

(6) Calls arrival rate at the network = Poisson process (the holding time or duration of the call has exponential distribution).

Equations for probability of Blocking, $P(B)$, depend on assumption that we make about what happens to calls that are blocked.

Case 1. Lost Calls Cleared:

➤ Assume that blocked calls are cleared (lost from the system). This assumption is valid for systems that can overflow blocked calls onto another trunk (*e.g.*, a high usage trunk).

Equation: Offered Traffic A = Carried Traffic $\frac{A_C}{(1 - P(B))}$

Case 2. Lost Calls Returning:

➤ Assume that blocked calls are retried until they are successfully carried. This assumption is valid for PBXs and corporate tie lines.

Equation: Offered Traffic A > or = Carried Traffic A_C

$P(B)$ for the case of Lost Calls Cleared also known as the Erlang B formula is given by:

$$P(B) = \frac{\frac{A^C}{C!}}{\sum_{k=0}^{C} \frac{A^k}{k!}} \quad \text{...(2.19)}$$

where, A is the traffic intensity and C is the number of channels by Carried traffic and A_{ca} is given as

$$A_{ca} = A[1 - p(B)] \quad \text{...(2.20)}$$

The efficiency of the channel usage is

$$\eta = \frac{A_{ca}}{C} \quad \text{...(2.21)}$$

The start-up system usually begins with a GOS of 0.02 (2% of the blocking probability) rising up to 0.5 as the system grows. If more subscribers are allowed in the system the blocking probability may reach unacceptable values.

The capacity of a trunked radio system where blocked calls are lost is tabulated for various values of GOS and numbers of channels in following table.

Capacity of an Erlang B System
(Number of Channels C Capacity (Erlangs) for GOS)

Erlang B Table

Number of channels C	Traffic Intensity (Erlangs)/Capacity for GOS			
	GOS = 0.01	GOS = 0.005	GOS = 0.002	GOS = 0.001
2	0.153	0.105	0.065	0.046
4	0,869	0.701	0.535	0.439
5	1.36	1.13	0.9	0.762
10	4.46	3.96	3.43	3.09
20	12	11.1	10.1	9.41
24	15.3	14.2	13	12.2
40	29	27.3	25.7	24.5
70	56.1	53.7	51	49.2
100	84.1	80.9	77.4	75.2

The Erlang B formula is plotted in Fig. 2.27. This chart is useful for determining GOS in rapid fashion, although computer simulations are often used to determine transient behaviours experienced by particular users in a mobile system.

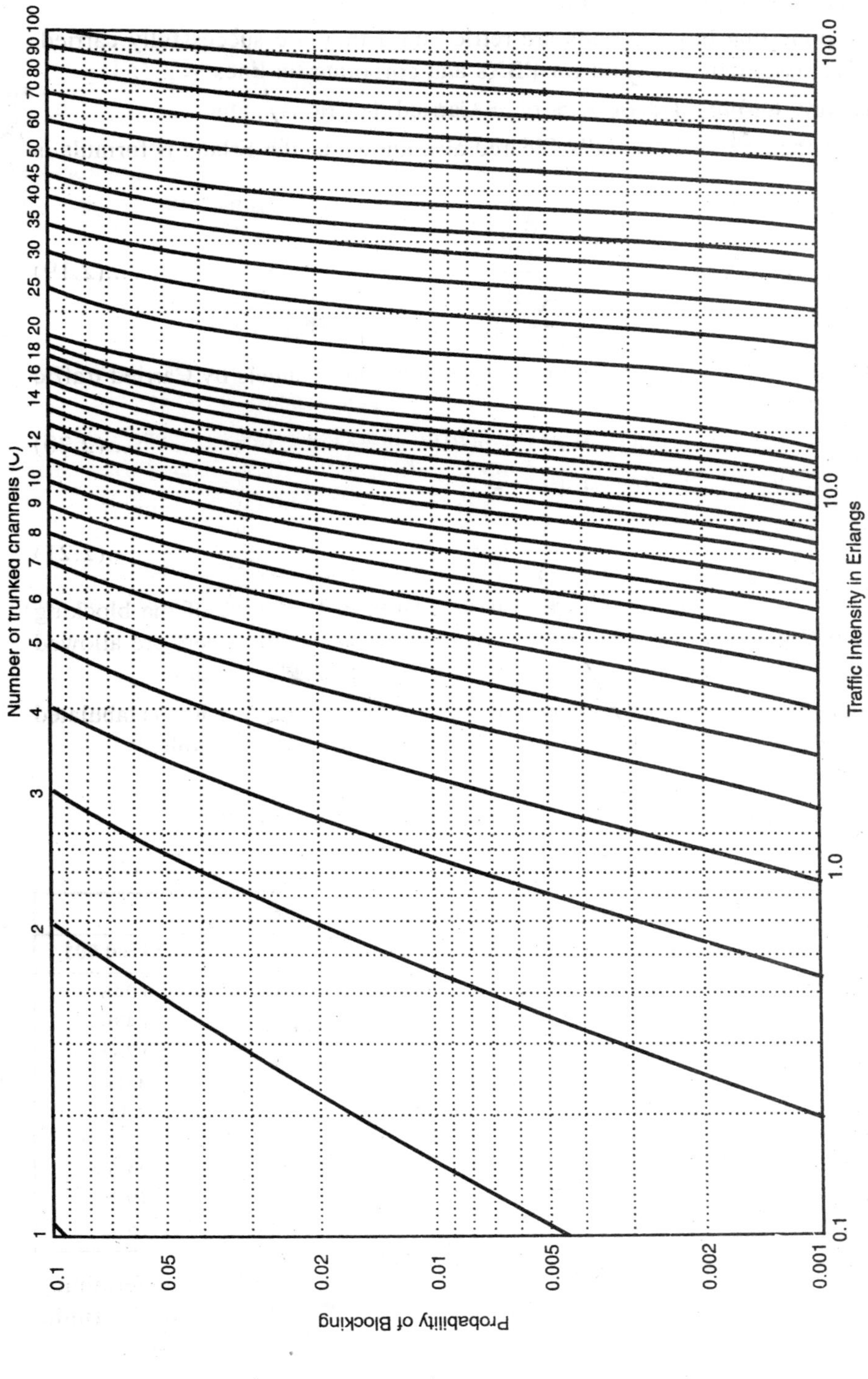

Fig. 2.27. The Erlang B chart showing the probability of blocking as functions of the number of channels and traffic intensity in Erlangs. (Refer to www.erlang.com)

2.7.5. Trunking Efficiency and Grade of Service

Trunking means to accommodate a large number of subscriber to share the relatively small number of channels in a cell by providing access to each user, on demand, from a pool of available channels.

In a trunked radio system, each user is allocated a channel on a per call basis, and upon termination of the call, the previously occupied channel is immediately returned to the pool of available channels. Trunk is a Latin word derived from TRUNCUS, which means, having to with the main routes of a transport or communication network. But in a communication network trunk word is used for long distance call in fixed telephone line.

The telephone company uses trunking theory to determine the number of telephone circuits that need to be allocated for office buildings with hundreds of telephones and this same principle is used in designing cellular radio system. In a trunked mobile radio system, when a particular user requests service and all the radio channels are already in use, the user is blocked, or denied access to the system.

In some systems, a queue may be used to hold the requesting users untill a channel becomes available. To design trunked radio systems that can handle a specific capacity at a specific "grade of service" it is essential to understand trunking theory and queuing theory.

The trunking theory was developed by Erlang, a Danish mathematician who, in the late 19th century, developed a theory that how a large population could be accommodated by a limited number of services.

Today, the measure of traffic intensity bears his name.

One Erlang represents the amount of traffic intensity carried by a channel that is completely occupied (one call-hour per hour or one call-minute per minute).

For example, a radio channel that is occupied for thirty minutes during an hour carries 0.5 Erlangs of traffic.

Example 2.3. A single GSM service provider supports 10 digital speech channels. Assume the probability of blocking is 1.0%. From the Erlang B chart find the traffic intensity. How many 3 minutes of calls this represent.

Solution. From the Erlang B Chart the traffic intensity = 5 Erlangs.

$$A_l = \lambda H$$

$$\lambda = \frac{A_l}{H} = 5/(3\text{ mins}/60) = 100\text{ calls}.$$

Example 2.4. A telephone switching board can handle 120 phones. Assuming the following, determine the outgoing traffic intensity and the number of channels.

- **On average 5 calls/hour per phone,**
- **Average call duration time = 4 minutes,**
- **60% of all calls made are external,**
- **QoS = 0.9%**

Solution. $A_T = U\lambda H$

$$\lambda^*U = (120 \text{ call}^*5 \text{ calls/hour})^*60\% = 360 \text{ call/hour}$$
$$H = 4 \text{ mins/call}$$

Therefore, $A_T = 360 * 4 * 1$ (1 hour/60 mins) = 24 Erlangs.

Thus 24 hours of circuit talk time is required for every hour of elapsed time.

Number of channels C from Erlang B chart = 34.

Example 2.5. Consider a telephone switchboard with 120 phones. Assuming the number of call is 3/hour/line, the average call duration is 4 minutes, and 55% of all calls are made external via a T-1 trunk (24 channels) to the PSTN. Determine carried traffic and channel usage.

Solution.

- Offered traffic $A = \lambda \times H$ = (150 phones × 3 calls/hr × 58%) × (4 mins./call) × (1 hour/60 mins) = 17.4 Erlangs
- Blocking Probability $P(B)$, C = 24 and A = 17.4, therefore, from the Erlang B Chart or formula $P(B)$ = 0.03.
- Carried Traffic, $A_{co} = A(1 - P(B))$ = 17.4 (1 – 03) = 16.9 Erlangs
- Channel usage $\eta = A_{ca}/C$ = 16.9/24 = 0.7 or 70%

Note. 16.9 Erlangs of traffic attempts to go across the T-1 trunk and 0.5 Erlang is blocked.

Example 2.6. An urban area has a population of two million residents. Three competing trunked mobile networks (systems *A*, *B*, and *C*) provide cellular service in this area. System *A* has 394 cells with 19 channels each, system *B* has 98 cells with 57 channels each, and system *C* has 49 cells, each with 100 channels. Find the number of users that can be supported at 2% blocking if each user averages two calls per hour at an average call duration of three minutes. Assuming that all three trunked systems are operated at maximum capacity, compute the percentage market penetration of each cellular provider.

Solution. System *A*:

Given, probability of blocking = 2% = 0.02

Number of channels per cell used in the system, C = 19

Traffic intensity per user,

Traffic intensity per user, $A_u = \lambda H = 2 \times (3/60) = 0.1$ Erlangs

For GOS = 0.02 and C = 19, from the Erlang B chart, the total carried traffic, A, is obtained as 12 Erlangs. Therefore, the number of users that can be supported per cell is

$$U = \frac{A}{A_u} = \frac{12}{0.1} = 120.$$

Since there are 394 cells, the total number of subscribers that can be supported by System A is equal to 120 × 394 = 47280.

System *B*:

Given, Probability of blocking = 2% = 0.02

Number of channels per cell used in the system, C = 57

Traffic intensity per user. $A_u = \lambda H = 2 \times (3/60) = 0.1$ Erlangs

GOS = 0.02 and $C = 57$, from the Erlang B chart, the total carried traffic, A, is obtained as 45 Erlangs. Therefore, the number of users that can be supported per cell is

$$U = \frac{A}{A_u} = \frac{45}{0.1} = 450.$$

Since there are 98 cells, the total number of subscribers that can be supported by System B is equal to $450 \times 98 = 44{,}100$.

System *C*:

Given, probability of blocking = 2% = 0.02

Number of channels per cell used in the system, $C = 100$

Traffic intensity per user, $A_u = \lambda H = 2 \times (3/60) = 0.1$ Erlangs

For GOS = 0.02 and $C = 100$, from the Erlang B chart, the total carried traffic, A, is obtained as 88 Erlangs. Therefore, the number of users that can be supported per cell is

$$U = \frac{A}{A_u} = \frac{88}{0.1} = 880$$

Since there are 49 cells, the total number of subscribers that can be supported by System C is equal to $880 \times 49 = 43{,}120$.

Therefore, total number of cellular subscribers that can be supported by these three systems are:

$$47{,}280 + 44{,}100 + 43{,}120 = 134{,}500 \text{ users.}$$

Since there are two million residents in the given urban area and the total number of cellular subscribers in System A is equal to 47,280 the percentage market penetration is equal to

$$\frac{47{,}280}{2{,}00{,}000} = 2.36\%$$

Similarly, market penetration of System B is equal to $\frac{44{,}100}{2{,}00{,}000} = 2.205\%$ and the market penetration of System C is equal to $\frac{43{,}120}{2{,}00{,}000} = 2.156\%$. The market penetration of the three systems combined is equal to $\frac{134{,}500}{2{,}00{,}000} = 6.725\%$.

2.7.6. Increasing the Capacity of Cellular Systems

As the demand for wireless service increases, the number of channels assigned to a cell eventually becomes insufficient to support the required number of users. At this point, cellular design techniques are needed to provide more channels per unit coverage area.

Main investment of a cellular network is the cost of the infrastructure including the cost of base stations and switching equipment, property (land for setting up the cell sites), installation, and links connecting the base stations. Costs are proportional to the number of base stations. All network operators start their operation with the minimum number of cell sites to cover a geographic area that requires the least initial investment. As the number of subscribers increases, network operators must acquire new cell sites to increase the capacity of the network and to support new subscribers. Various methods of increasing the capacity are:

1. To obtain additional spectrum for new subscribers, simple but expensive.
2. Change the cellular topology.
   Cell splitting - cells in areas of high uses can be split into smaller cells.
   Cell sectoring - cells are divided using directional antennas into a number of sectors each with their own set of channels.
3. Micro cells - antennas move to buildings, lamp posts. Very small cells, possibly an antenna in every room.
4. Antenna and cluster.

2.8. CELL SPLITTING

Cell splitting is the process of subdividing a congested cell into smaller cells, each with its own base station and a corresponding reduction in antenna height and transmitter power. Cell splitting increases the capacity of a cellular system since it increases the number of times that channels are reused. By defining new cells which have a smaller radius than the original cells and by installing these smaller cells (called *microcells)* between the existing cells, capacity increases due to the additional number of channels per unit area. Figure 2.28 illustrates cell splitting from Radius R to $R/2$ to $R/4$.

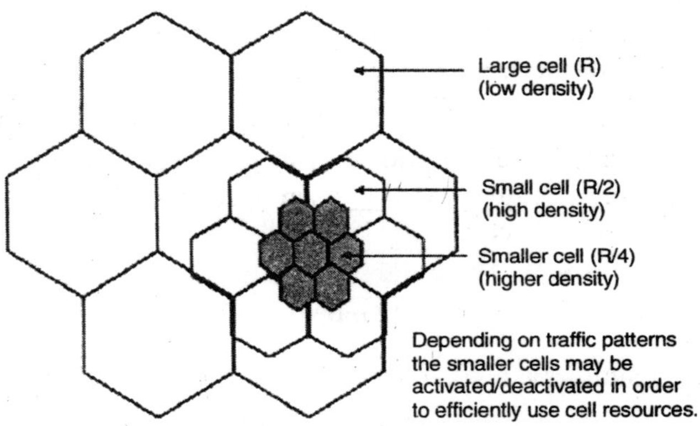

Fig. 2.28. An example of cell splitting.

If the radius of the new smaller cell is $R/2$, in order to cover the entire service area with smaller cells, approximately four times as many cells would be required. The increased number of cells would increase the number of clusters over the coverage region, which in turn would increase the number of channels, and thus capacity, in the coverage area. Cell splitting allows a system to grow by replacing large cells with smaller cells, while the channel allocation scheme required to maintain the minimum co-channel reuse ratio Q (see Eq. (2.6)) between co-channel cells.

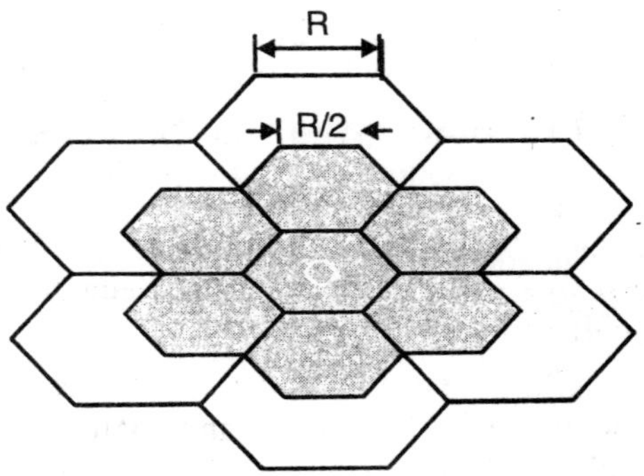

Fig. 2.29. Cell splitting with radius half of original cells.

For the new cells to be smaller in size, the transmit power of these cells must be reduced.

The transmit power of the new cells with radius half that of the original cells can be found by examining the received power P_r at the new and old cell boundaries and setting them equal to each other. This is necessary to ensure that the frequency reuse plan for the new microcells behaves exactly as for the original cells. The received power at the large (old) cell and smaller (new) cell is,

$$P_r[\text{at old cell boundary}] = P_{t1}(R)^{-\alpha} \quad ...(2.22)$$

$$\text{And } P_r[\text{at new cell boundary}] = P_{t2}\left(\frac{R}{2}\right)^{-\alpha} \quad ...(2.23)$$

where P_{t1} and P_{t2} are the transmit powers of the larger and smaller cell base stations, respectively, and α is the path loss exponent. On the basis of equal received power, we have

$$P_{t1}(R)^{-\alpha} = P_{t2}\left(\frac{R}{2}\right)^{-\alpha} \quad \text{or} \quad \frac{P_{t1}}{P_{t2}} = 2^{\alpha} \quad ...(2.24)$$

Taking the logarithm we can express this as:

$$10 \log 10\left(\frac{P_{t1}}{P_{t2}}\right) = \alpha\, 10 \log_{10}(2) = 3(\alpha)\text{ dB} \quad ...(2.25)$$

For $a = 4$, $\left(\frac{P_{t1}}{P_{t2}}\right) = 12$ dB.

Thus, with cell splitting where the radius of the new cell is one-half of that the old cell, we can achieve a 12 dB reduction in the transmit power. In other words, the transmit power must be reduced by 12 dB in order to fill in the original coverage area with microcells, while maintaining the S/I requirement.

Example 2.7. Suppose each cell is allocated 60 channels regardless of the cell size. Find the number of channels contained in a 3 × 3 km^2 area without cell splitting, *i.e.*, R = 1 km and with cell splitting, $\frac{R}{2}$ = 0.5 km.

Solution. The number of cells for R = 1 km.

(*a*) Each large cell can cover 2.6 km^2, for 9 km^2 approximately need $\frac{9}{2.56} = 3.5$ cells. A better approximation is 4 cells.

So, the number of channels is 4 × 60 = 240.

(*b*) With small cells, the number of cells is approximately $\left(\frac{1}{0.5}\right)2 \times 4 = 16$.

Then the number of channels is 16 × 60 = 960.

In practice, not all cells are split at the same time. It is often difficult for service providers to find real estate that is perfectly situated for cell splitting. Therefore, different cell sizes will exist simultaneously. In such situations, special care needs to be taken to keep the distance between co-channel cells at the required minimum, and hence channel assignments become more complicated. Also, handoff issues must be addressed so that high speed and low speed traffic can be simultaneously accommodated (the umbrella cell approach is commonly used). When there are two cell sizes in the same region. Equation above shows that one cannot simply use the original transmit power for all new cells or the new transmit power for all the original cells. If the larger transmit power is used for all cells, some channels used by the smaller cells would not be sufficiently separated from co-channel cells. On the other hand, if the smaller transmit power is used for all the cells, there would be parts of the larger cells left unserved. For this reason, channels in the old cell must be broken down into two channel groups, one that corresponds to the smaller cell reuse requirements and the other that corresponds to the larger cell reuse requirements. The larger cell is usually dedicated to high speed traffic so that handoffs occur less frequently.

The two channel group sizes depend on the stage of the splitting process. At the beginning of the cell splitting process, there will be fewer channels in the small power groups. However, as demand grows, more channels will be required, and thus the smaller groups will require more channels. This splitting process continues until all the channels in an area are used in the lower power group, at which point cell splitting is complete within the region, and the entire system is reseated to have a smaller radius per cell. *Antenna down tilting,* which deliberately focuses radiated energy from the base station toward the ground (rather than toward the horizon), is often used to limit the radio coverage of newly formed microcells.

2.8.1. Sectoring

As shown earlier cell splitting achieves capacity improvement by essentially rescaling the system. By decreasing the cell radius R and keeping the co-channel reuse ratio D/R unchanged, cell splitting increases the number of channels per unit area. However, another way to increase capacity is to keep the cell radius unchanged and seek methods to decrease the D/R ratio. As we now show, sectoring increases S/I so that the cluster size may be reduced. In this approach, first the S/I is improved using directional antennas, then capacity improvement is achieved by reducing the

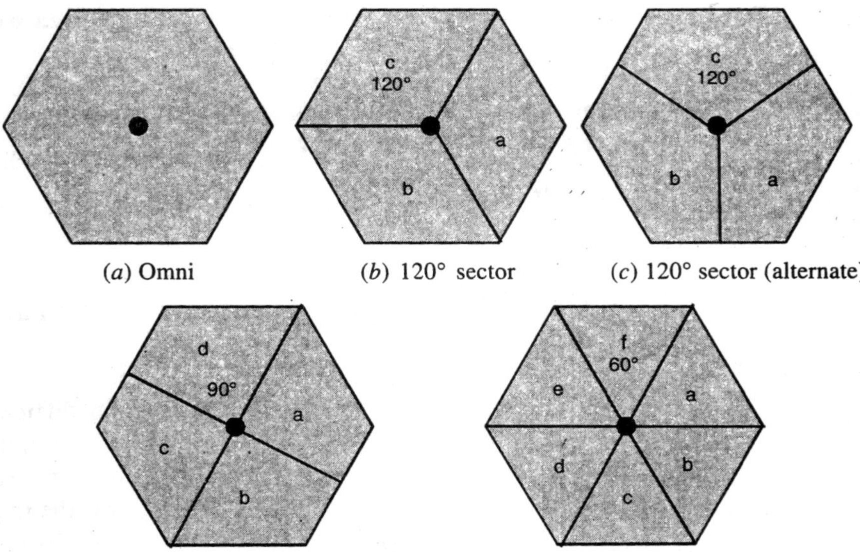

(*a*) Omni (*b*) 120° sector (*c*) 120° sector (alternate)

(*d*) 90° sector (*e*) 60° sector.

Fig. 2.30. Several sector examples.

number of cells in a cluster, thus increasing the frequency reuse. However, in order to do this successfully, it is necessary to reduce the relative interference without decreasing the transmit power. The co-channel interference in a cellular systeı may be decreased by replacing a single Omni-directional antenna at the base statio by several directional antennas, each radiating within a specified sector. By usin directional antennas, a given cell will receive interference and transmit with only fraction of the available co-channel cells. The technique for decreasing co-chann interference and thus increasing system performance by using directional antenna is called sectoring. The factor by which the co-channel interference is reduce depends on the amount of sectoring used. A cell is normally partitioned into thre 120° sectors or six 60° sectors as shown in Fig. 2.30.

When sectoring is employed, the channels used in a particular cell are broke down into sectored groups and are used only within a particular sector, as illustrate in Fig. 2.31.

Assuming seven-cell reuse, for the case of 120° sectors, the number of interferer in the first tier is reduced from six to two. This is because only two of the six cc channel cells receive interference with a particular sectored channel group. Referrin to Fig. 2.32, consider the interference experienced by a mobile located in th right-most sector in the centre cell labeled A_0.

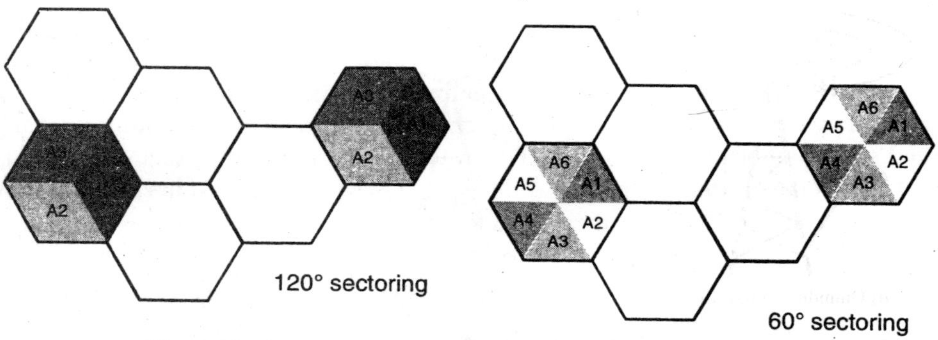

Fig. 2.31. 120° and 60° sectoring with directional antennas.

There are three co-channel cell sectors labeled A_1, A_2 and A_3 to the right of the center cell, and three co-channel cells sectors labeled A_4, A_5 and A_6 to the left of the center cell. Out of these six co-channel cells, only two cells A_5 and A_6 have sectors with antenna patterns which radiate into the center cell, and hence a mobile in the center cell will experience interference on the forward link from only these two sectors. Figure 2.34 shows the various sectored cells.

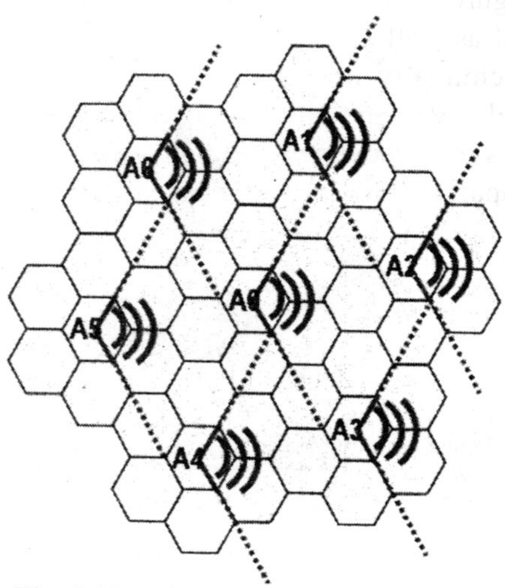

Fig. 2.32. Reduction of interference from co-channels due to 120° sectoring.

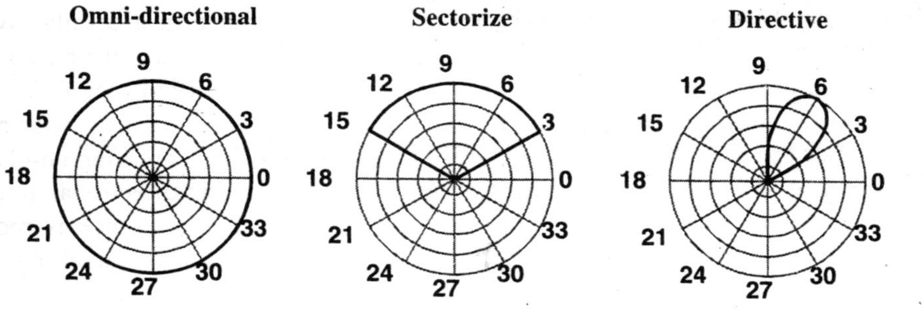

Fig. 2.33. Various antenna patterns.

Omnidirection antenna provides same gain in every direction but it is suitable for one cell pattern so it is rarely used. Figure 2.34(*a*) shows that there are three frequencies associated with antenna.

While Fig. 2.34(*b*) is having sectorised antenna, it is most commonly used on highways. It shows that one frequency per cell is assigned by operator.

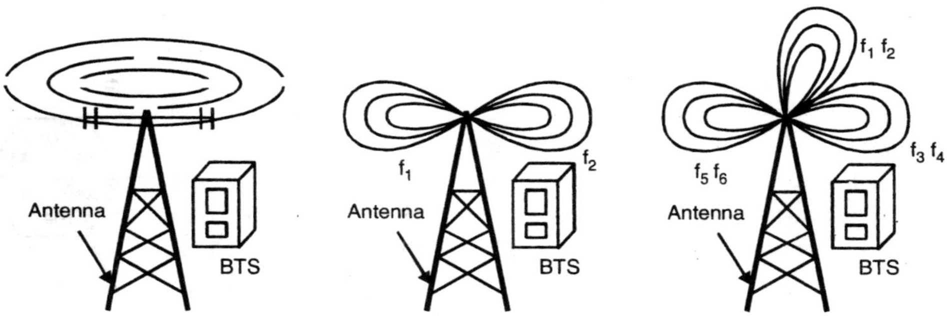

(*a*) Omnidirectional Antenna (*b*) 2 Sectorised BTS (*c*) 3 Sectorised BTS

Fig. 2.34. Omnidirectional and sectorised cells.

Figure 2.34(*c*) is a three sectored antenna, it is most commonly used by operator in rural as well as in urban areas. It shows that 2 frequencies are assigned per cell and technically it is named (2 + 2 + 2) mean one BTS have 6 frequencies or channel (TRx) and 2 frequencies assigned per cell. It may be (3 + 3 + 3) mean BTS consists of 9 frequencies and 3 frequencies per cell is assigned. And each antenna is 120° apart. While in 2 sectorised (Fig. 2.34(*b*)) each antenna is 180° apart.

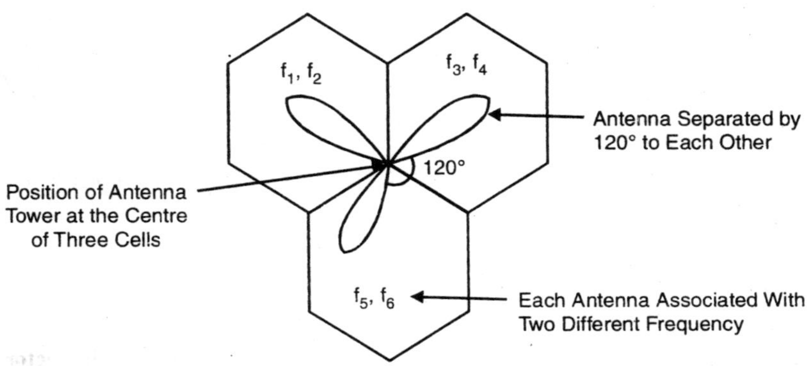

Fig. 2.35. Most practical pattern of cell used by network operators.

Worst-Case Scenario in 120° Sectoring

Let D be the distance between the adjacent co-channel cells.

With the distance approximation and use path loss component for the Worst-Case Scenario of 120° Sectoring (Fig. 2.36).

The resulting S/I for this case can be found using the Equation (2.26)

$$\left(\frac{S}{I}\right)_{120} = \frac{R^{-\alpha}}{D^{-\alpha} + (D+0.7R)^{-\alpha}}$$

$$= \frac{1}{q^{-\alpha} + (q+0.7)^{-\alpha}} \quad ...(2.26)$$

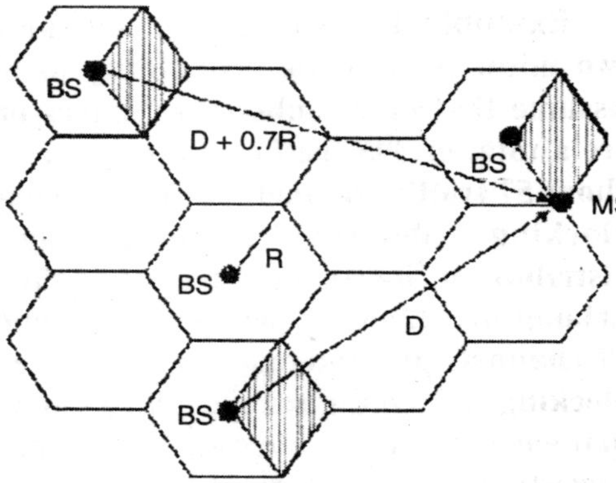

Fig. 2.36. Worst case scenario, 120°.

Using q = 4.6 and α = 4 in the last equation we get $\frac{S}{I}$ = 285 or 24.5 dB.

The S/I for a mobile unit served by a cell site with a 120 degrees directional antenna is a significant improvement over the omni-directional case where the worst case S/I was shown to be 17 dB. In real life scenario under heavy traffic, the S/I could be 6 dB weaker than in the equation due to irregular terrain configurations and imperfect locations. The resulting 18 dB S/I is still adequate. This S/I improvement allows the wireless engineer to then decrease the cluster size N in order to improve the frequency reuse, and thus the system capacity. In practical systems, further improvement in S/I is achieved by down tilting the sector antennas such that the radiation pattern in the vertical (elevation) plane has a notch at the nearest co-channel cell distance.

The improvement in S/I implies that with 120° sectoring, the minimum required S/I of 18 dB can be easily achieved with seven-cell reuse, as compared to 12-cell reuse for the worst possible situation in the unsectored case. Thus, sectoring reduces interference, which amounts to an increase in capacity by a factor of 12/7, or 1.714. In practice, the reduction in interference offered by sectoring enable planners to reduce the cluster size N, and provides an additional degree of freedom in assigning channels. The penalty for improved S/I and the resulting capacity improvement from the shrinking cluster size is an increased number of antennas at each base station, and a decrease in trunking efficiency due to channel sectoring at the base station. Since sectoring reduces the coverage area of a particular group of channels, the number of handoffs increases, as well. Fortunately, many modern base stations support sectorization and allow mobiles to be handed off from sector to sector within the same cell without intervention from the MSC, so the handoff problem is often not a major concern.

It is the loss of traffic due to decreased trunking efficiency that causes some operators to shy away from the sectoring approach, particularly in dense urban areas where the directional antenna patterns are somewhat ineffective in controlling radio propagation. Because sectoring uses more than one antenna per base station, the available channels in the cell must be subdivided and dedicated to a specific antenna. This breaks up the available trunked channel pool into several smaller pools, and decreases trunking efficiency.

Example 2.8. Consider a cellular system in which an average call lasts two minutes, and the probability of blocking is to be not more than 1%. Assume that every subscriber makes one call per hour, on average. If there are a total of 395 traffic channels for a seven-cell reuse system, there will be about 57 traffic channels per cell. Assume that blocked calls are cleared so the blocking is described by the Erlang B distribution. From the Erlang B distribution, it can be found that the unsectored system may handle 44.2 Erlangs or 1326 calls per hour. Now employing 120° sectoring, there are only 19 channels per antenna sector (57/3 antennas). For the same probability of blocking and average call length, it can be found from the Erlang B distribution that each sector can handle 11.2 Erlangs or 336 calls per hour. Since each cell consists of three sectors, this provides a cell capacity of 3 × 336 = 1008 calls per hour, which amounts to a 24% decrease when compared to the unsectored case. Thus, sectoring decreases the trunking efficiency while improving the S/I for each user in the system.

Solution. Worst-Case Scenario in 60° Sectoring:

S/I for a worst case scenario in 60° is given by (Fig. 2.30):

$$\left(\frac{S}{I}\right)_{60} = \frac{R^{-\alpha}}{(D+0.7R)^{-\alpha}} = (q+0.7)^{\alpha} \qquad \ldots(2.27)$$

For q = 4.6 and α = 4, the above equation gives S/I = 789 or 29 dB.

This indicates a further reduction of co-channel interference. Using the argument as was used for the three sector case and subtracting 6 dB from 29 dB the remaining 23 dB is still more than adequate. For heavy traffic the 60° sector configuration can be used to reduce the co-channel interference. However, with the six sector configuration reduces the trunking efficiency and increases the number of necessary handoffs even more. If the unsectored system is compared to the six sector case, the degradation in trunking efficiency can be shown to be 44%.

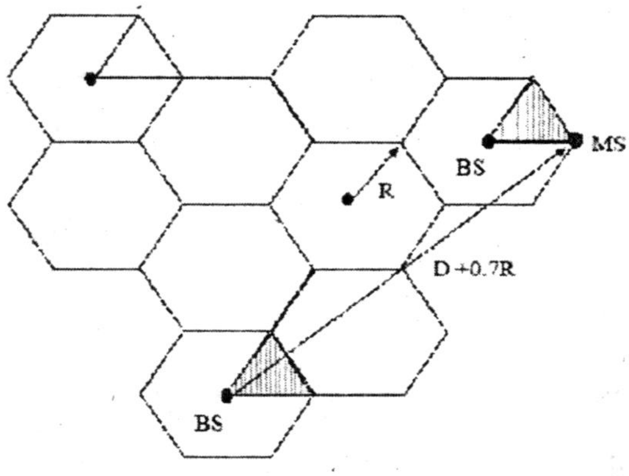

Fig. 2.37. Worst case scenario, 60° sectoring.

2.8.2. Microcell Zone

When we employed the sectoring concept, handoff is increased which increases the load on the switching and control link elements of the mobile system.

A solution to this problem was presented by lee. This proposal is based on a microcell concept for $N = 7$ cell reuse. Following figure explains clearly this concept, in the figure we are considering a three zone site [may be more or less] connected to a single base station and share the same radio equipment.

The zones are connected by co-axial cable, fiberoptic cable, or microwave link to the base station. As a mobile travels within the cell, it is served by the zone with the strongest signal. This approach is superior to sectoring since Antennas are placed at the outer edges of the cell, and any base station channel may be assigned to any zone by the base station.

As a mobile travels frame one zone to another within the cell, it retains the same channel.

Thus, like in a sectoring, a handoff is not required at the MSC when the mobile travels between zones within the cell. The base station simply switches the channel to a different zone site. In this way, a given channel active only in the particular zone in which the mobile is travelling and hence the base station radiation is localized and interference is reduced.

The channels are distributed in time and space by all three zones and are also re-used in co-channel cells in the normal fashion. This technique is use-full along highways or along urban traffic corridors.

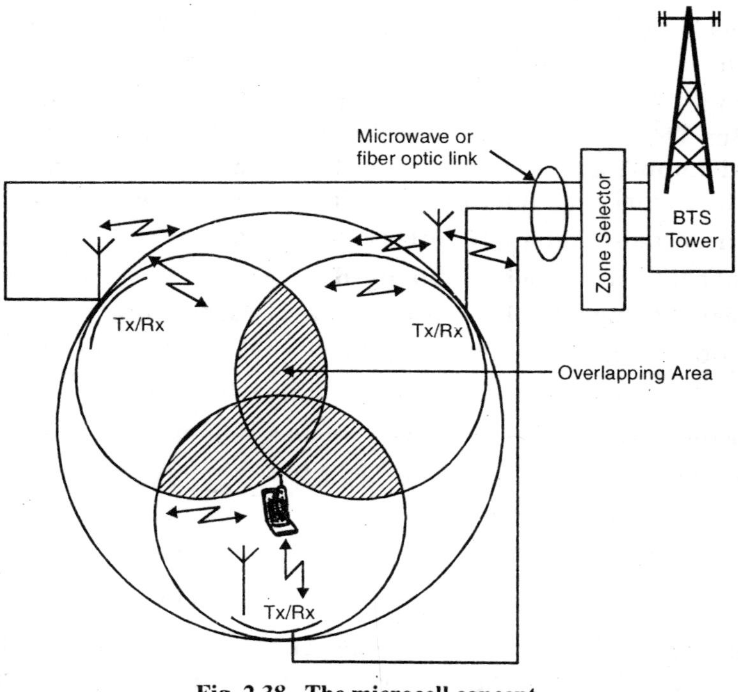

Fig. 2.38. The microcell concept.

2.8.3. Antenna and Clusters

Generally in hexagonal model base station transmitters are placed in centre. And maximally, an omnidirectional antennas are used in centre excited cells. The factor N is called the cluster size. If the cluster size N is reduced, while the cell size is kept constant, more clusters are required to cover a given area and hence more capacity is achieved.

A larger size of N causes the ratio between the cell radius and the distance between co-channel cells to decrease which results weaker co-channel Interference and small cluster size indicated that co-channel cells are located much closer together.

2.8.4. MicroCells

As cells become smaller, antennas height become smaller. It moves from the tops of tall building or hills to the tops of small buildings or the sides of large buildings, where they form microcells. Decrease in cell size is accompanied by a reduction in the rediated power levels from the base stations and the mobile units.

Microcells are useful in city streets or along highways and inside large public buildings.

Table 2.3. Typical Parameters for Macrocells and Microcells

	Macrocell	Microcell
Cell radius	1 to 20 km	0.1 to 1 km
Transmission power	1 to 10 W	0.1 to 1 W
Average delay spread	0.1 to 10 μs	10 to 100 ns
Maximum bit rate	0.3 Mbps	1 Mbps

No loss in trunking efficiency is experienced. Zone cell architectures are being adopted in many cellular personal communication system.

Example 2.9. In a cellular system, the received power at the mobile unit is –98 dBm. The cell structure is based on a 7-cell reuse pattern. If the thermal noise and amplifier noise powers are –120 dBm and –110 dBm, respectively and the CCI from each interfering stations is –121 dBm, then calculate

(*i*) the signal-to-total interference ratio in dB

(*ii*) signal-to-signal noise ratio in dB

(*iii*) signal-to-signal noise and interference ratio in dB

Solution. Converting dBm to mW:

$$\text{Signal power } P_s = 10_{-9.8} = 1.58 \times 10_{-10} \text{ mW}$$

$$\text{Thermal noise power } N_{th} = 10_{-12} = 1.0 \times 10_{-12} \text{ mW}$$

$$\text{Amplifier noise power } N_{amp} = 10_{-11} = 1.0 \times 10_{-11} \text{ mW}$$

$$\text{CCI power } I_T = 10_{-12.1} = 7.94 \times 10_{-13} \text{ mW}$$

$$\text{Signal-to-CCI ratio SIR} = 1.58 \times 10_{-10}/6.\ (7.94 \times 10_{-13})$$

$$= 33.2, \text{ in dB} = 15.22 \text{ dB}$$

Example 2.10. The transmitter power of a base station is 10 W. If the coverage of this area is to be split half to that minicells (half the size) can be created to accommodate additional users in the area, what must be the transmitter power at the base station at the minicell in order to keep the CCI at the same level as that of the unsplit cell :

Assume α = 3.5.

Solution. To keep CCI from the interfering stations unchanged, the power at the boundary needs to be keep the same, regardless of whether the cell has been split or not. Thus we have

$$10\, R^{-\alpha} = P\left(\frac{R}{2}\right)^{-\alpha} = P = \frac{10}{(2)_{3.5}} = 0.86 \text{ W.}$$

Example 2.11. For a mobile system with acceptable performance, the signal-to-CCI ratio must be at least 20 dB. What must be the value of $\frac{R}{D}$? Assume α = 3.

Solution. Signal-to-CCI ratio

$$\text{SIR} = \left(\frac{1}{6}\right)\left(\frac{D}{R}\right)_{\alpha} = 20 \text{ dB}$$

$$\Rightarrow \qquad \frac{1}{6}\left(\frac{D}{R}\right)^{\alpha} = 100$$

$$\left(\frac{D}{R}\right) = (600)^{1/3} = 8.4.$$

Example 2.12. If a mobile service provider is planning to offer 20 voice channels per cell for grade of service of 10%, calculate the offered traffic and the carried traffic. If the service provider has 10 cells, what is the total carried traffic in Erlangs ?

Solution. Offered Load (From Table) = 17.6 Erlang/cell;

$$\text{Carried Load} = 17.6 \times 0.9 = 15.84$$

Erlang/cell

$$\text{Total carried traffic} = 15.84 \times 20 = 316.8 \text{ Erlang.}$$

Example 2.13. Consider a cellular system with the following data :

- **Average call duration: 2.5 minutes**
- **Probability of blocking: 1%**
- **Average number of call per subscriber: 1 call/hour**
- **Number of traffic channel: 400 for a 7-cell reuse system**

Calculate the number of calls per hour. If 120 degree sectoring per cell has been adopted, then determine the total calls per hour.

Solution. No. of channel per cell = $\frac{400}{7} = 57$

From the Erlang B Chart the traffic intensity A_I = ~46 Erlangs

$$A_I = \lambda \text{H},$$

thus

$$\lambda = \frac{A_I}{H} = \frac{46}{(2 \text{ mins}/60)} = 1380 \text{ call.}$$

With sectoring, the number of channel = $\frac{56}{3} = 19.$

From the Erlang B Chart the traffic intensity A_I = 12 Erlangs, and thus $\lambda = \frac{12}{0.0333}$ = 360 calls per hour.

Finally, total calls/hour = 3 × 360 = 1080.

Example 2.14. A cellular mobile radio system has the following characteristics:

- **Transmitted signal power at a base station = –90 dBm**
- **Cluster size = 7 cells**

  The loss exponent factor = 4
- **The signal-to-(noise + the total interference ratio) = 16 dB.**

Assuming that all the interfering base stations are at equidistant from the desired base station, and thermal and amplifier noise at the mobile unit are negligible, determine:

(*i*) The co-channel interference (CCI) in W and dBm,

(*ii*) The adjacent channel interference (ACI) in W and dBm.

Solution. Given $D = R\sqrt{(3N)}$, $N = 7$,

therefore, 6 interfering channels, $\alpha = 4$

(*i*) The signal to the co-channel interference

$$= \text{SIR} = [R\sqrt{(3N)}]_{\alpha/io} = 441/6 \text{ m} = 73.4$$

Thus the co-channel interference is:

$$\text{CCI} = \frac{S}{73.4} = 10_{-90/10}/73.4 = 1.36 \times 10_{-11} = 13.6 \text{ Pw},$$

In dBm = –78.6

(*ii*) Since signal-to noise + interference ratio is

$$= \frac{S}{(N + I_T)} = 16 \text{ dB} = 10_{-1.6} = 39.8$$

Since thermal noise is assumed to be negligible, Thus

$$I_T = \frac{S}{39.8} = \frac{10_{-90/10}}{39.8} = 2.5 \times 10_{-11} \text{ W}$$

And the total interference

$$I_T = CCI + ACI$$

Therefore, $ACI = IT - CCI = 25 \text{ pW} - 13.6 \text{ pW} = 11.4 \text{ pW},$

In dBm = –79.4

Example 2.15. Assume each user of a base station mobile radio system averages three calls per hour, each call lasting an average of 5 minutes.

(*a*) What is the traffic intensity of each user ?

(*b*) Find the number of users that could use the system with 1% blocking if only one channel is available.

(*c*) Find the number of users that could use the system with 1% blocking if five trunked channels are available.

(*d*) If the number of users you found in (*c*) is suddenly doubled, what is the new blocking probability of the five channel trunked mobile radio system ? Would this be acceptable performance ? Justify why or why not ?

Solution. (*a*) Traffic Intensity

$$H = 5 \text{ minutes}, \lambda = \frac{\text{Number of calls}}{\text{Unit time}}$$

$$\lambda = \frac{3}{\text{Hour}}$$

$$A_v = \lambda H = \frac{3 \text{ calls}}{\text{Hour}} \times \frac{5 \text{ minutes} \times 14\ R}{60 \text{ minutes}}$$

$$A_v = 0.25 \text{ Erlangs}$$

(*b*) $\text{GoS} = 0.01,\ C = 1$; How many users ?

$$A = UA_v$$

$$\text{GoS} = \frac{A}{1+A} = \frac{UA_v}{1+UA_v}$$

$$\Rightarrow \quad U = 0.04$$

⇒ User, since one user could be supported on one channel.

(*c*) 1% blocking probability, 5 channels of users ?

From Erlang Chart $A = 1.45$

So, $$U = \frac{A}{A_v} = \frac{1.45}{0.25} = 5.8 \cong 5 \text{ users}$$

(*d*) $$U = 10$$

$$(\text{GoS})_{\text{New}} = \frac{A^5}{\sum_{i=0}^{5} \frac{A^i}{i!}}$$

Since $$A = UA_v = (10)\,(0.25) = 2.5$$

Solve for GoS, we get GoS = 0.0697.

Example 2.16. **(*a*) Consider a cellular system with 630 full duplex channels and a 7-cell reuse pattern so that each cell is allocated 90 channels. Assume the system has been designed so that at peak traffic intensity; the blocking probability is 2%. What is the SIR assuming propagation exponents $\alpha = 3$, 4, and 5 ? That is, find the SIR for each of the indicated propagation exponents.**

(*b*) It is proposed to improve the signal-to-interference ratio by employing 120 degree antenna sectoring. Since there are three sectors per cell, each sector will now have 30 channels. What is the peak traffic intensity per sector and per cell that can be supported by this system ? What is the SIR assuming propagation exponent $\alpha = 3$, 4, and 5 ?

(*c*) Since sectoring improves the SIR, it is also proposed to change to a 4-cell reuse pattern. Specify the number of channels per cell, the number of channels per sector, the peak traffic intensity per cell and per sector, and the SIR for the different propagation exponent of $\alpha = 3$, 4 and 5.

Solution. $$\text{GoS} = 2\%$$

$$C = 90$$

Erlang chart gives $$A \cong 80 \frac{\text{Erlangs}}{\text{Cell}}$$

$$\frac{S}{I} = \frac{P_{\text{Signal}}}{\sum P_{\text{Interference}}}$$

$$P_{\text{Signal}} = Po\left(\frac{d_{\text{Signal}}}{do}\right)^{-\alpha}$$

$$P_{\text{Interference}} = Po\left(\frac{d_{\text{Interference}}}{do}\right)^{-\alpha}$$

Since Interference is coming from 6 equidistant sources

$$\frac{S}{I} = \frac{1}{6}\left(\frac{d_{\text{Signal}}}{d_{\text{Interference}}}\right)^{-\alpha}$$

$$q = \frac{d_{\text{Interference}}}{d_{\text{Signal}}} = \sqrt{3N}$$

So, $$\frac{S}{I} = \frac{1}{6}(\sqrt{3N})^{+\alpha}$$

Thus

α	$\frac{S}{I}$(dB)
3	12.05
4	18.66
5	25.27

(*b*) By employing a 120° antenna we are eliminating a third of the interference. We can now use only two interfering sources thus

$$C = \frac{90}{3} = 30$$

Since $G = 2\%$

Erlang Chart gives ⇒ $A = 22 \frac{\text{Erlang}}{\text{Cell}}$

$$\frac{S}{I} = \frac{1}{2}\left(\sqrt{3N}\right)^{\alpha}$$

Thus

α	$\frac{S}{I}$(dB)
3	16.82
4	23.43
5	30.04

(*c*) If we use a 4-cell reuse pattern we get 157 $\frac{\text{Channel}}{\text{Cell}}\left(\frac{630}{4} = 157.5\right)$ and 52 channel per section. The Peak Traffic Intensity (Since $C = 52$, GoS = 2) is from Erlang Chart

$$A = 42\frac{\text{Erlang}}{\text{Section}} \quad \text{or} \quad A = 126\frac{\text{Erlang}}{\text{Cell}}$$

Thus

α	$\frac{S}{I}$(dB)
3	13.17
4	18.57
5	23.97

2.9. ANTENNA

As we know Antenna is very important device to transport wireless messages from one point to another point in air media.

We do not need any physical media for the transport of electromagnetic waves. Somehow, we have to couple the energy from the transmitter to the outside world and in reverse, from the outside world to the receiver. This is exactly what antennas do.

Antenna couple electromagnetic energy to and from space to and from a wire or coaxial cable.

A theoretical reference antenna is the isotropic radiator, a point in space radiating equal power in all direction. All points with equal power are located on a sphere with the antenna as its center. The radiation pattern is symmetrical in all directions as shown in following Fig. 2.39.

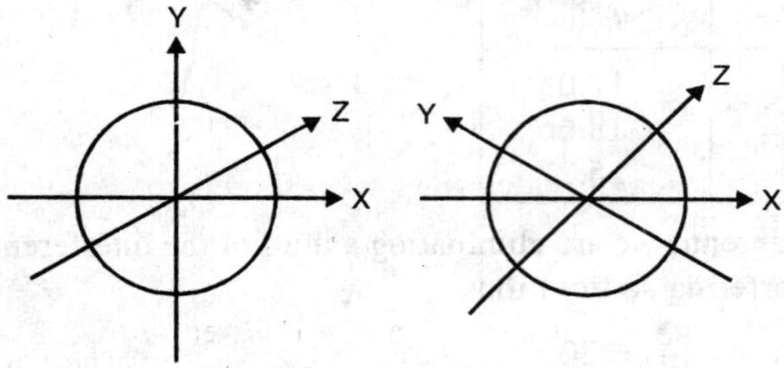

Fig. 2.39. Isotropic pattern of antenna.

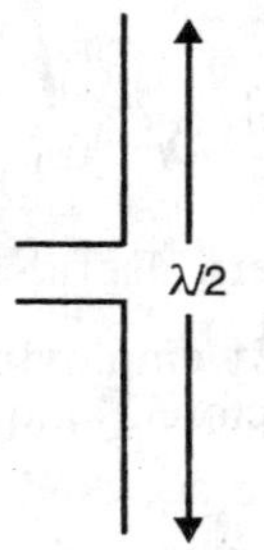

Fig. 2.40.

However, such an antenna does not exist in reality. Real antennas all exhibit directive effects, the intensity of radiation is not the same in all directions from the antenna.

The simplest real antenna is a thin, center feed dipole also called Hertzian dipole as shown in following Fig. 2.40.

The dipole consists of two collinear conductors of equal length separated by a small feeding a gap. The length of the dipole is not arbitrary but for example half the wavelength λ of the signal to transmit results in a very efficient radiation of energy. If mounted on the roof of a car the length of λ/4 is efficient.

This is also known as marconi antenna.

A λ/2 dipole has a uniform or omnidirectional radiation pattern in one plane and a figure eight pattern in the other two planes.

This type of antenna can only overcome environmental challenges by boosting the power level of the signal. Challenges could be mountains, valleys, buildings etc. If an antenna is positioned in a valley or between buildings an omnidirectional radiation pattern is not very useful. In this case, directional antennas with certain fixed preferential transmission and reception directions can be used.

Directed antennas are typically applied in cellular systems.

Several directed antennas can be combined on a single pole to construct a sectorized antenna. Two or more antennas can also be combined to improve reception by counteracting the negative effects of multipath propagation. (Fig. 2.41)

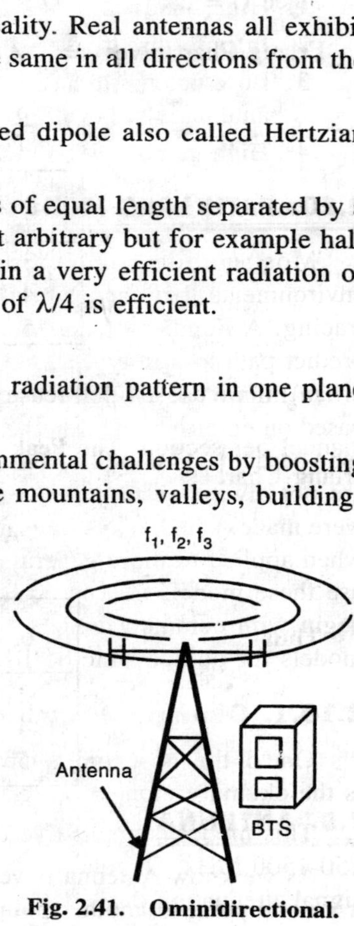

Fig. 2.41. Ominidirectional.

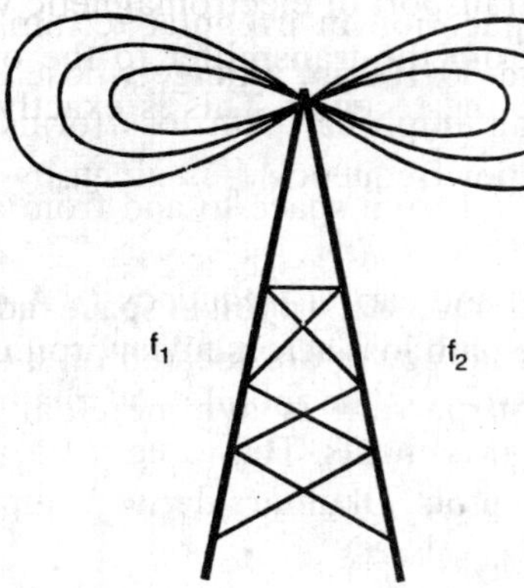

Fig. 2.42.

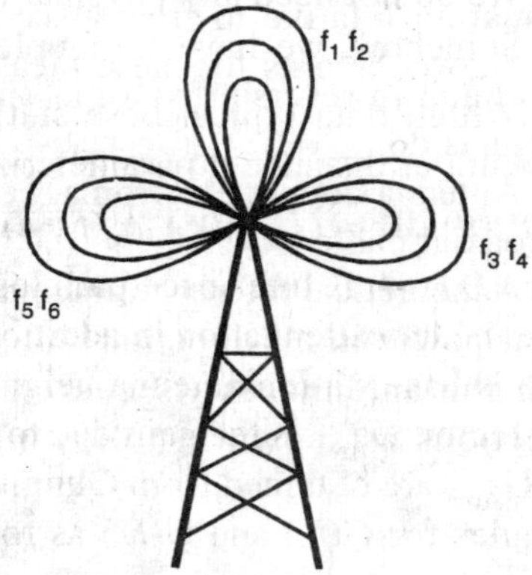

Fig. 2.43.

Pattern useful for plane area but provides low gain, therefore, not useful. (Fig. 2.42)

2 sectored antenna pattern useful for highways. (Fig. 2.43)

Three sectored antenna pattern useful for plane area, hill area, crowded area etc. Following are ideal characteristics of an antenna.

1. High bandwidth.
2. Infinite directivity.
3. Low beamwidth for directional antenna and high beamwidth for omni antenna.
4. High gain.

2.10. PATH LOSS MODELS (OKUMARA MODEL, HATA MODEL)

Most mobile communication systems operate in complex propagation environments that cannot be accurately modelled by free space. Path loss or ray tracing. A number of path loss models have been developed over the years to predict path loss in typical wireless environments such as large urban macrocells, urban microcells, and more recently, inside buildings. These models are mainly based on empirical measurements over a given distance in a given frequency range for a particular geographical area or buildings. However applications of these models are not always restricted to environments in which the empirical measurements were made, which may compromise the accuracy of such empirically based models when applied to more general environments. Nevertheless, many wireless systems use these models as a basis for performance analysis. In our discussion we will begin with common models for urban macrocells and then describe more recent models for outdoor microcells and indoor propagation.

2.10.1. Okumura Model

One of the most common models for signal prediction in large urban macrocells is the okumura model.

This model is applicable over distance of 1-100 km and frequency ranges of 150-1500 MHz. Okumura used extensive measurements of base station to mobile signal attenuation throughout Tokyo City to develope a set of curves giving median attenuation relative to free space of signal propagation in irregular terrain. The base station heights for these measurements were 30-100 m, a range whose upper end is high than typical base stations today. The empirical path loss formula of Okumura at distance d parameterized by the carrier frequency f_c is given by

$$P_L(d)\text{ dB} = L(f_c, d) + A_\mu(f_c, d) - G(h_t) - G(h_r) - G_{AREA} \quad ...(1)$$

where $L(f_c, d)$ is free space path loss at distance d and carrier frequency f_c, $A_\mu(f_c, d)$ is the median attenuation in addition to free space path loss across all environments, $G(h_t)$ is base station antenna height gain factor, $G(h_r)$ is the mobile antenna height gain factor. G_{Area} is the gain due to the type of environments. The values of $A_\mu(f_c, d)$ and G_{Area} are obtained from Okumura's empirical plots. Okumura derived empirical formulas for $G(h_t)$ and $G(h_r)$ as follows :

$$G(h_r) = 20 \log (h_t/200),\ 30\text{ m} < h_t < 1000\text{ m}$$

$$G(h_r) = \begin{cases} 10 \log_{10} (hr/3) & h_r \le 3\text{ m} \\ 20 \log_{10} (hr/3) & 3\text{ m} < h_r < 10\text{ m} \end{cases}$$

Okumura's model has a 10-14 dB standard deviation between the path loss predicted by the model and the path loss associated with one of the measurements used to develop the model.

2.10.2. Hata Model

The Hata model is an empirical formulation of the graphical path loss data provided by Okumura and is valid over roughly the same range of frequencies, 150-1500 MHz. This model simplifies calculation of path loss because it is a closed form formula and is not based on empirical curves for the different parameters. The standard formula for empirical path loss in urban areas under the Hata model is

$$P_{L,\text{ urban}}(d)\text{ dB} = 69.55 + 26.16 \log_{10} (f_c) - 13.82 \log_{10} (h_t) - a(h_r) + (44.9 - 6.55 \log_{10} (h_t)) \log_{10} (d)$$

The parameters in this model area the same as under the Okumura model, and $a(h_r)$ is a correction factor for the mobile antenna height based on the size of the coverage area.

For small to medium size cities, this factor is given by

$$a(h_r) = (1.1 \log_{10} (f_c) - 0.7)\, h_r - (1.56 \log_{10} (f_c) - 0.8)\text{ dB}$$

For larger cities at frequencies $f_c > 300$ MHz by

$$a(h_r) = 3.2 (\log_{10} (11.75\, h_r))^2 - 4.97\text{ dB}$$

Corrections to the urban model are made for suburban and rural propagation, so that these models are (respectively)

$$P_{L,\text{ suburban}}(d)\text{ dB} = P_{L,\text{ urban}}(d)\text{ dB} - 2[\log_{10} (f_c/28)]^2 - 5.4$$

and

$$P_{L,\text{ rural}}(d)\text{ dB} = P_{L,\text{ urban}}(d)\text{ dB} - 4.78 [\log_{10} (f_c)]^2 + 18.33 \log_{10} (f_c) - K$$

where K range from 35.94 to 40.94. Unlike the Okumura model, the Hata model does not provide for any path specific correction factors. Hata model well approximates the Okumura model for distance $d > 1$ km. Hence, it is good model for first generation (AMPS) but it does not model propagation well in current cellular systems with smaller cell size and higher frequencies.

2.11. BREWSTER ANGLE

Brewster angle is the angle at which no reflection occurs in the medium of origin. It occurs when the incident angle θ_B is such that the reflection coefficient R is equal to zero. Brewester angle is given by the value of θ_B which satisfies

$$\sin \theta_B = \sqrt{\frac{\varepsilon_1}{\varepsilon_1 + \varepsilon_2}}$$

For the case when the first medium is free space and the second medium has a relative permittivity ε_r then above equation can be expressed as

$$\sin \theta_B = \frac{\sqrt{\varepsilon_r - 1}}{\sqrt{\varepsilon_r^2 - 1}}$$

Note that it occur only for vertical polarization.

2.12. SCATTERING

Scattering is a general physical process where some forms of radiation, such as light, sound or moving particles, are forced to deviate from a straight trajectory by one or more localized non uniformities in the medium through which they pass in. The actual received signal in a mobile radio environment is often stronger than what is predicted by reflection and diffraction models alone. This is because when a radio wave impinges on a rough surface, the reflected energy is spread out (diffused) in all directions due to scattering. Objects such as lamp posts and trees tend to scatter energy in all directions, thereby providing additional radio energy at a receiver. Flat surfaces that have much larger dimensions than a wavelength may be modelled as reflective surfaces. However, the roughness of such surfaces often induces propagation effects different from the specular reflection described earlier in this chapter. Surface roughness is often tested using the Rayleigh criterion which defines a critical height (h_c) of surface protuberances for a given angle of incidence θ_i, given by

$$h_c = \frac{\lambda}{8 \sin \theta_i}$$

A surface is considered smooth if its minimum to maximum protuberance h is less than h_c, and is considered rough if the protuberance is greater than h_c. For rough surfaces, the flat surface reflection coefficient needs to be multiplied by a scattering loss factor f_s, to account for the diminished reflected field. If a surface height h is a Gaussian distributed random variable with a local mean and found f_s to be given by

$$\rho_s = \exp\left[-8\left(\frac{\pi\sigma_n \sin \theta_i}{\lambda}\right)^2\right]$$

where σ_n is the standard deviation of the surface height about the mean surface height. The scattering loss factor derived by Ament and it got modified by Boithias to give better agreement with measured results and is given by

$$\rho_s = \exp\left[-8\left(\frac{\pi\sigma_h \sin\theta_i}{\lambda}\right)^2\right] I_0\left[8\left(\frac{\pi\sigma_h \sin\theta_i}{\lambda}\right)^2\right]$$

where I_0 is the Bessel function of the first kind and zero order.

2.13. DIRECTIVITY

Directivity is a figure of merit for an antenna. It measures the power density of antenna that radiates in the direction of its strongest emission, versus the power density radiated by an ideal isotropic radiator (which emits uniformly in all directions) radiating the same total power.

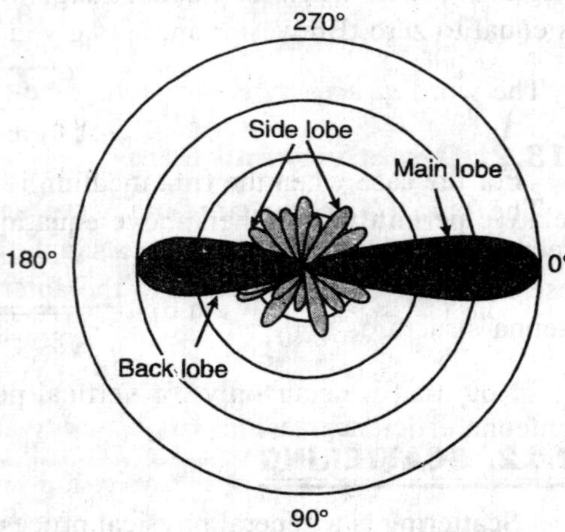

Fig. 2.44. Antenna focus to 0°.

An antenna's directivity is a component of its gain, the other component is its (electrical) efficiency. Directivity is an important measure because most emissions are intended to go in a particular direction or at least in a particular plane (horizontal or vertical), emissions in other directions or planes are wasteful.

The directivity of an actual antenna can vary from 1.76 dBi for a short dipole, to as much as 50 dBi for a large dish antenna.

2.13.1. Definition

The *directivity, D*, of an antenna is the maximum value of its *directive gain.* Directive gain is represented as $D(\theta, \phi)$, and compares the *radiation intensity (power per unit solid angle)* $U(\theta, \phi)$ that an antenna creates in a particular direction against the average value over all directions:

$$D(\theta, \phi) = \frac{U(\theta, \phi)}{\frac{P_{tot}}{(4\pi)}}$$

Here θ and ϕ are the standard spherical coordinate angles, $U(\theta, \phi)$ is the *radiation intensity,* which is the power density per unit solid angle, and P_{tot} is the *total radiated power* (TRP). The quantities $U(\theta, \phi)$ and P_{tot} satisfy the relation:

$$P_{tot} = \int_{\phi=0}^{\phi=2\pi} \int_{\theta=0}^{\theta=\pi} U \sin\theta \, d\theta \, d\phi;$$

that is, the total radiated power P_{tot} is the power per unit solid angle $U(\theta, \phi)$ integrated over a spherical surface. Since there are 4π *steradians* on the surface of a sphere, the quantity $\frac{P_{tot}}{4\pi}$ represents the *average* power per unit solid angle.

In other words, directive gain is the radiation intensity of an antenna at a particular (θ, ϕ) coordinate combination divided by what the radiation intensity would have been had the antenna been an isotropic antenna radiating the same amount of total power into space.

Directivity, then, is the maximum directive gain value found among all possible solid angles:

$$D = \max\left(\frac{U}{\frac{P_{tot}}{(4\pi)}}\right) = \frac{U(\theta, \phi)|_{max}}{\frac{1}{4\pi}\int_0^{2\pi}\int_0^{\pi} U(\theta, \phi)\sin\theta\, d\theta\, d\phi}$$

The word *directivity* is also sometimes used as a synonym for directive gain.

2.13.2. Directivity and Gain

The gain (G) of an antenna is an actual quantity which is less than the directivity (D) due to the ohmic losses in the antenna or in its radome: in transmitting, these losses involve power fed to the antenna which is not radiated but heats the antenna structure.

$$G = kD\ (0 < k < 1)$$

k antenna efficiency factor (k is usually very close to 1).

REVIEW QUESTIONS

1. Explain the concept of frequency reuse.
2. To increase the cellular system capacity, what are the different strategies that you can apply.
3. What is the relation of dB with Neper ?
4. Define the following terms in brief :
   (*a*) Trunking (*b*) Grade of Service (*c*) Sectoring
5. How Q and S/I can effect the system capacity. Explain with one example ?
6. What do you mean by interference, how it affects the performance of a cellular system ?
7. In question 2.9, which device will have shorter battery life and why ?
8. What are the different factors that are responsible for interference.
9. What is the frequency reuse distance ? What are the parameters on which it depends. Show that frequency reuse factor for a cellular system is given by K/S where K is average number of channels per cell and S is the total number of channels available to service provider.
10. Explain different performance criterion of a cellular mobile system with special emphasis on grade of service.
11. Give all difference between fixed channel assignment and nonfixed channel assignment.
12. Explain co-channel interference and how it can be minimized ?
13. Prove that for a hexagonal geometry, the co-channel reuse distance is $Q = \sqrt{3N}$ where $N = i^2 + ij + j^2$.
14. What is the effect of antenna height on cell area coverage ? Explain in detail.
15. Obtain the relation for signal to interference ratio for a 7-cell cluster in cellular architecture.
16. Calculate the Brewster angle for a wave impiging on ground having a permittivity of $\varepsilon_r = 4$.
17. Explain the effect of scattering phenomenon in mobile radio environment.

□□□

CHAPTER 3

Cell Coverage for Signal and Antenna Structure

INSIDE THIS CHAPTER

3.0. Introduction
3.1. Mobile Point to Point Model
3.2. Propagation Over Water and in Flat Area or in Free Space
3.3. Foliage Loss
3.4. Long Distance and Near Distance Propagation of Mobile Signal
3.5. Point to Point Prediction Model
3.6. Cell Site Antenna Height
3.7. Cell Coverage Area
3.8. Mobile to Mobile Propagation
3.9. Characteristics of Basic Antenna Structures
3.10. Antenna at Cell Site (Omnidirectional antenna)
3.11. Mobile Antenna

3.0. INTRODUCTION

As we know from previous chapter that cell is a hexagonal shape geometrical area and it's coverage depends upon frequency of signal and signal strength as well as geographical area and subscriber density. If it is plane area like Delhi then cell area will be very small because subscriber density is very much high but if it is plane area of rural side or hill side then cell area may be very big. When operator designs network then he wants that whole geographical area should be covered with large number of subscribers but with less number of cell site.

But technically it is impossible to cover 100 per cent area. So when network planner designs a network then he should consider following conditions.

1. Urban area
2. Rural area
3. Subscriber density
4. Hill area
5. Highway
6. Sea shore
7. Foliage areas.

There are so many models available to design a network to get maximum coverage but one model called production model is more suitable because it provides less than 3 dB deviation than practical value.

3.0.1. Predection Model Considers Following Conditions

(a) Incident Angle and Elevation Angle with Ground

Let ground incident angle is ϕ and elevation angle is θ.

Incident angle ϕ is the angle of signal arrival at the ground and elevation angle θ is angle of signal arrival at mobile unit as shown in Fig. 3.1.

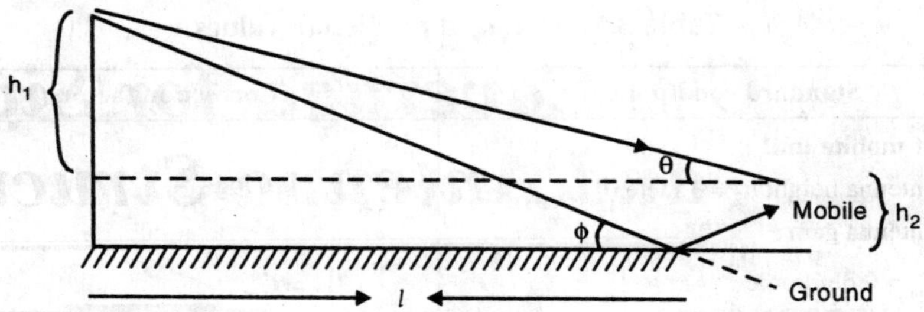

Fig. 3.1.

$$\text{Incident angle } \phi = \tan^{-1} \frac{h_1 + h_2}{l}$$

(b) Reflection Angle and Reflection Point

As per Snell's law incident angle and reflection angle are same.

But it is applicable if reflection point is plane surface. As per practical calculation if slope is less than 10 with respect to ground then reflection angle can be obtained by Snell's law.

Example 3.1. Find out incident angle for a mobile communication if cell site antenna height is 100 m and mobile antenna height is 2 meter and communication path is 10 meter.

Solution. From Fig. 3.1

$$\tan \phi = \frac{h_1 + h_2}{l}$$

$$\Rightarrow \quad \phi = \tan^{-1} \frac{h_1 + h_2}{l}$$

$$\Rightarrow \quad \tan^{-1} \frac{100 + 2}{10}$$

$$\phi = \tan^{-1} \frac{102}{10}$$

$$\phi = 84.4°$$

3.1. MOBILE POINT TO POINT MODEL

This model is also known as Lee model to obtain efficient design of a cell area. This model can be obtained in following steps :

(1) Generate a standard condition.

(2) An area to area prediction model.

3.1.1. Generate a Standard Condition

It is very difficult to generate any standard condition for point to point model but based on some practical data some correction factors are obtained on basis of standard conditions at cell site and at mobile site.

During design of model we will have to use these data as shown in Table 3.1.

Table 3.1. Standard prediction values.

Standard conditions	Correction factor
(*a*) **At mobile unit :**	
Antenna height $h_2 = 3$ meter	$\beta_1 = 10 \log \frac{h_2'}{h_2}$
Antenna gain	
$G_m = 0$ db/dipole	$\beta_2 = G_m'$
(*b*) **At Cell Site :**	
Transmitted power	
$P_t = 10$ W	$\beta_3 = 10 \log \frac{P_t'}{10}$
Antenna height $h_1 = 30$ meter	$\beta_4 = 20 \log \frac{h_1'}{h_1}$
Antenna gain $G_t = 6$ dB/Dipole	$\beta_5 = g_m'$

3.1.2. Area to Area Prediction Model

As we know area structure is different for a cell place to place, somewhere area is very flat and somewhere it is not flat and may be some area flat and some area non flat.

If we consider Delhi then we find that some area is flat with low density of subscribers and some area is non flat with high density of subscribers. Therefore, it is very difficult to generate standard deviations. So network planner collects data from whole geographical area and considered that some area is flat even some area is not flat.

Reason is to get average value for prediction model and the average value of standard deviation indicates the degree of roughness.

As we know structure of city varied place to place within city due to man made structures (buildings, offices etc.)

Therefore, prediction area curve is different due to different structures, so we take mean value of the collected data at different places and use to design network in optimized way.

Area to area prediction model can be represented by two parameters.

(1) Kilometer intercept point.

(2) Path loss point.

1 kilometer intercept point is the point at which power will be received from transmitter of cell site. See Fig. 3.2.

Following are methods to find value of above parameter experimentally.

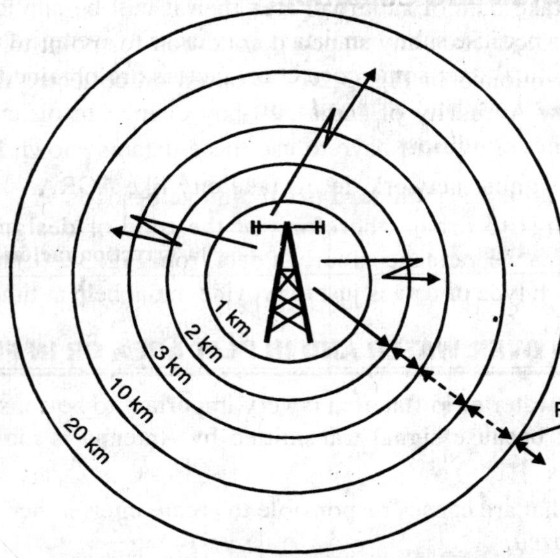

Fig. 3.2(*a*). Propagation path of man made structure.

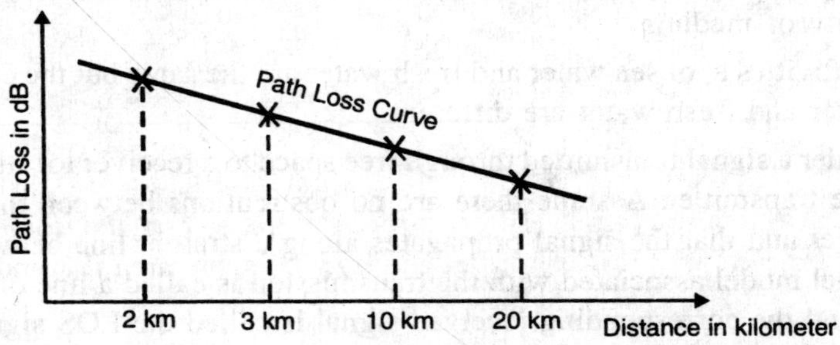

Fig. 3.2(*b*).

1. Select an area of interest and then compare with man made area of standard type shown in Fig. 3.2(*c*).

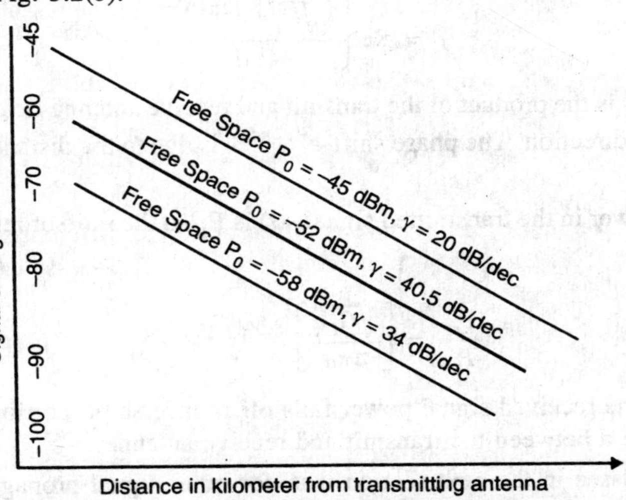

Fig. 3.2(*c*). Propagation loss in different cities.

Note that there are three types of area in India—Urban area, suburban area and rural area. But if we take data of suburban area then it will be applicable more or less for all types of area because suburban data is very close to average data of all areas.

Note that data mentioned in Fig. 3.2(*c*) is based on collected data by TRAI (Telephony Regulatory Authority of India). It may change in different countries and in different weather conditions.

At the time of designing network, let us take city like AGRA.

AGRA city belongs to urban, therefore, at the time of designing cell site, antenna height, cell area we can take suburban data from Fig. 3.2 (*c*) but practically it may be different. Such type of data is just to provide initial help at time of planning.

3.2. PROPAGATION OVER WATER AND IN FLAT AREA OR IN FREE SPACE

Propagation over water or in flat area is very important to be considered during designing of network because signal transmitted by Antenna is more affected in such areas.

So, if we know what are causes responsible to create interference then it can be controlled and optimized.

Frequency of radio signal is very much affected by die-electric constants and conductivity of medium.

Permittivities ε_r of sea water and fresh water are the same but the conductivities of sea water and fresh water are different.

Consider a signal transmitted through free space to a receiver located at distance d from the transmitter. Assume there are no obstrcutions between the transmitter and receiver and that the signal propagates along a straight line between the two. The channel model associated with the transmission is called a line of sight (LOS) channel, and the corresponding received signal is called the LOS signal or ray.

Free space path loss introduces a complex scale factor resulting in the received signal

$$r(t) = \text{Re}\left\{\frac{\lambda\sqrt{G_l}\,e^{-j2\pi d/\lambda}}{4\pi d}\,u(t)\,e^{J2\pi fct}\right\} \quad ...(3.1)$$

where $\sqrt{G_l}$ is the product of the transmit and receive antenna field radiation pattern in the LOS direction. The phase shift $e^{-j2\pi d/\lambda}$ is due to the distance d that the wave travels.

The power in the transmitted signal $s(t)$ is P_t, so the ratio of received to transmit power is

$$\frac{P_r}{P_t} = \left[\frac{\sqrt{G_l}\,\lambda}{4\pi d}\right]^2 \quad ...(3.2)$$

Thus, the received signal power falls off in inverse proportion to the square of the distance d between the transmit and receive antennas.

We will see in the next section that, for other signal propagation models, the received signal power falls of more quickly relative to this distance.

The received signal power is also proportional to the square of the signal wave length, so as the carrier frequency increases the received power decreases.

This dependence of received power on the signal wavelength λ is due to the effective area of the receive antenna. However, directional antenna can be designed so that received power is an increasing function of frequency for highly directional links.

The received power can be expressed in dBm.

$$P_r \text{ dBm} = P_t \text{ dBm} + 10 \log_{10}(G_l) + 20 \log_{10}(\lambda) - 20 \log_{10}(4\pi) - 20 \log_{10}(d) \quad ...(3.3)$$

Free space path loss is defined as the path loss of free space model

$$P_L \text{ dB} = 10 \log_{10} \frac{P_t}{P_r} = -10 \log_{10} \frac{G_l \lambda^2}{(4\pi d)^2} \quad ...(3.4)$$

The free space path gain is thus

$$P_G = -P_L = 10 \log_{10} \frac{G_l \lambda^2}{(4\pi d)^2} \quad ...(3.5)$$

3.2.1. Two Ray Model

Two ray model is used when a single ground reflection dominates the multipath effect as shown in Fig. 3.3.

The received signal consists of two components.

The LOS component or ray, which is just the transmitted signal propagating through free space and a reflected component or ray, which is the transmitted signal reflected of the ground.

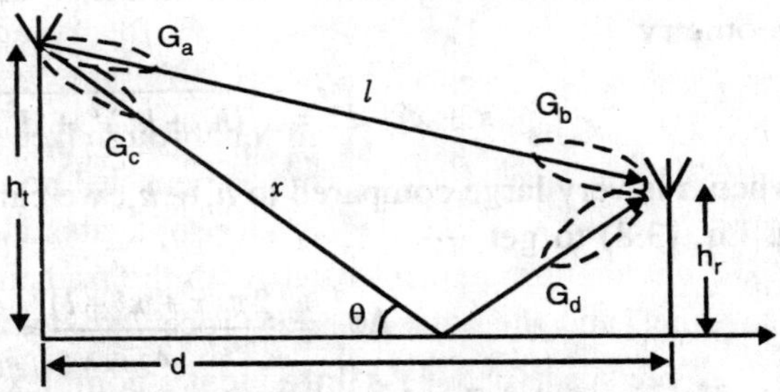

Fig. 3.3. Two ray-model for propagation in open area.

The received LOS ray is given by the free space propagation loss formula of Eq. (3.3).

The reflected ray is shown in Fig. 3.3 by the segment x and x'.

If we ignore the effect of surface wave attenuation then by superposition, the received signal for the two ray model is

$$r_{2\text{-ray}}(t) = \text{Re}\left\{\frac{\lambda}{4\pi}\left[\frac{\sqrt{G_l}\, v(t)\, e^{-j2\pi l/\lambda}}{l} + \frac{R\sqrt{G_r}\, v(t-\tau)\, e^{\frac{-j2\pi(x+x')}{\lambda}}}{x+x'}\right]\right\} e^{j2\pi f_c t} \quad ...(3.6)$$

where $\tau = \dfrac{(x + x' - l)}{c}$ is the time delay of the ground reflection relative to the LOS ray,

$$\sqrt{G_l} = \sqrt{G_a \cdot G_b}$$

is the product of the transmit and receive antenna field radiation patterns in the LOS direction, R is the ground reflection co-efficient, and

$$\sqrt{G_r} = \sqrt{G_c G_d}$$

is the product of the transmit and receive antenna field radiation patterns corresponding to the rays of length x and x' respectively.

The delay spread of the two ray model equals the delay between the LOS ray and the reflected ray $\frac{(x+x'-l)}{C}$.

If the transmitted signal is narrowband relative to the delay spread $\tau << B_0^{-1}$ then $U(t) \approx U(t-\tau)$.

With this approximation the received power of the two ray model for narrowband transmission is

$$P_r = P_t \left[\frac{\lambda}{4\pi}\right]^2 \left| \frac{\sqrt{G_l}}{l} + \frac{R\sqrt{G_r}\, e^{-j\Delta\phi}}{x+x'} \right|^2 \quad \text{...(3.7)}$$

where $\Delta\phi = 2\pi(x+x'-l)/\lambda$ is the phase difference between the two received signal components. Equation (3.7) is very close to practical data. If d denotes the horizontal separation of the antennas, h_t the transmitter, and h_r the receiver height then from geometry

$$x + x' - l = \sqrt{(h_t+h_r)^2 + d^2} - \sqrt{(h_t-h_r)^2 + d^2} \quad \text{...(3.8)}$$

when d is very large compared to $h_t + h_r$, we can use a Taylor series approximation in Eq. (3.8) to get

$$\Delta\phi = \frac{2\pi(x+x'-l)}{\lambda} \approx \frac{4\pi h_t h_r}{\lambda d} \quad \text{...(3.9)}$$

The ground reflection coefficient is given by

$$R = \frac{\sin\theta - z}{\sin\theta + z} \quad \text{...(3.10)}$$

where $z = \sqrt{\varepsilon_r - \frac{\cos^2\theta}{\varepsilon_r}}$ for vertical polarization

$= \sqrt{\varepsilon_x - \cos^2\theta}$ for horizontal polarization

and ε_r is the die-electric constant of ground.

We see from Fig. 3.3 and Eq. (3.10), for asymptotically large d, $x + x' \approx l \approx d$, $\theta = 0$, $G_l \approx G_r$ and $R \approx -1$ then Eq. (3.7) yields

$$P_r \approx \left[\frac{\lambda\sqrt{G_l}}{4\pi d}\right]^2 \left[\frac{4\pi h_l h_r}{\lambda d}\right]^2 P_t$$

$$= \left[\frac{\sqrt{G_l}\, h_t h_r}{d^2}\right]^2 \cdot P_t \quad \text{...(3.11)}$$

or in dB

$$P_r \text{ dBm} = P_t \text{ dBm} + 10 \log_{10} G_l + 20 \log_{10} (h_t h_r) - 40 \log_{10} (d) \qquad ...(3.12)$$

Above approximation is valid for water surface, free space, flat area.

Thus, in the limit of asymptotically large d, the received power falls off inversely with the 4th power of d and is independent of the wavelength λ. The received signal becomes independent of λ because directional antenna arrays have a received power that does not necessarily decrease with frequency, and combining the direct path and reflected signal effectively forms an antenna array.

3.3. FOLIAGE LOSS

As we know electromagnetic signal is having basic characteristics that whenever it will incident on any object, it will reflect and form standing waves and these standing waves may interfere input signal from antenna.

There are lot of objects between cell antenna and mobile like tree, buildings, etc.

Reflected wave depends upon size of tree, branches and their distribution as well as size of building. To consider foliage loss every such parameter will have to be taken into consideration.

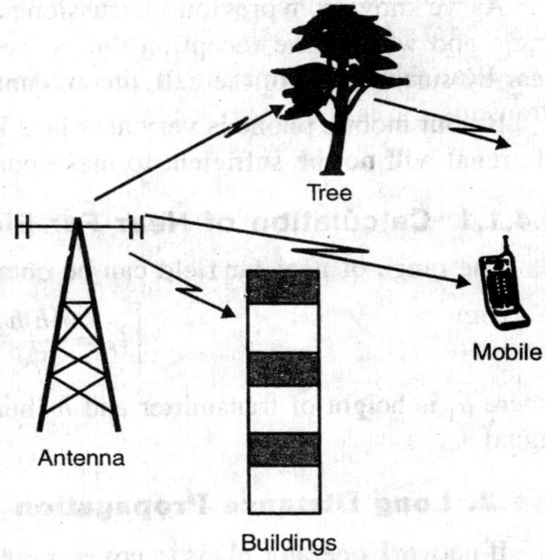

Fig. 3.4. Propagation of mobile signal.

However, a rough estimate should be sufficient for the purpose of system design. In tropic area size of tree is very big and in metropolation cities buildings size are also very big and it is very difficult to penetrate by mobile signals.

In this case, the signal will propagate from the top of the tree and deflect to the mobile receiver. Note that as frequency increases of a electromagnetic signal then penetration will decrease.

In summer the foliage is very heavy but in winter it is less.

Sometime the foliage loss can be treated as a wire-line loss in dBm.

Generally foliage loss occurs with respect to the frequency to –ve fourth power (f^{-4}). Foliage loss is around 40 dB/decade at frequency of 800 MHz which is double of free space loss.

3.4. LONG DISTANCE AND NEAR DISTANCE PROPAGATION OF MOBILE SIGNAL

It is difficult to differentiate between near distance and long distance but technically it is assumed that if distance is less than 70 km then it is assumed near distance otherwise it is assumed to be long distance.

3.4.1. Propagation in Near-in Distance

When antenna transmits signal then due to electromagnetic field a near field and far field is created.

Near field is very near to antenna and field available in near field may interfere propagation of electromagnetic signal.

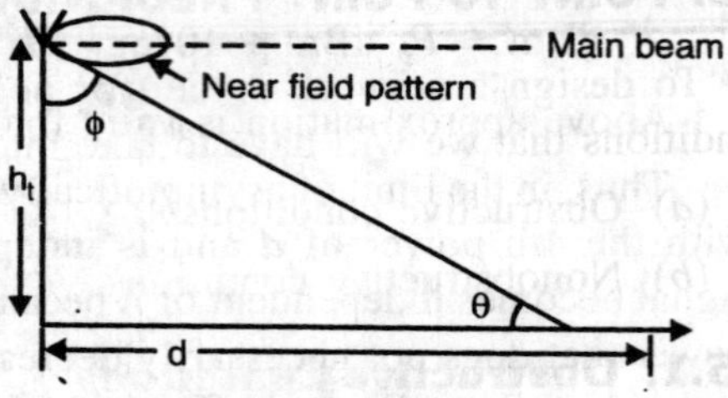

Fig. 3.5. Near field pattern.

Within 2 km radius the antenna beamwidth is very narrow and in the vertical plane particularly due to omnidirectional antenna.

So if mobile unit is within 2 km then signal reception at mobile may be very weak.

As we know from previous discussion nearer the mobile unit larger the elevation angle and weaker the reception due to vertical pattern of near field. Therefore, near by surrounding of the cell site antenna can affect the reception.

If your mobile phone is very near [< 2 km] or very far [>50 km] then reception of signal will not be sufficient to make conversation.

3.4.1.1. Calculation of Near Far Field Propagation

The range of near far field can be obtained by following formula

$$\left[l_F = \frac{4h_1 h_2}{\lambda} \right] \quad ...(3.13)$$

where h_1 is height of transmitter and h_2 height of mobile unit, λ is wavelength of signal.

3.4.2. Long Distance Propagation

If network operator plans to cover more area then height of cell antenna should be high particularly in high density area of subscribers.

But at the time of designing it should be kept in mind that in large area there is large interference due to objects like tree, buildings, etc.

For a high site, the low atmospheric phenomenon would cause the ground wave path to propagate in non straight line fashion.

If distance is large then it becomes impossible that signal will directly reach to mobile. If frequency of operation is 800 MHz and distance is more than 200 km then during travel of such big distance wave will undergo a lot of dielectric constant variation. At 12 km above the ground dielectric constant will change with temperature which decreases with height at a rate of about 6.5 °C/km and reaches –50°C at the upper boundary of the troposphere.

If wave travels under tropospheric propagation then it may undergo lot of reflection and refraction.

(a) Tropospheric Reflection

This type of reflection will occur when there are abrupt changes in the dielectric constant of the atmosphere.

(b) Tropospheric Refraction

This refraction is a gradual bending of the rays due to the changing effective dielectric constant of the atmosphere through which wave is passing.

3.5. POINT TO POINT PREDICTION MODEL

To design a network model for point to point communication there are two conditions that we will have to take into consideration.

(*a*) Obstructive conditions

(*b*) Nonobstructive conditions.

3.5.1. Obstructive Conditions

In obstructive conditions the direct path between cell site and mobile unit is obstructed by the terrain counter. Obstructive conditions can be treated as follows.

(a) Apply Area to Area Prediction

Area to area prediction is same as in non obstructive conditions. Analysis will be done in next section.

(b) Diffraction Loss

Diffraction loss can be obtained from single knife edge or multiple knife edge surfaces as per following Figs. 3.6 and 3.7.

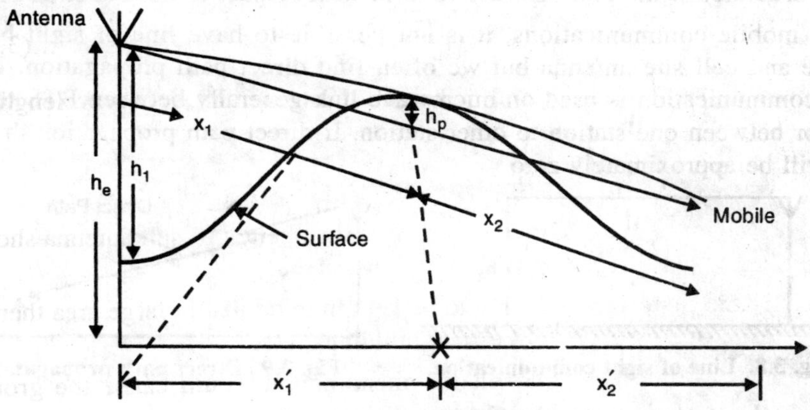

Fig. 3.6. Single knife edge.

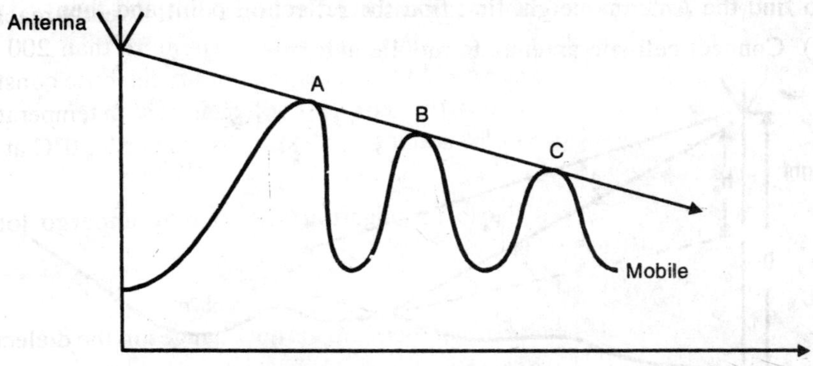

Fig. 3.7. Multiple knife edge.

First find out the following parameter for a single knife edge.

Distance x_1 and x_2 from the knife edge to the cell site and to mobile unit and the height of the knife edge to find the new parameter v.

$$v = -h_p \sqrt{\frac{2}{\lambda}\left(\frac{1}{x_1} + \frac{1}{x_2}\right)} \quad \text{...(3.14)}$$

As soon as we will find v the diffraction loss can be obtained from following relations

if $v \geq 1$, $L = 0$ dB ...(3.15)

$0 \leq v < 1$ then $L = 20 \log (0.5 + 0.62v)$ ...(3.16)

$-2.4 \leq v < -1$ then $L = 20 \log (0.4 - \sqrt{0.118 - (0.1v + 0.38)^2}$ ...(3.17)

If height of knife edge is zero ($h_P = 0$) and $v = 0$ and loss will be around 6 dB and path becomes tangential to knife edge.

3.5.2. In Nonobstructive Conditions

In this case direct path from cell site to the mobile unit is not obstructed by any obstacle. Note that nonobstructive term is different from line of sight communication. Line of site communications term is defined with respect to man made structures while non-obstructive term with respect to natural structure.

In mobile communications, it is not possible to have line of sight between mobile and cell site antenna but we often find direct path propagation. Line of sight communication is used on microwave link generally between BTS tower to BSC or between one station to other station. If direct path propagation then path loss will be approximately zero.

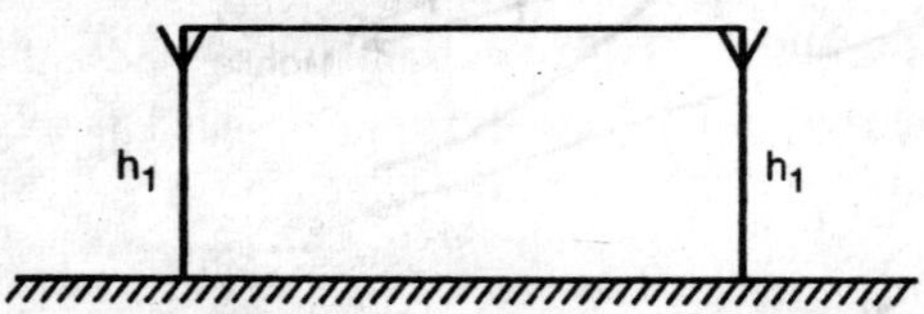

Fig. 3.8. Line of sight communication.

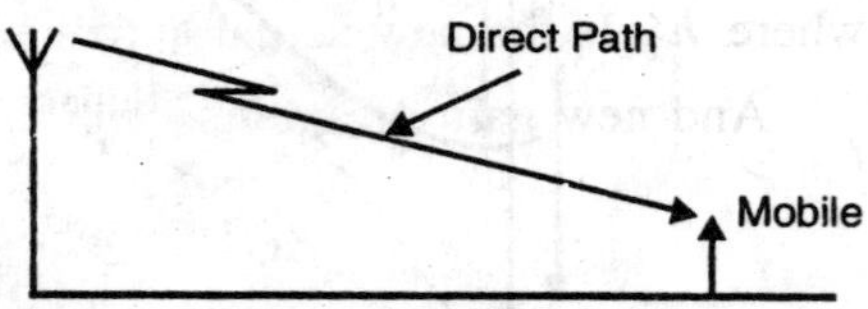

Fig. 3.9. Direct path propagation.

Finding the Antenna Height Gain

To find the Antenna height first find the reflection point and then

(*a*) Connect cell site antenna to mobile antenna.

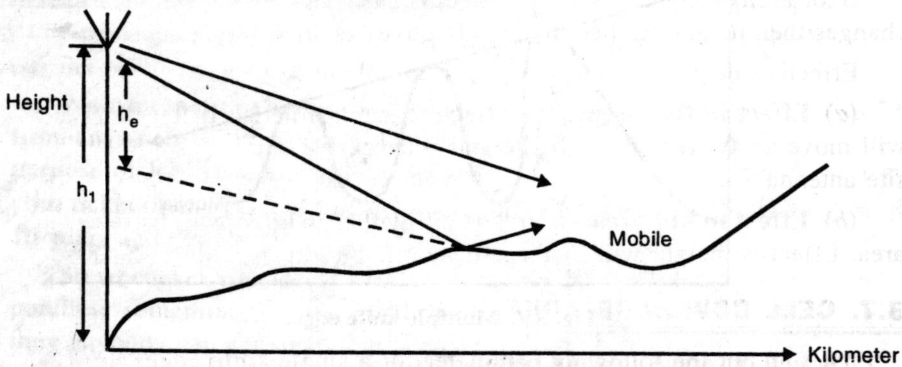

Fig. 3.10. Distance.

(*b*) Extend the reflected ground plane.

(*c*) Measure the effective antenna height. The effective antenna height is measured from the point where the reflected ground plane and the cell site antenna location meet.

(*d*) Calculate the antenna height gain ΔG.

Formula of ΔG can be represented as follows

$$\left[G = 20 \log \frac{h_e}{h_1} \right] \quad \text{...(3.18)}$$

where h_e is effective height of antenna and h_1 is physical height of antenna.

As mobile unit changes his position then effective height of antenna will change, result gain will change.

As effective height of antenna will increase gain of antenna will increase and reception of signal strength will increase.

3.6. CELL SITE ANTENNA HEIGHT

There are several parameters which need to be consider, during cell site antenna height. As we know if mobile subscriber will change his position then effective height of antenna and gain both will change.

If antenna height changes by ΔH then whole signal strength map will get change. New gain will be

$$\Delta G = 20 \log \frac{h_1'}{h_1} \quad \text{...(3.19)}$$

where h_1' is the new actual antenna height and h_1 is the old Antenna height.

And new gain of antenna due to effective height will become

$$\Delta G' = 20 \log \frac{h_e'}{h_e} \quad \text{...(3.20)}$$

where $h_e' = h_e \pm \Delta H$

h_e is old effective height of antenna and h_e' is new effective height of antenna.

Location of the Antenna Changed

If location of antenna changes then effective gain will change. Because if location changes then height of antenna and effective height will get change.

Effective height will change due to following conditions.

(*a*) Effect in flat area: The effective antenna height increases if mobile unit will move away from the cell site and it decreases if mobile unit approaches cell site antenna.

(*b*) Effect in hill area: If area is hill type then effect will be reverse as in flat area. Effective height will increase if mobile unit approaches cell site antenna.

3.7. CELL COVERAGE AREA

The cell coverage area in a cellular system is defined as the expected percentage of locations within a cell where the received power at these locations is above a given minimum power level.

Consider a base station inside a circular cell of a given radius R. All mobiles within the cell require some minimum received SNR for acceptable performance. Assuming a given model for noise, the SNR requirement translates to a minimum received power P_{min} throughout the cell.

The transmit power at the base station is designed for an average received power P_R at the cell boundary which is average over the shadowing variations.

However, shadowing will cause some locations within the cell to have received power below P_R, and others will have received power exceeding P_R.

As shown in Fig. 3.11, where we show contours of constant received power based on a fixed transmit power at the base station for path loss and average shadowing for path loss and random shadowing.

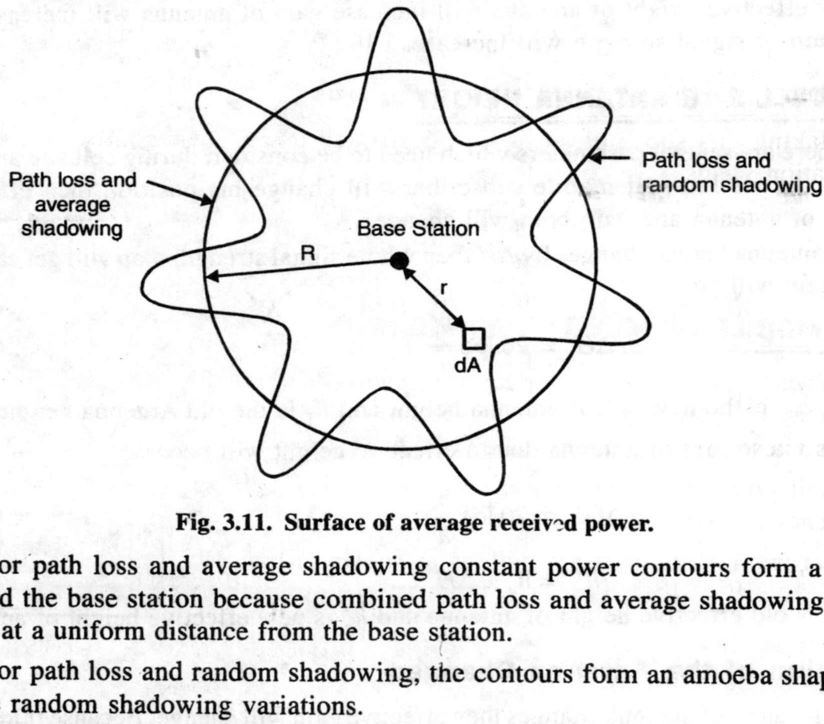

Fig. 3.11. Surface of average received power.

For path loss and average shadowing constant power contours form a circle around the base station because combined path loss and average shadowing is the same at a uniform distance from the base station.

For path loss and random shadowing, the contours form an amoeba shape due to the random shadowing variations.

The constant power controus for combined path loss and random shadowing indicate the challenge that shadowing passes in cellular system design. Specifically, it is not possible for all users at the cell boundary to receive the same power level. Thus, either the base station must transmit extra power to ensure users affected by shadowing receive their minimum required power P_{min}, which causes excessive interference to neighbouring cells, or some users within the cell will found their minimum received power requirement.

In fact, since the Gaussian distribution has infinite tails, any mobile in the cell has a nonzero probability of experiencing received power below its required minimum power, even if the mobile is close to the base station.

This makes sense intuitively because a mobile may be in a tunnel or blocked by a large building, regardless of its proximity to the base station.

The percentage of area within a cell where the received power exceeds the minimum required power $P_{\min}$ is obtained by taking an incremental area dA at radius r from the base station in the cell as shown in Fig. 3.11. Let $P_r(r)$ be the received power in dA from combined path loss and shadowing. Then the total area within the cell where the minimum power requirement is exceeded is obtained by integrating overall incremental areas where this minimum power is exceeded.

$$C = E\left[\frac{1}{\pi R^2}\int_{\text{cell area}} 1[P_r(r) > P_{\min} \text{ in } dA]\, dA\right]$$

$$= \frac{1}{\pi R^2}\int_{\text{cell area}} E[1]\, P_r(r) > P_{\min} \text{ in } dA]\, dA \qquad \text{...(3.20)}$$

Let $\qquad P_A = P(P_r(r) > P_{\min}) \text{ in dA}$

Then $\qquad P_A = E[1(P_r(r) > P_{\min} \text{ in } dA]]$

Making this substitution in Eq. (3.20) and using polar coordinates for the integration yields.

$$C = \frac{1}{\pi R^2}\int_{\text{Cell area}} P_A\, dA = \frac{1}{\pi R^2}\int_0^{2\pi}\int_0^{R} P_A r\, dr\, d\theta \qquad \text{...(3.21)}$$

3.8. MOBILE TO MOBILE PROPAGATION

Whenever mobile to mobile communication occurs then transmitter and receiver both may be in motion or may not be but in both conditions signal propagation passes via so many obstacles called buildings, trees, etc.

And propgation channel acts like a filter with a time varying function of frequency and time as $H(f, t)$.

Let two mobile units are moving as shown in Fig. 3.12.

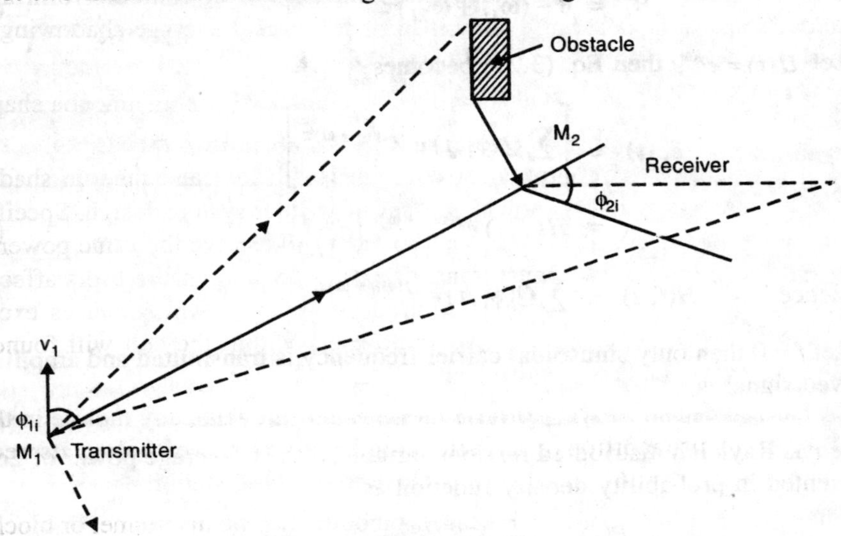

Fig. 3.12.

Let M_1 is transmitting a signal

$$S(t) = U(t)\, e^{j\omega t} \quad \text{...(3.22)}$$

and received signal at M_2 mobile unit is

$$S_i(t)\, U(t-\tau)\, e^{j[(\omega_0 + \omega_1 i + \omega_{2i})(t-\tau)+\theta]} \quad \text{...(3.23)}$$

where $U(t)$ is signal and

ω_0 is carrier frequency

θ is uniform phase but randomly distributed

τ time delay at receiver due to path propagation

ω_{1i} is Doppler shift of transmitting mobile unit

ω_{i2} is Doppler shift of receiving mobile unit

$$\omega_{1i} = \frac{2\pi}{\lambda} V_1 \cos \phi_{1i} \quad \text{...(3.24)}$$

$$\omega_{i2} = \frac{2\pi}{\lambda} V_2 \cos \phi_{2i} \quad \text{...(3.25)}$$

where ϕ_{i1} and ϕ_{2i} are random angles due to distant receiver unit.

Let us consider that the received signal is summation of received signal from n different path.

Then resultant signal at the receiver is

$$S_r(t) = \sum_{i=1}^{n} S_i(t) = \sum_{i=1}^{n} r_i U(t-\tau) \exp\{j[(\omega_0 + \omega_{1i} + \omega_{2i})](t-\tau) + \theta_i\} \quad \text{...(3.26)}$$

$$\Rightarrow \sum_{i=1}^{n} Q(\phi_{i,t})\, U(t-\tau)\, e^{j\omega_0 (t-\tau)} \quad \text{...(3.27)}$$

where

$$Q(\phi_{i,t}) = r_i \exp\{j[\omega_{1i} + \omega_{2i})t + \theta_i']\} \quad \text{...(3.28)}$$

$$\theta_i' = \theta - (\omega_{1i} + \omega_{2i})\tau \quad \text{...(3.29)}$$

Let $U(t) = e^{j\omega t}$ then Eq. (3.27) becomes

$$e_r(t) = \left[\sum_{i=1}^{n} Q(\phi_i, t)\, e^{-j(\omega_0 + \omega)\tau}\right]$$

$$= H(f, t)\, e^{j(\omega_0 + \omega)t} \quad \text{...(3.30)}$$

Hence

$$H(f, t) = \sum_{i=1}^{n} Q(\phi_i, t)\, e^{-j(\omega_0 + \omega)\tau} \quad \text{...(3.31)}$$

Let $f = 0$ then only sinusoidal carrier frequency is transmitted and amplitude of received signal is

$$r = H(0, t) \quad \text{...(3.32)}$$

where r is Rayleigh distributed random variable with its average power of $2\sigma^2$ and represented in probability density function as

$$P(r) = \frac{r}{\sigma^2} e^{-r^2/2\sigma^2} \quad \text{...(3.33)}$$

Equation (3.33) represents distribution of power of mobile unit with respect to other mobile unit.

3.9. CHARACTERISTICS OF BASIC ANTENNA STRUCTURES

An antenna is a vital component of any communication system. An antenna is basically a length of conductor and acts as a conversion device.

It serves a link between the transmitter and receiver. The first type of conversion takes place when an antenna is used for transmission where electrical energy is converted into electromagnetic waves.

The second conversion occurs at the receiving end where EM waves are transformed into electrical signal that is applied to the input of the receiver.

In two way communication, the same antenna can be and often is used for both transmission and reception. This is possible because any antenna transfers energy from the surrounding environment to its input receiver terminals with the same efficiency that it transfers energy from the output transmitter terminals into the surrounding environment, assuming that the same frequency is used in both directions.

Antenna characteristics are essentially the same whether an antenna is sending or receiving electromagnetic energy.

3.9.1. Radiation Patterns

An antenna will radiate power in all directions but typically does not perform equally in all directions. A common way to characterize the performance of an antenna is the radiation pattern, which is a graphical representation of the radiation properties of an antenna as a function of space coordinates.

The simplest pattern is produced by an idealized antenna known as the isotropic antenna.

An isotropic antenna is a point in space that radiates power in all directions equally.

The actual radiation pattern for isotropic antenna is a sphere with the antenna at the center.

The actual size of a radiation pattern is arbitrary. What is important is the relative distance from the antenna position in each direction. The relative distance determines the relative power.

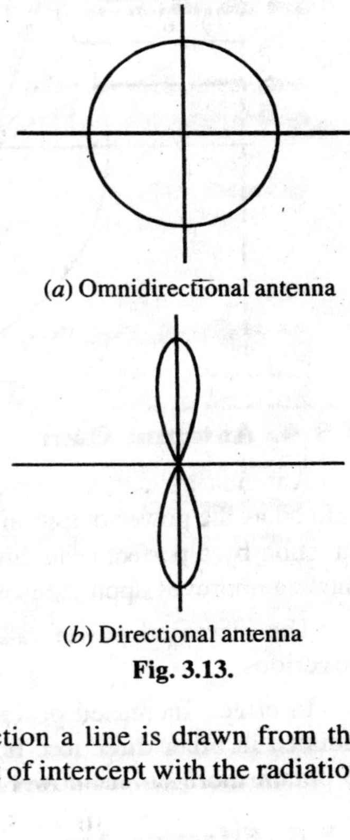

(*a*) Omnidirectional antenna

(*b*) Directional antenna

Fig. 3.13.

To determine the relative power in a given direction a line is drawn from the antenna position at the appropriate angle and the point of intercept with the radiation pattern is determined.

3.9.2. Antenna Types

There are two types of antennas. The simplest and most basic antennas are the half wave dipole or Hertz antenna and the quarter wave vertical or marconi antenna.

A half wave dipole has a uniform or omnidirectional radiation pattern in one dimension and a figure eight pattern in the other two dimensions.

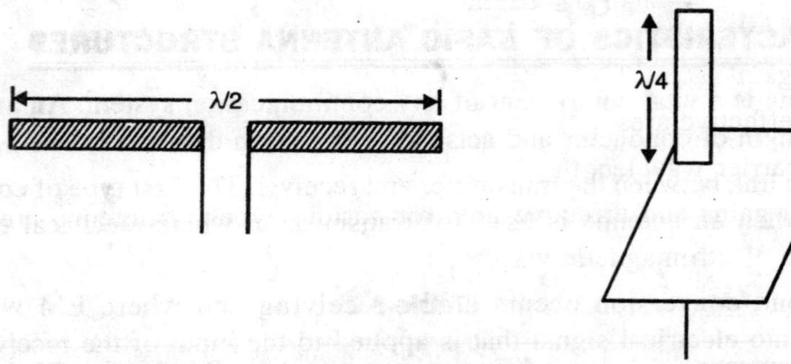

Fig. 3.14.

3.9.3. Parabolic Antenna

Parabolic reflective antenna is used in terrestrial micro and satellite applications. In mobile radio applications parabolic antenna is used for cell to cell communications.

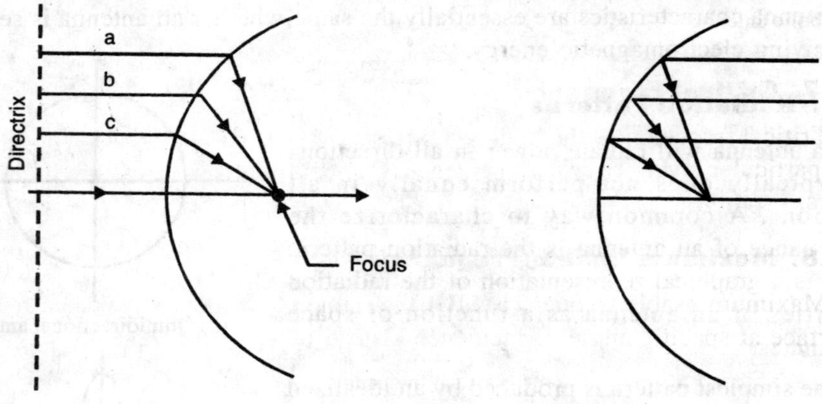

Fig. 3.15. Parabolic antenna.

3.9.4. Antenna Gain

Antenna gain is a measure of the directivity of an antenna. Antenna gain is defined as the power output in a particular direction compared to that produced in any direction by a perfect omnidirectional antenna. If an antenna has a gain of 3 dB that antenna improves upon the isotropic antenna in the directivity by 3 dB or a factor of 2.

The increased power radiated in a given direction is at the expense of other directions.

In effect, increased power is radiated in one direction by reducing the power radiated in other direction. It is important to note that antenna gain does not refer to obtain more output power than input power but rather to directivity.

3.9.5. Effective Area

Effective area of an antenna is related to the physical size of the antenna and to its shape.

3.9.6. Relation Between Gain and Effective Area

$$G = \frac{4\pi Ae}{\lambda^2} \quad \text{...(3.34)}$$

where G is gain

A_e is effective area

λ is carrier wavelength.

Antenna gains and effective area for various types of antenna are shown in Table 3.2.

Table 3.2.

Type of antenna	Effective area	Power gain with respect to isotropic antenna
Isotropic	$\frac{\lambda^2}{4\pi}$	1
Half wave dipole	$\frac{1.6\lambda^2}{4\pi}$	1.64
Parabolic antenna	0.56 A	$\frac{7A}{\lambda^2}$

3.9.7. Critical Frequency

Critical frequency is the highest frequency which can be reflected back to earth by a particular layer at vertical incidence. Critical frequency is different for various layers.

3.9.8. Maximum Usable Frequency

Maximum usable frequency (MUF) is frequency which is reflected back from a surface at specific angle of incidence instead of vertical frequency.

$$\text{MUF} = \frac{\text{Critical frequency}}{\cos\theta} \Rightarrow f_c \sec\theta \quad \text{...(3.35)}$$

3.10. ANTENNA AT CELL SITE (OMNIDIRECTIONAL ANTENNA)

Antenna at cell site may be directional or may be omnidirectional. Let us consider omnidirectional antenna. Generally, antenna of gain 6 dB and 9 dB is used in mobile radio environment.

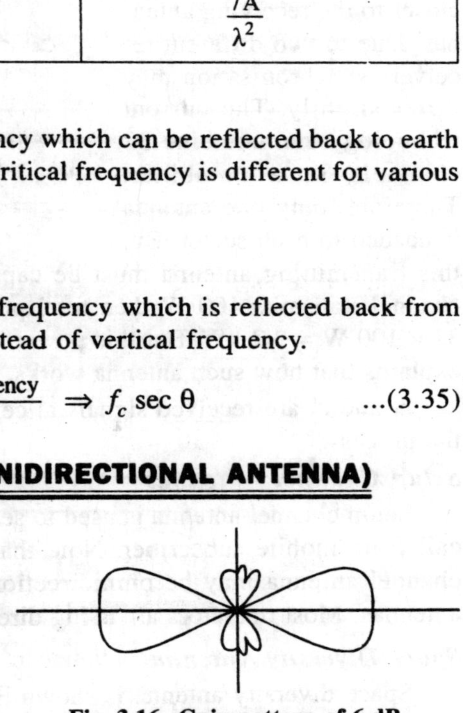

Fig. 3.16. Gain pattern of 6 dB omnidirectional antenna.

Figure 3.16 shows the pattern of gain of antenna. Currently network operators are not using omnidirectional antenna for mobile wave propagations.

3.10.1. Antenna at Cell Site (Directional Antenna)

If network operator will use directional antenna then gain will be more and radiation pattern will be different.

In mobile radio environment we generally use directional antenna with frequency re-use concept. If $K = 7$ cell pattern, very frequently use frequency pattern and if 210 channels will get used then each cell will get 30° frequency. Each 120° sector would have one transmitting antenna and two receiving antenna and serve 15 radio channels. Network operator using 2 receiving antenna for diversity application.

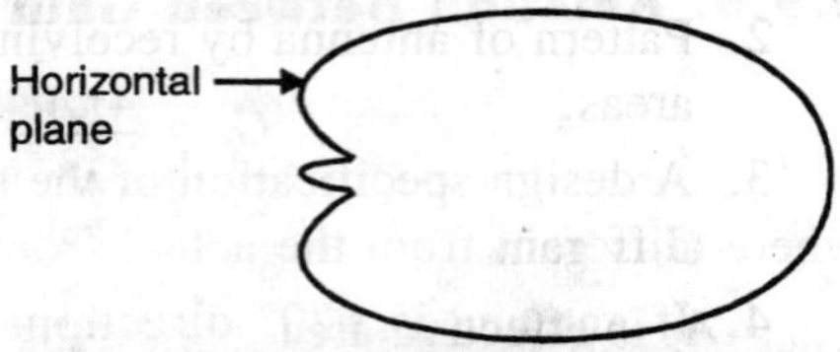

Fig. 3.17. 8 dB directional pattern of 120° sectored antenna.

If number of subscribers increasing gradually then re-use frequency concept is more economical than cell splitting concept. If we use frequency re-use concept then every cell will get more channels. Two transmitting antennas in each sector are placed relatively closer to the receiving antennas. Due to two different receivers synchronisation may differ slightly. The current technology can combine 32 channels in a combiner. Therefore, only one antenna is needed in each sector. But this transmitting antenna must be capable to deliver more power. If each radio channel transmits 100 W then total power that can be transmitted by antenna is 32 × 100 W = 3.2 kW. But this power is determined by manufacturer. Figure 3.18 explains that how such antenna works.

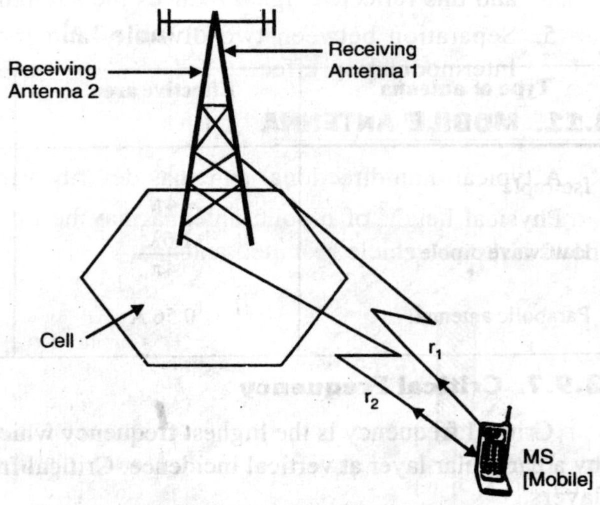

Fig. 3.18. Space diversity antenna.

r_1 and r_2 are received signals at cell antenna and then combined together and the process.

Setup Channel Antennas

Setup channel antenna is used to search a mobile subscriber unit or to access a call from mobile subscriber. Note that setup channel is carry only data. Setup channel antenna may be omnidirectional antenna or consist several directional antennas. Most operators are using directional antennas.

Space Diversity Antenna

Space diversity antenna is shown in Fig. 3.18 and it is used at the cell site to receive the same signal. Degree of synchronisation between two antennas depends upon separation between two antennas. Generally this separation is 45°. When both receiving signal strength combined then resultant signal becomes strong and are easy to process.

Some important points regarding cell site antenna:

1. If cell site antenna height is lower than height of surrounding buildings and other obstacles then mobile unit receive pattern is different as to mobile receive in free space area without obstacle.

2. Pattern of antenna by receiving mobile unit is different in rural and urban areas.
3. A design specification of the front to back ratio of a directional antenna is different from the actual front to back ratio in mobile radio applications.
4. If antenna is 120° directional antenna then front to back ratio is about 10 dB less than the front lobe regardless of whether a weak side lobe pattern or no sidelobe pattern is designed in free space. This occur because the strong signal radiates in front and reflects from the surrounding objects and this reflected signal reduces the strength of resultant signal.
5. Separation between two diversity antennas should be small to reduce intermodulation effect.

3.11. MOBILE ANTENNA

A typical omnidirectional antennas described in Fig. 3.19.

Physical height of mobile antenna may be different cell to cell. Figure 3.19 shows a motor vehicle mounted antenna.

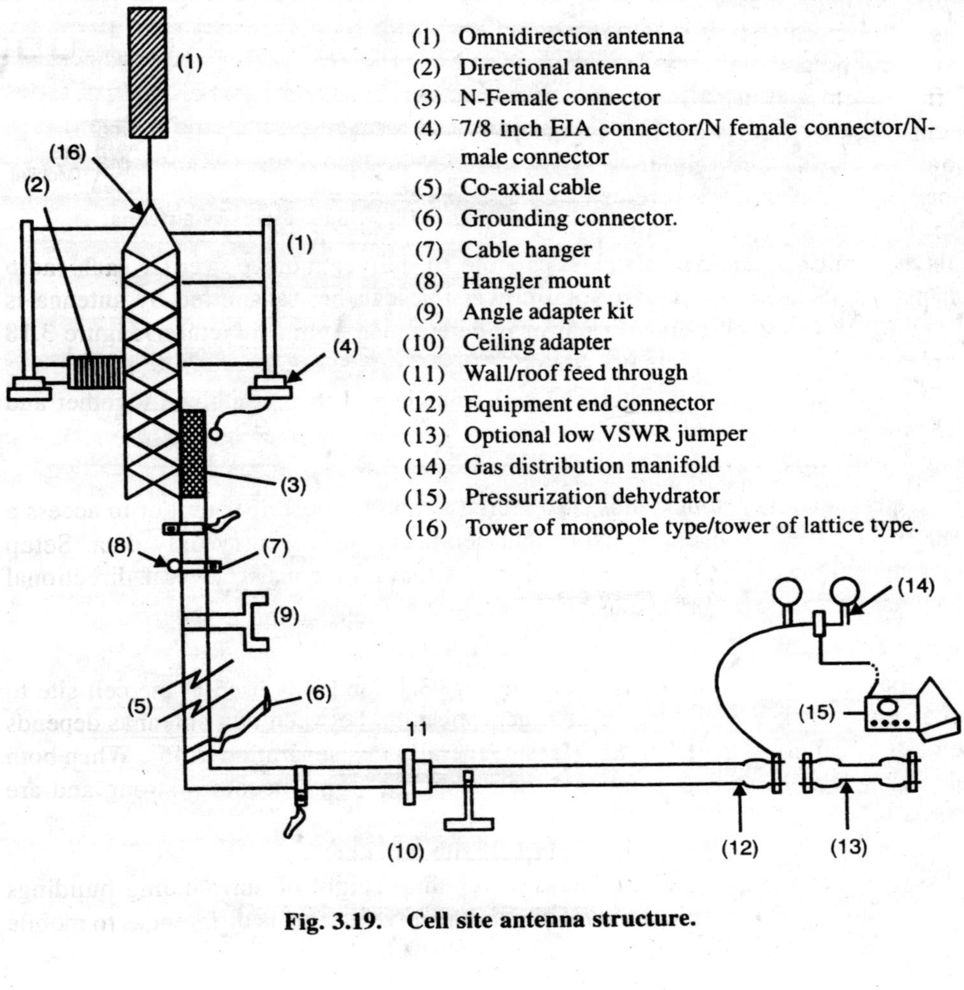

Fig. 3.19. Cell site antenna structure.

3.11.1. Room Mounted Antenna

Physical structure will be same of a room mounted antenna but physical height and gain may be different. It depends on whether the antenna will be placed in rural area or urban area.

REVIEW QUESTIONS

1. What do you mean by antenna ?
2. What are basic characteristics of antenna ?
3. Explain Foliage loss.
4. Explain point to point model.
5. Discuss about the features of antenna structure used for cell coverage in cellular architecture based system.
6. Why hexagonal shape is preferred over other cell shapes in cellular network.
7. What are the difference between omni and directive antenna.
8. Explain mobile to mobile propagation.
9. Discuss about the features of antenna structures used for cell coverage in cellular architecture based systems.
10. What is effect of antenna height ?

❑❑❑

4 CHAPTER

Frequency Management & Channel Assignment, Handoff & Dropped Calls

4.0. INTRODUCTION

INSIDE THIS CHAPTER

4.0. Introduction
4.1. Handoff
4.2. Channel Assignment Schemes and Handoff
4.3. GSM (Global System for Mobile Communication)
4.4. GPRS (General Packet Radio Service)
4.5. Edge Technology
4.6. Dropped Call Rates and Their Evaluations

Frequency management is having very important role in designing of cellular network. As we know, we need two frequency for connecting a call. Frequency is very important resource as it is allocated by Government authorities. 2 to 3 frequency allocated in one telecom circle for one operator and one frequency can be share with eight subscribers only. So, it become a big challenge for operator to use available frequency with optimum application. Handoff is also very important because halfoff is a technique for reducing the call drop during conversation going on in movement of a subscriber.

4.1. HANDOFF

In cellular telecommunications, the term **handoff** refers to the process of transferring an ongoing call or data session from one channel connected to the core network to another. In satellite communications it is the process of transferring satellite control responsibility from one earth station to another without loss or interruption of service. The British English term for transferring a cellular call is **handover**, which is the terminology standardised within such European originated technologies as GSM and UMTS.

In telecommunications there are two reasons why a handoff (handover) might be conducted: if the phone has moved out of range from one cell site (base station) and can get a better radio link from a stronger transmitter, or if one base station is full the connection can be transferred to another nearby base station.

The most basic form of handoff is that used in GSM and analog cellular networks, where a phone call in progress is redirected from one cell site and its transmit/receive frequency pair to another base station (or sector within the same cell) using a different frequency pair without interrupting the call. As the phone can be connected to only one base station at a time and therefore, needs to drop the radio link for a brief period of time before being connected to a different, stronger transmitter, this is referred to as a **hard handoff**. This type of handoff is described as "break before make" (referring to the radio link).

In CDMA systems the phone can be connected to several cell sites simultaneously, combining the signalling from nearby transmitters into one signal using a rake receiver. Each cell is made up of one to three (or more) sectors of coverage, produced by a cell site's independent transmitters outputting through antennas pointed in different directions. The set of sectors the phone is currently linked to is referred to as the "active set". A soft handoff occurs when a CDMA phone adds a new sufficiently-strong sector to its active set. It is so called because the radio link with the previous sector(s) is not broken before a link is established with a new sector; this type of handoff is described as "make before break". In the case where two sectors in the active set are transmitted from the same cell site, they are said to be in **softer handoff** with each other.

There are also **inter-technology handoffs** where a call's connection is transferred from one access technology to another, *e.g.*, a call being transferred from GSM to W-CDMA. But in existing technology it is not possible.

4.2. CHANNEL ASSIGNMENT SCHEMES AND HANDOFF

4.2.1. Channel Assignment Strategy

For efficient utilization of the radio spectrum, a frequency reuse scheme that is consistent with the objectives of increasing capacity and minimizing interference is required. A variety of channel assignment strategies have been developed to achieve these objectives. Channel assignment strategies can be classified as either *fixed* or *dynamic*. The choice of channel assignment strategy impacts the performance of the system, particularly as to how calls are managed when a mobile user is handed off from one cell to another.

4.2.2. Fixed Channel Assignment Strategy

In a fixed channel assignment strategy, each cell is allocated a predetermined set of voice channels. Any call attempt within the cell can only be served by the unused channels in that particular cell. If all the channels in that cell are occupied, the call is *blocked* and the subscriber does not receive service. Several variations of the fixed assignment strategy exist. In one approach, called the *borrowing strategy,* a cell is allowed to borrow channels from a neighbouring cell if all of its own channels are already occupied. The mobile switching center (MSC) supervises such borrowing procedures and ensures that the borrowing of a channel does not disrupt or interfere with any of the calls in progress in the donor cell.

4.2.3. Dynamic Channel Assignment Strategy

In a dynamic channel assignment strategy, voice channels are not allocated to different cells permanently. Instead, each time a call request is made, the serving base station requests a channel from the MSC. The switch then allocates a channel to the requested cell following an algorithm that takes into account the likelihood of future blocking within the cell, the frequency of use of the candidate channel, the reuse distance of the channel, and other cost functions. Accordingly, the MSC only allocates a given frequency if that frequency is not presently in use in the cell or any other cell which falls within the minimum restricted distance of frequency reuse to avoid co-channel interference. Dynamic channel assignment reduces the likelihood of blocking, which increases the trunking capacity of the system, since all the available channels in a market are accessible to all of the cells. Dynamic channel assignment strategies require the MSC to collect real-time data on channel occupancy, traffic distribution, and *radio signal strength indications* (RSSI) of all channels on a continuous basis. This increases the storage and computational load on the system but provides the advantage of increased channel utilization and decreased probability of a blocked call.

4.2.4. Handoff Strategies

When a mobile moves into a different cell while a conversation is in progress, the MSC automatically transfers the call to a new channel belonging to the new base station. This handoff operation not only involves identifying a new base station, but also requires that the voice and control signals be allocated to channels associated with the new base station. Processing handoffs is an important task in any cellular radio system. Many handoff strategies prioritize handoff requests over call initiation requests when allocating unused channels in a cell site. Handoffs must be performed successfully and as infrequently as possible, and be imperceptible to the users. In order to meet these requirements, system designers must specify an optimum signal level at which to initiate a handoff. Once a particular signal level is specified as the minimum usable signal for acceptable voice quality at the base station receiver (normally taken as between –90 dBm and –100 dBm), a slightly stronger signal level is used as a threshold at which a handoff is made. This margin, given by, $\Delta = P_{r\ \text{handoff}} - Pr_{\text{minimum usable}}$ cannot be too large or too small. If Δ is too large, unnecesary handoffs which burden the MSC may occur, and if Δ is too small, there may be insufficient time to complete a handoff before a call is lost due to weak signal conditions. Therefore, Δ is chosen carefully to meet these conflicting requirements. Figure 4.1 illustrates a handoff situation.

In deciding cases of handoff, it is important to ensure that the drop in the measured signal level is not due to momentary fading and that the mobile is actually moving away from the serving base station. In order to ensure this, the base station monitors the signal level for a certain period of time before a handoff is initiated. This running average measurement of signal strength should be optimized so that unnecessary handoffs are avoided, while ensuring that necessary handoffs are completed before a call is terminated due to poor signal level. The length of time needed to decide if a handoff is necessary depends on the speed at which the vehicle is moving. If the slope of the short-term average received signal level in a given time interval is steep, the handoff should be made quickly. Information about the vehicle speed, which can be useful in handoff decisions, can also be computed from the statistics of the received short-term fading signal at the base station.

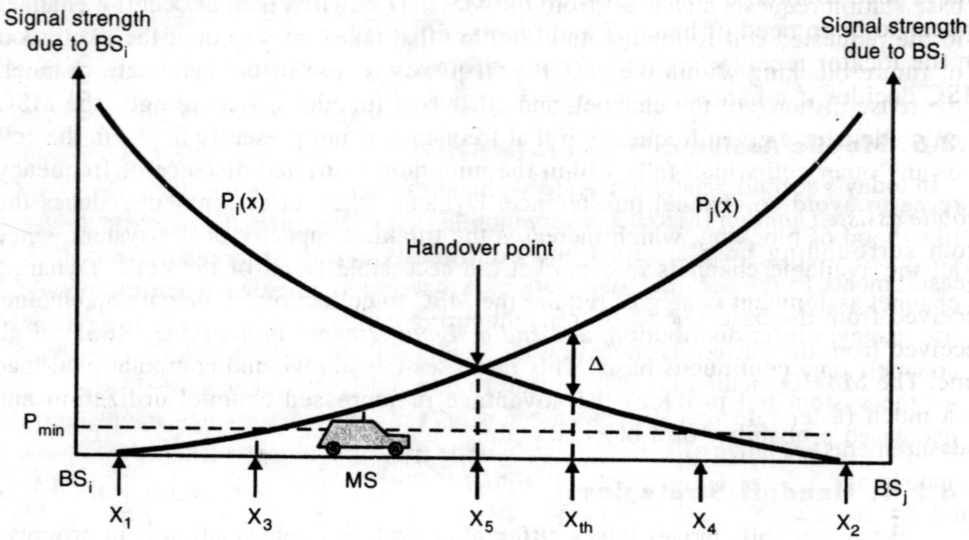

Fig. 4.1. By looking at the variation of signal strength from either base station it is possible to decide on the optimum area where handoff can take place.

The time over which a call may be maintained within a cell, without handoff, is called the *dwell time.* The dwell time of a particular user is governed by a number of factors, including propagation, interference, distance between the subscriber and the base station, and other time varying effects. Even when a mobile user is stationary, ambient motion in the vicinity of the base station and the mobile can produce fading; thus, even a stationary subscriber may have a random and finite dwell time. The statistics of dwell time vary greatly, depending on the speed of the user and the type of radio coverage. For example, in mature cells which for vehicular highway users, most users tend to have a relatively constant speed and travel along fixed and well-defined paths with good radio coverage. In such instances, the dwell time for an arbitrary user is a random variable with a distribution that is highly concentrated about the mean dwell time. On the other hand, for users in dense, cluttered microcell environments, there is typically a large variation of dwell time about the mean, and the dwell times are typically shorter than the cell geometry would otherwise suggest. It is apparent that the statistics of dwell time are important in the practical design of handoff algorithms.

4.2.5. Network Initiated Handoff

Base stations make measurements of signal strength and report back to MSC, which makes the decision to handoff. In first generation analog cellular systems, signal strength measurements are made by the base stations and supervised by the MSC. Each base station constantly monitors the signal strengths of all of its reverse voice channels to determine the relative location of each mobile user with respect to the base station tower. In addition to measuring the RSSI of calls in progress within the cell, a spare receiver in each base station, called the locator receiver, is used to scan and determine signal strengths of mobile users which are in neighbouring cells. The *locator receiver* is controlled by the MSC and is used to monitor the signal strength of users in neighbouring cells which appear to be in need of handoff and reports all RSSI values to the MSC. Based on the locator receiver signal strength information from each base station, the MSC decides if a handoff is necessary or not.

4.2.6. Mobile Assisted Handoff (MAHO)

In today's second generation systems, handoff decisions are *mobile assisted.* In *mobile assisted handoff* (MAHO), every mobile station measures the received power from surrounding base stations and continually reports the results of these measurements to the serving base station. A handoff is initiated when the power received from the base station of a neighbouring cell begins to exceed the power received from the current base station by a certain level or for a certain period of time. The MAHO method enables the call to be handed over between base stations at a much faster rate than in first generation analog systems since the handoff measurements are made by each mobile, and the MSC no longer constantly monitors signal strengths. MAHO is particularly suited for microcellular environments where handoffs are more frequent.

During the course of a call, if a mobile moves from one cellular system to a different cellular system controlled by a different MSC, an *inter MSC handoff* becomes necessary. An MSC engages in an intersystem handoff when a mobile signal becomes weak in a given cell and the MSC cannot find another cell within its system to which it can transfer the call in progress. There are many issues that must be addressed when implementing an inter MSC handoff. For instance, a local call may become a long-distance call as the mobile moves out of its home system and becomes a roamer in a neighbouring system. Also, compatibility between the two MSCs must be determined before implementing an intersystem handoff.

4.2.7. Prioritizing Handoffs

Guard Channels : One method for giving priority to handoffs is called the *guard channel concept,* whereby a fraction of the total available channels in a cell is reserved exclusively for handoff requests from ongoing calls which may be handed off into the cell. This method has the disadvantage of reducing the total carried traffic, as fewer channels are allocated to originating calls. Guard channels, however, offer efficient spectrum utilization when dynamic channel assignment strategies, which minimize the number of required guard channels by efficient demand-based allocation, are used.

4.2.8. Queuing of Calls

Queuing of handoff requests is another method to decrease the probability of forced termination of a call due to lack of available channels. There is a tradeoff between the decrease in probability of forced termination and total carried traffic. Queuing of handoffs is possible due to the fact that there is a finite time interval between the time the received signal level drops below the handoff threshold and the time the call is terminated due to insufficient signal level. The delay time and size of the queue is determined from the traffic pattern of the particular service area. It should be noted that queuing does not guarantee a zero probability of forced termination, since large delays will cause the received signal level to drop below the minimum required level to maintain communication and hence lead to forced termination.

4.2.9. Practical Handoff Considerations

In practical cellular systems, several problems arise when attempting to design for a wide range of mobile velocities. High speed vehicles pass through the coverage region of a cell within a matter of seconds, whereas pedestrian users may never need a handoff during a call. Particularly with the addition of microcells to provide capacity, the MSC can quickly become burdened if high speed users are constantly being passed between very small cells. Several schemes have been devised to handle the simultaneous traffic of high speed and low speed users while minimizing the handoff intervention from the MSC. Another practical limitation is the ability to obtain new cell sites.

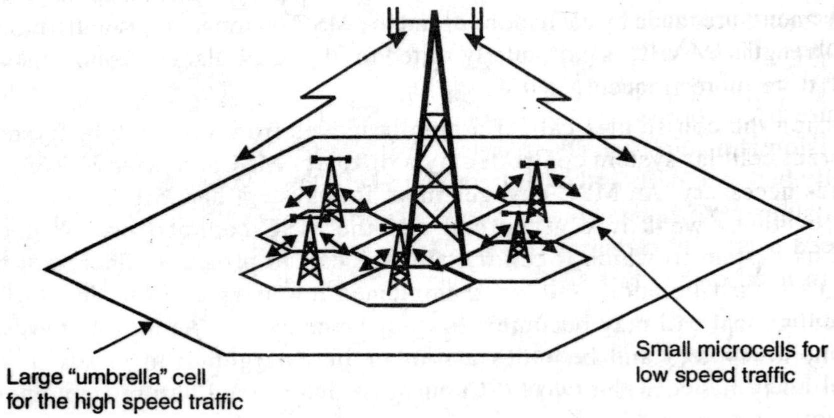

Fig. 4.2. Umbrella cell configuration.

Although the cellular concept clearly provides additional capacity through the addition of cell sites, in practice it is difficult for cellular service providers to obtain new physical cell site locations in urban areas. Zoning laws, ordinances, and other non-technical barriers often make it more attractive for a cellular provider to install additional channels and base stations at the same physical location of an existing cell, rather than find new site locations. By using different antenna heights (often on the same building or tower) and different power levels, it is possible to provide "large" and "small" cells which are co-located at a single location. This technique is called the *umbrella cell* approach and is used to provide large area coverage to high speed users while providing small area coverage to users travelling at low speeds. Figure 4.2 illustrates an umbrella cell which is collocated with some smaller microcells. The umbrella cell approach ensures that the number of handoffs is minimized for high speed users and provides additional microcell channels for pedestrian users. The speed of each user may be estimated by the base station or MSC by evaluating how rapidly the short-term average signal strength on the Reverse Voice Channel (RVC) changes over time. If a high speed user in the large umbrella cell is approaching the base station, and its velocity is rapidly decreasing, the base station may decide to hand the user into the co-located microcell, without MSC intervention.

Another practical handoff problem in microcell systems is known as *cell dragging*. Cell dragging results from pedestrian users that provide a very strong signal to the base station. Such a situation occurs in an urban environment when there is a line-of-sight (LOS) radio path between the subscriber and the base station. As the user travels away from the base station at a very slow speed, the average signal strength does not decay rapidly. Even when the user has travelled well the designed range of the cell, the received signal at the base station may be above the handoff threshold, thus a handoff may not be made. This creates a potential interference and traffic management problem, since the user has meanwhile travelled deep within a neighbouring cell. To solve the cell dragging problem, handoff thresholds and radio coverage parameters must be adjusted carefully.

In first generation analog cellular systems, the typical time to make a handoff, once the signal level is deemed to be below the handoff threshold, is about 10 seconds. This requires that the value for Δ be of the order of 6 dB to 12 dB. In digital cellular systems such as GSM, the mobile assists with the handoff procedure by determining the best handoff candidates, and the handoff, once the decision is made, typically requires only 20 milli seconds. Consequently, Δ is usually between 0 dB and 6 dB in modern cellular systems. The faster handoff process supports a much greater range of options for handling high speed and low speed users and provides the MSC with substantial time to “rescue” a call that is in need of handoff.

Another feature of newer cellular systems is the ability to make handoff decisions based on a wide range of metrics other than signal strength. The co-channel and adjacent channel interference levels may be measured at the base station or the mobile, and this information may be used with conventional signal strength data to provide a multi-dimensional algorithm for determining when a handoff is needed.

The IS-95 code division multiple access (CDMA) spread spectrum cellular system provides a unique handoff capability that cannot be provided with other wireless systems. Unlike channelized wireless systems that assign different radio channels during a handoff (called a *hard handoff*), spread spectrum mobiles share the same channel in every cell. Thus, the term *handoff does* not mean a physical change in the assigned channel, but rather that a different base station handles the radio communication task. By simultaneously evaluating the received signals from a single subscriber at several neighbouring base stations, the MSC may actually decide which version of the user's signal is best at any moment in time. This technique exploits macroscopic space diversity provided by the different physical locations of the base stations and allows the MSC to make a “soft” decision as to which version of the user's signal to pass along to the PSTN at any instance. The ability to select between the instantaneous received signals from a variety of base stations is called *soft handoff*.

4.2.10. Type of Handoff (Hand Over)

From previous discussion as we know handover is a provision to make your call continuing when you have moved from one cell to another or when you have moved far away from serving cell. It is important to note that the decision to perform a handover is always made by the BSC that is currently serving the subscriber, except for the handover for traffic reasons. In the latter case the MSC makes the decision. There are 4 different types of handover and the best way to analyse them is to follow the subscriber as he moves. But handover may be initiated by mobile or by network.

4.2.10.1. Intra-cell–Intra BSC Handover

The smallest of the handover is the intra-cell handover where the subscriber is handed over to another traffic channel (generally in another frequency) within the same cell. In this case, the BSC controlling the cell makes the decision to perform handover. (Fig. 4.3)

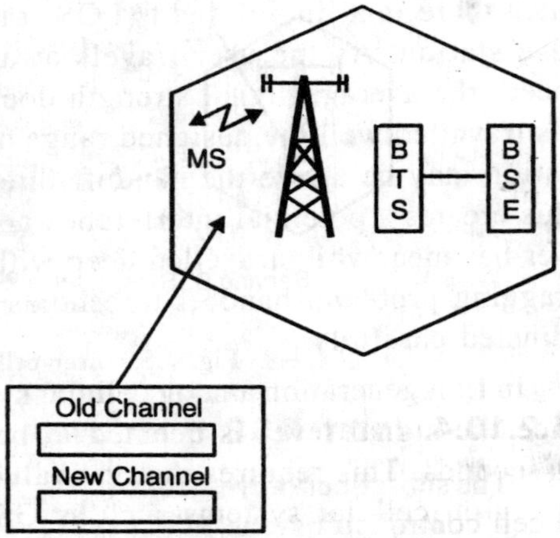

Fig. 4.3. Intra-cell–Intra BSC handover.

4.2.10.2. Inter-cell–Intra BSC Handover

The subscriber moves from cell 1 to cell 2 but within BSC. In this case, handover process is controlled by the BSC. Traffic connection with cell 1 is released when the connection with cell 2 is setup successfully. (Fig. 4.4)

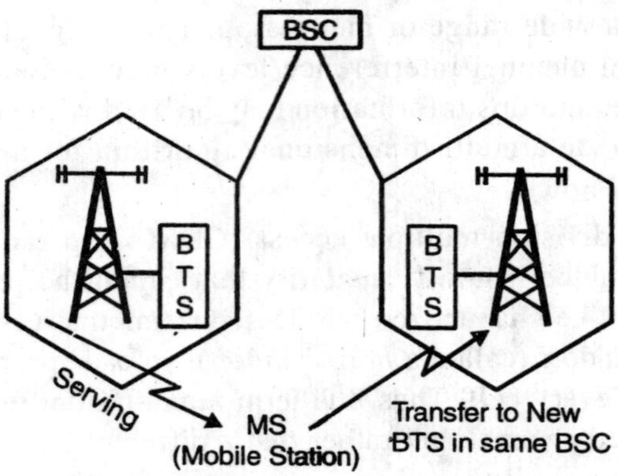

Fig. 4.4. Inter-cell–Intra BSC handover.

4.2.10.3. Inter-cell–Inter BSC Handover

The subscriber moves from cell 2 to cell 3, which is served by another BSC. In this case, the handover process is carried out by the MSC, but the decision to make the handover is still done by the first BSC. The connection with the first BSC and BTS is released when the connection with the new BSC and BTS is setup successfully.

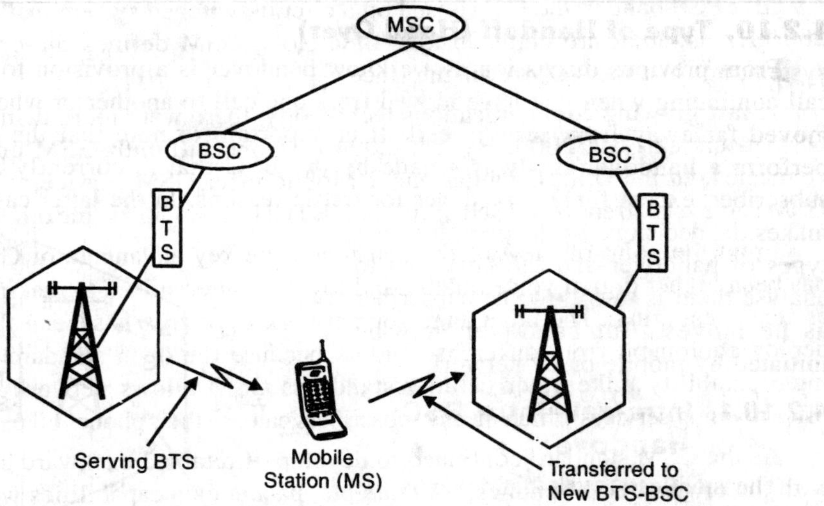

Fig. 4.5. Inter-cell–Inter BSC handover.

4.2.10.4. Inter MSC Handover

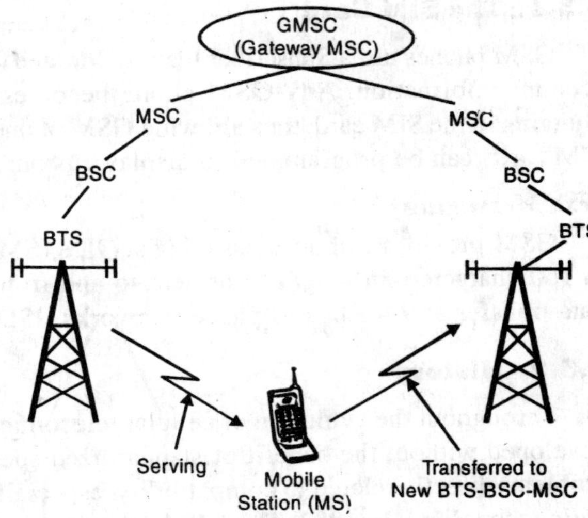

Fig. 4.6. Inter MSC handover.

The subscriber moves from a cell controlled by one MSC/VLR to a cell in the domain of another MSC/VLR. This case is more complicated. The MSC/VLR currently serving the subscriber contacts the target MSC/VLR and the traffic connection is transferred to the target MSC/VLR. As both MSC are part of the same network, the connection is established smoothly. It is important to note, however, that the target MSC and the source MSC are two telephone exchanges. The call can be transferred between two exchanges only if there is a telephone number identifying the target MSC.

Note: For more detail about Handoff in CDMA/WCDMA system refer to chapter seven.

4.3. GSM (GLOBAL SYSTEM FOR MOBILE COMMUNICATION)

GSM (Global System for Mobile Communications) : A digital cellular phone technology based on TDMA that is the predominant system in Europe, but also used worldwide. Developed in the 1980s, GSM was first deployed in seven European countries in 1992. It operates in the 900 MHz and 1.8 GHz bands in Europe and the

1.9 GHz PCS band in the U.S. Based on a circuit-switched system that divides each 200 kHz channel into eight 25 kHz time slots, GSM defines the entire cellular system, not just the TDMA air interface.

A fast-growing communications technology, there were more than 250 million GSM users early in 2000. By mid-2004, the one billionth GSM customer was connected. In the U.S., Cingular and T-Mobile offer GSM service, while in India GSM service is offered by BSNL, AirTel, HUTCH, IDEA and some other companies.

From the point of view of the consumer, the key advantage of GSM systems has been higher digital voice quality and low cost alternatives to making calls such as text messaging. The advantage for network operators has been the ability to deploy equipment from different vendors because the open standard allows easy interoperability. Like other cellular standards GSM allows network operators to offer roaming services which means subscribers can use their phones all over the world.

As the GSM standard continued to develop, it retained backward compatibility with the original GSM phones; for example, packet data capabilities were added in the Release '97 version of the standard, by means of GPRS. Higher speed data transmission has also been introduced with EDGE in the Release '99 version of the standard.

4.3.1. The SIM Card

GSM phones use a Subscriber Identity Module (SIM) smart card that contains user account information. Any GSM phone becomes immediately programmed after plugging in the SIM card, thus allowing GSM phones to be easily rented or borrowed. SIM cards can be programmed to display custom menus for personalized services.

Text Messaging

GSM provides a short messaging service (SMS) that enables text messages up to 160 characters in length to be sent to and from a GSM phone. It also supports data transfer at 9.6 Kbps to packet networks, ISDN users.

4.3.2. History

Throughout the evolution of cellular telecommunications, various systems were developed without the benefit of standardized specifications. This presented many problems directly related to compatibility, especially with the development of digital radio technology. In 1982, The GSM group (*"Groupe Special Mobile"* (French) 1, 2, 3 and 4) was formed to address these problems. The name of the system comes from the name of this group, though later the decision was made to keep the initials but to change what they stood for. Originally the group was hosted by European Conference of Postal and Telecommunications Administrations (CEPT).

From 1982 to 1985 discussions were held to decide between building an analog or digital system. After multiple field tests, a digital system was adopted for GSM. The next task was to decide between a narrow or broadband solution. In May 1987, the narrowband time division multiple access (TDMA) solution was chosen.

By the mid-1980s, many of the European countries had developed their own systems. This led to disagreement on what system to use. The conflict almost stopped the project. However, the European Union intervened and all 15 countries decided to choose the standard recommended by CEPT. In February 1987, eight different systems were tested in a competition held in Paris. A system developed by researchers Torleiv Maseng and Odd Trandem at the Norwegian University of Science and Technology won and was chosen. This was particularly surprising as they had no industry backing. Their system won by its ability to counter signal reflection in urban areas and in mountainous terrain.

The technical fundamentals of the GSM system were defined in 1987. In 1989, the European Telecommunications Standards Institute (ETSI) took over control and by 1990 the first GSM specification was completed, amounting to over 6,000 pages of text. Commercial operation began in 1991 with Radiolinja in Finland.

In 1998, the 3rd Generation Partnership Project (3GPP) was formed. Originally, it was intended only to produce the specifications of the next (third, 3G) generation of mobile networks. However, 3GPP also took over the maintenance and development of the GSM specification. ETSI is a partner in 3GPP.

GSM provides recommendations, not requirements. The GSM specifications define the functions and interface requirements in detail but do not address the hardware. The reason for this is to not limit the designers yet still make it possible for the operators to buy equipment from multiple suppliers.

4.3.3. Radio Interface

GSM is a cellular network, which means that mobile phones connect to it by searching for cells in the immediate vicinity. GSM networks operate in four different frequency ranges. Most GSM networks operate in the 900 MHz or 1800 MHz bands. Some countries in the Americas (including the United States and Canada) use the 850 MHz and 1900 MHz bands because the 900 and 1800 MHz frequency bands were already allocated.

The rarer 400 and 450 MHz frequency bands are assigned in some countries, notably Scandinavia, where these frequencies were previously used for first-generation systems.

In the 900 MHz band the uplink frequency band is 890-915 MHz, and the downlink frequency band is 935-960 MHz. This 25 MHz bandwidth is subdivided into 124 carrier frequency channels, each spaced 200 kHz apart. Time division multiplexing is used to allow eight speech channels per radio frequency channel. There are eight radio timeslots (giving eight burst periods) grouped into what is called a TDMA frame. The channel data rate is 270.833 kbit/s, and the frame duration is 4.615 ms.

The transmission power in the handset is limited to a maximum of 2 watts in GSM850/900 and 1 watt in GSM1800/1900.

GSM originally used a variant of linear predictive coding (LPC). The purpose of LPC is to reduce the bit rate. The LPC provides parameters for a filter that mimics the vocal tract. The signal passes through this filter, leaving behind a residual signal. This was encoded at 13 kbit/s. However, updates to the standard allowed for GSM-EFR, and AMR-Narrowband—which produce wireline like quality at 12.2 or 13 kbit/s.

There are four different cell sizes in a GSM network macro, micro, pico and umbrella cells. The coverage area of each cell varies according to the implementation environment. Macro cells can be regarded as cells where the base station antenna is installed on a mast or a building above average rooftop level. Micro cells are cells whose antenna height is under average rooftop level; they are typically used in urban areas. Picocells are small cells whose diameter is a few dozen meters; they are mainly used indoors. On the other hand, umbrella cells are used to cover shadowed regions of smaller cells and fill in gaps in coverage between those cells.

Cell radius varies depending on antenna height, antenna gain and propagation conditions from a couple of hundred meters to several tens of kilometers. The longest distance the GSM specification supports in practical use is 35 km or 22 miles. There are also several implementations of the concept of an extended cell, where the cell radius could be double or even more, depending on the antenna system, the type of terrain and the timing advance.

Indoor coverage is also supported by GSM and may be achieved by using an indoor picocell base station, or an indoor repeater with distributed indoor antennas fed through power splitters, to deliver the radio signals from an antenna outdoors to the separate indoor distributed antenna system. These are typically deployed when a lot of call capacity is needed indoors, for example in shopping centers or airports. However, this is not a prerequisite, since indoor coverage is also provided by in-building penetration of the radio signals from nearby cells.

The modulation used in GSM is Gaussian minimum shift keying (GMSK), a kind of continuous-phase frequency shift keying. In GMSK, the signal to be modulated onto the carrier is first smoothed with a Gaussian low-pass filter prior to being fed to a frequency modulator, which greatly reduces the interference to neighbouring channels (adjacent channel interference).

The GSM network is the source of the mysterious "dit dit dit, dit dit dit, dit dit dit" signal that can be heard intermittently on home stereo systems, televisions, computers, and personal music devices such as Apple's iPod. This can be attributed to the Time Division multiplexing method, as well as RFI.

4.4. GPRS (GENERAL PACKET RADIO SERVICE)

Wireless telecommunication infrastructure is evolving fast to accommodate high speed data traffic while simultaneously handling voice traffic. First generation mobile phones had only voice facility. These were replaced by second generation (2G) digital phones with added fax, data and messaging services. The third generation (3G) technology has added multimedia facilities to 2G phones.

To narrow the gap between 2G and 3G communication systems, **General Packet Radio Services (GPRS),** or the so called **2.5G,** has been introduced as an intermediary. GPRS allows high speed data transfer over conventional GSM and TDMA networks. It is the outcome of continued efforts to push high speed data over existing mobile networks. Instead, GSM network is used to obtain GPRS user-profile data from location register databases.

GPRS utilizes 1 to 8 radio time slots for multiple users to share. The user data is put into packets and transported over Public Land and Mobile Network (PLMN) using the internet protocol (IP) backbone. GPRS offers high speed of 14400 bps to 115000 bps for wireless internet access and interfaces to the public data networks (PDNs). It allows short bursts of traffic, such as e-mail, but can also handle large volumes of data.

General Packet Radio Service (GPRS) is a standardized packet switched data service for GSM based networks. The GPRS service provides :

➤ Efficient use of radio resources
➤ A flexible service at low cost to the users
➤ Fast setup/access time
➤ Efficient transport of packets in the GSM network, up to 115 kbit/s
➤ Simultaneous GSM and GPRS, co-existence without disturbance
➤ Connectivity to other data networks, for example IP and X. 25

GPRS introduces packet data transmission to the mobile subscriber. GPRS is being designed to work with existing GSM infrastructure with additional packet switching nodes. GPRS will ensure direct connectivity from mobile to almost any data source in the world. GPRS radio resources will be shared by several users, thus giving better channel utilization. The user may remain connected to the accessed data network as long as desired, but only need to be charged for the data volume received and/or sent. Each packet of data in GPRS contains the destination address, thus there is no need to reserve and establish a path for the packet to follow the receiver. GPRS is user friendly—providing fast call setup times and allowing the user to be online without charging for more than the data transferred.

4.4.1. GPRS Architecture

It should be noted that little or no hardware upgrades are required in existing GSM nodes and that the same transmission links can be used. Standard GSM functions like authentication should also be used for GPRS. The ciphering functionality will be new.

There will be three new logical nodes introduced to handle the GPRS applications in the GSM networks.

➤ **SGSN**—Serving GPRS Support Node
➤ **GGSN**—Gateway GPRS Support Node
➤ **PTM-SC**—Point To Multipoint Service Center

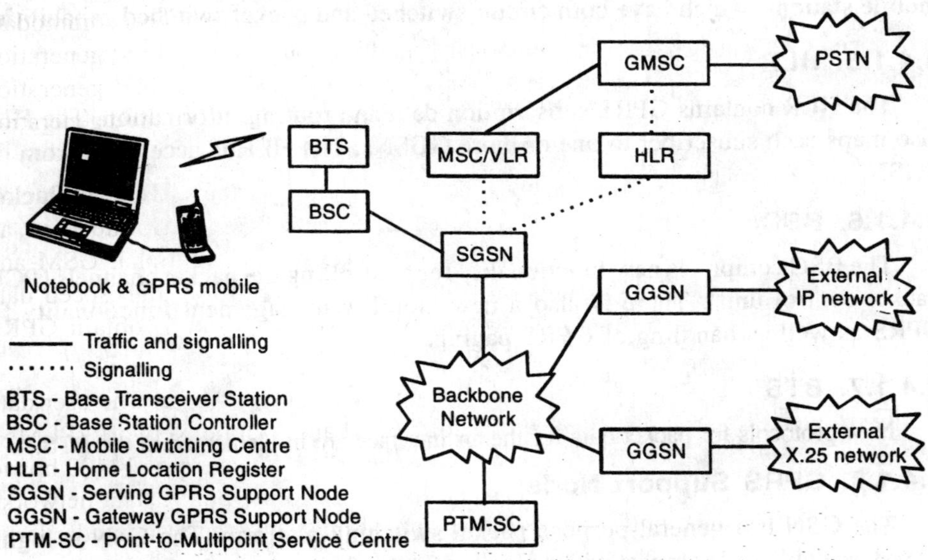

Fig. 4.7.

The basic idea of a GPRS network is to offer a logical channel for packet transmission between GGSN and the GPRS subscriber via the SGSN serving the geographical area the subscriber is currently located. The nodes are interconnected via a backbone network, that is an Internet Protocol (IP) network.

4.4.1.1. SGSN

It provides packet routing to and from the SGSN service area. It is serving all the subscribers that are physically located in the SGSN service area. The traffic is routed from the SGSN to the BSC via the BTS to the mobile terminal.

The SGSN provides :

➤ Authentication, ciphering and IMEI check
➤ Mobility management
➤ Logical link management
➤ Charging data
➤ Connection to the HLR, all GPRS subscription information in the traffic system will be handled within the current logical node HLR
➤ Connection to the MSC

4.4.1.2. GGSN

GGSN provides the interface towards the external IP packet networks and X. 25 networks. It translates data formats, signalling protocols and address information in order to permit communication between different networks. The functionality for routing to the right SGSN and protocol conversion is also provided by the GGSN.

4.4.1.3. PTM-SC

The PTM-SC handles the PTM traffic. It has connections to GPRS backbone network and the HLR.

4.4.1.4. MSC/VLR

An interface is provided between MSC and SGSN to coordinate signalling for mobile stations, which have both circuit switched and packet switched capabilities.

4.4.1.5. HLR

The HLR contains GPRS subscription data and routing information. The HLR also maps each subscriber to one or more GGSNs. The HLR is accessible from the SGSN.

4.4.1.6. BSC

The BSC comprises new functionality for controlling the packet channels (PCU, packet control unit). There is also a new mobility management functionality for GPRS as well as handling of GPRS paging.

4.4.1.7. BTS

New protocols for packet data for the air interface are implemented in the BTS.

4.4.1.8. GPRS Support Node

The GSN is a general-purpose packet switching platform. The GSN is highly robust, scalable and compact with advanced functions capturing faults and protecting overloads. A GSN consists mainly of the PSS core platform on which the SGSN and/or GGSN software executes. The PSS core platform consists of hardware, including back plane switch, interfaces and processors as well as software components such as OS, high availability middleware, and *O* and *M* support. The SGSN and GGSN applications include GPRS protocols and session and mobility management as well as *O* and *M*.

4.4.2. GPRS Terminals

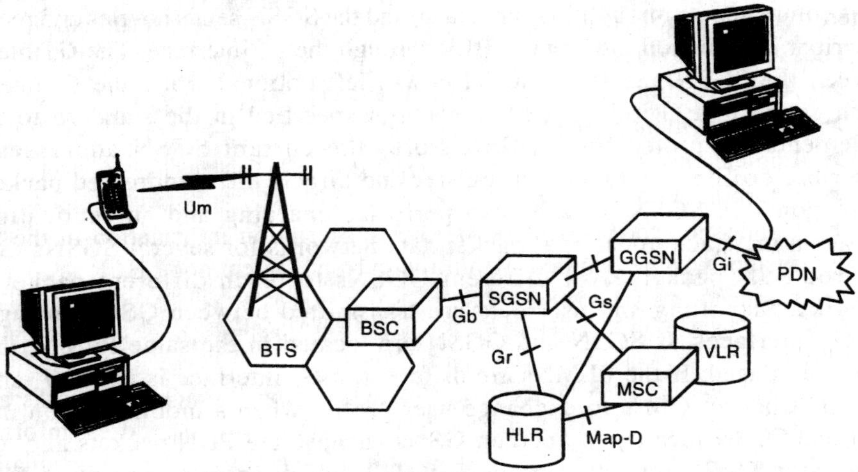

Fig. 4.8.

4.4.3. GPRS BSS

GPRS has minor impact on the existing GSM BSS making it easy to reuse existing component and links without major modifications. This is possible because GPRS uses the same frequency bands and hopping techniques, the same TDMA frame structure, the same radio modulation and burst structure as GSM.

A new functional component, called PCU, (Packet Control Unit) was added to the BSS in the GPRS standard to support the handling of data packets. The PCU (not shown above) is placed logically between the BSS and the GPRS NSS. Unlike the voice circuit connections, however, connections in GPRS have to be established and released between the BSS and the MS only when data needs to be transported over the air interface.

4.4.4. GPRS NSS

The GPRS NSS can be viewed as an overlay network ensuring the link between mobile users and data networks. GPRS introduces a new functional element to the GSM infrastructure GSN (GPRS Support Node) which can be either a SGSN (Serving-GSN) or a GGSN (Gateway-GSN). This addition is necessary for the GSM network in order to support packet data services. The network is generally divided into several service areas controlled by separate SGSNs.

The serving GPRS support node (SGSN) serves to deliver data packets to mobile stations within the service area. It performs mobility management, logical link control, packet routing and transfer, authentication, and charging. SGSN refers to a location register that stores subscriber profile and location information.

The gateway GPRS support node (GGSN) converts GPRS packets coming from SGSN into packet data protocol for transmission through packet data networks. It acts as an interface for the GPRS backbone to the packet data networks. The packet data protocol (PDP) addresses of incoming data packets are then converted into GSM addresses of the proper destinations. Several interfaces have been introduced in GPRS to define entity-to-entity interactions.

For instance, the G_b interface is required between the BSC and the SGSN. Two GSNs communicate through a G_n interface, and the SGSN sends queries and receives subscriber information to/from the HLR through the G_r interface. The G_s interface between the SGSN and the MSC/VLR was left optional while the G_i interface which connects a GGSN to a PDN was not specified in the standard to allow implementation preferences. GGSN stores the current SGSN addresses and subscriber profile in its location register and directs the readdressed packets to the responsible SGSN. GGSN also performs charging and authentication. It acts as an interface to external packet data networks for several SGSNs. SGSN may route its packets over different GGSNs to reach different packet data networks. Signalling and user data are transmitted between GSNs through G_n and G_p interfaces. If SGSN and GGSN are located in the same PLMN, the G_n interface is used. If the PLMNs are different, a G_p interface is used. G_n and G_p interfaces permit GSNs to exchange user profile when a mobile station moves from one SGSN locality to another. GSNs encapsulate PDN packets and tunnel them using GPRS tunneling protocol (GTP), which operates on top of TCP/IP. All GSNs are connected through IP-based GPRS backbone.

Connection between SGSN and base station controller (BSC) is achieved through the G_b interface. G_i interfaces connected PLMN with PDNs. Home location register (HLR) stores user profile, current SGSN address, and PDF address of each subscriber and communicates with SGSN through G_r interface. GGSN connects HLR through G_c interface. In order to update its location register, GGSN queries HLR of user location and profile through G_c interface. To permit coordination between GPRS and the existing GSM service, GSM networks must be functionally upgraded to suit the requirements.

4.4.5. Classes of GPRS Terminals

Three classes of GPRS terminals have been defined. The class of the device determines the speed at which GPRS can be used.

For example, the majority of GPRS terminals will be able to download data at speeds of up to 24 kbps (kilobytes per second). At the higher end, speeds are theoretically possible up to 171.2 kbit/sec when 8 slots are assigned at the same time to a single user. In reality 40-50 kbps. The **uplink frequencies** used are in the range of 890-915 MHz, these are used to carry the data from mobile station to BSC (Base Station Controller).

While the **downlink frequencies** are in the range of 935-960 MHz, these are used to carry the data from BSC (Base Station Controller) to mobile station.

The 25 MHz band is divided into 125 channels which are 200 kHz apart as shown in Fig. 4.9.

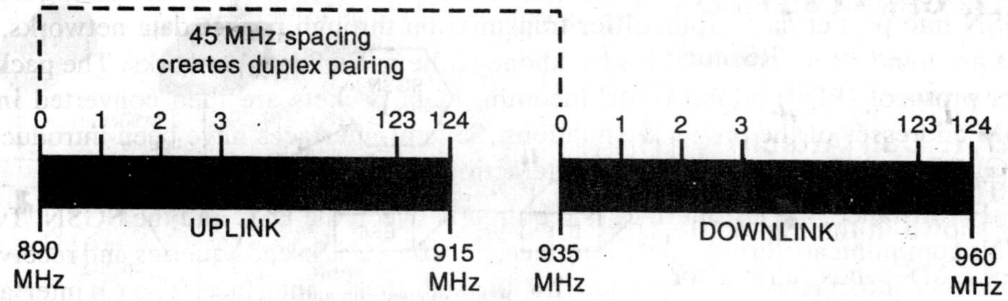

Fig. 4.9.

The uplink and downlink speed rates for various networks are shown in the Table 4.1.

Table 4.1.

Type	Uplink (Sending)	Downlink (Receiving)
GPRS	14 kbps	28-64 kbps
GSM CSD	9.6-14 kbps	9.6-14 kbps
HSCSD	28 kbps	28 kbps
Dial-UP	56 kbps	56 kbps
ISDN Standard	64 kbps	64 kbps
ADSL	256 kbps	512 kbps
Broadband	2 Mbps	2 Mbps

4.4.6. GPRS Multislot Classes

Multislot classes are product dependant, and determine the maximum achievable data rates in both the uplink and downlink directions. Written as (for example) 3 + 1 or 2 + 2, the first number indicates the amount of downlink timeslots (what the mobile phone is able to receive from the network). The second number indicates the amount of uplink timeslots (how many timeslots the mobile phone is able to transmit).

The class indicates the mobile phone capabilities.

Class A

Class A mobile phones can be connected to both GPRS and GSM services simultaneously.

Class B

Class B mobile phones can be attached to both GPRS and GSM services, using one service at a time. Class B enables making or receiving a voice call, or sending/ receiving an SMS during a GPRS connection. During voice calls or SMS, GPRS services are suspended and then resumed automatically after the call or SMS session has ended.

Class C

Class C mobile phones are attached to either GPRS or GSM voice service. You need to switch manually between services.

4.4.7. GPRS Call Flow

Call Flow in GPRS consists of two steps :

4.4.7.1. Call Attach

The process for call attach is as shown in the Fig. 4.10.

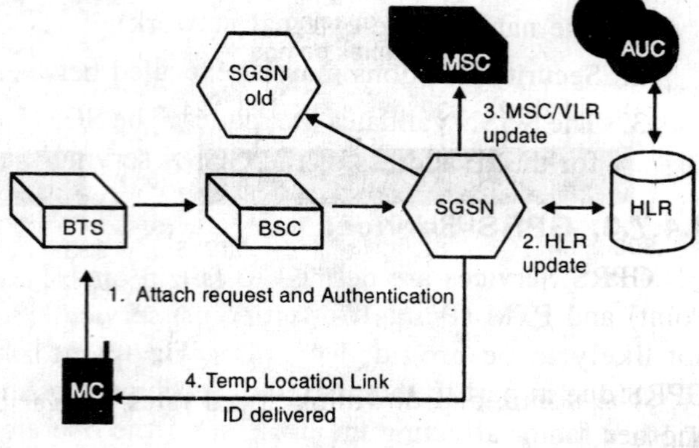

Fig. 4.10.

1. MS sends request for a call attach to the new SGSN via the BTS and the BSC.
2. The new SGSN identifies old SGSN and asks the old SGSN for the IMSI number.
3. If the MS is known in the old SGSN, it sends the IMSI and authentication triplets to the new SGSN.
4. If the MS is unknown in the old SGSN, the new SGSN order the MS to send its IMSI to the new SGSN.
5. The new SGSN fetches the authentication triplets for the corresponding IMSI, from the HLR.
6. The new SGSN authenticates the MS.
7. The new SGSN sends an Update Location (IMSI, new SGSN address) to notify the HLR that MS has changed the SGSN.
8. The HLR sends a Cancel_Location to the old SGSN in order to purge information regarding this MS.
9. The HLR sends subscriber data to the new SGSN.
10. If required, the SGSN informs the MS about new temporary identities.

4.4.7.2. PDP Context Activation

To send and receive GPRS data, the MS must perform PDP context activation after the GPRS attach, which means that the MS must specify the external data network it wants to reach. The process for PDP context activation is as shown in the Fig. 4.11 :

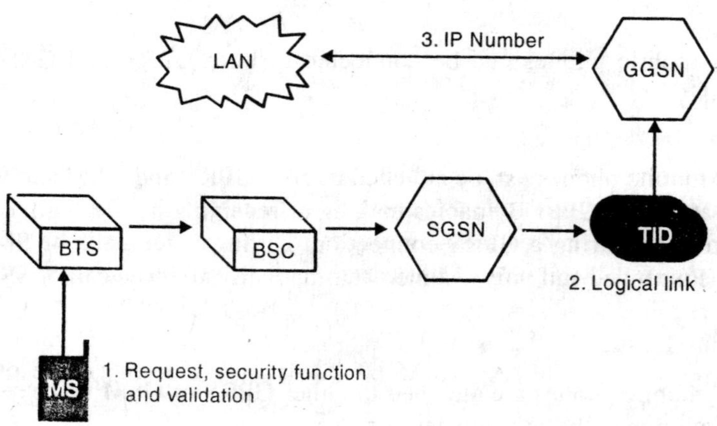

Fig. 4.11.

1. The MS sends a PDF context activation request to the SGSN with the APN (the name of the external network).
2. Security functions may be executed between the MS and the SGSN.
3. The SGSN validates the request/The SGSN asks the DNS in GPRS network for the IP address of the GGSN serving the APN in question.

4.4.7.3. GPRS Services

GPRS Services are defined to fall in one of two categories : PTP (Point-To-Point) and PTM (Point-To-Multipoint) services. Some of the GPRS services are not likely to be provided by network operators during early deployment of GPRS due in part to the phased development of the standard. Market demand is another factor affecting the decision of the operators regarding which services to offer first.

PTP (Point-To-Point) Services

GPRS will support applications based on IP. Applications based on the Connection Oriented Network Protocols are also defined to be supported. The X.25 protocol was initially mentioned but has been dropped in recent standard developments. Table 4.2 illustrates the general description of the PTP services and some possible applications.

Table 4.2.

Service	Description	Applications
PTP-CONS Point-to-point Connection Oriented Network Service	➤ Bursty transactive or interactive applications. ➤ A logical relation is established between users. ➤ Multiple packets are sent between a single source and a single destination.	Credit card validations electronic monitoring. Telnet applications Data base access and information retrieval.
PTP-CLNS Point-to-point Connectionless Network Service	➤ Datagram type service for bursty applications. ➤ No logical link required between users. ➤ Packets are sent between a single source and a single destination. ➤ Each packet is independent of its predecessor and successor.	Electronic mail Internet's World Wide Web

PTM (Point-To-Multipoint) Services

The PTM services provide the subscribers with the capability to send data to multiple destinations within one single service request. Table 4.4 shows a general description of these services and some possible applications. With the exception of PTM-M (Point-To-Multipoint Multicast) services, groups must be defined and members are required to join an ongoing call to become participants. A PTM-G (Point-to-Multipoint Group) call is usually restricted to members located within a specific geographical area. An IP-M (IP-Multicast) call is on the other hand independent of the geographical area of the participants and can be internal to the network or distributed across the internet. Table 4.3 illustrates the general description of the FTP services and some possible applications.

Table 4.3.

Service	Description	Applications
PTM-M Point-to-Multipoint Multicast	➤ Messages are transmitted to a specific geographical area and optionally to a specified group within that area. ➤ The recipients are anonymous. ➤ Delivery time is scheduled. ➤ Unidirectional transmission.	News Weather and traffic reports
PTM-G Point-To-Multipoint Group Call	➤ Messages are transmitted to a specific group within a specific geographical area. ➤ Group members must join the PTM-G call to become participants. ➤ Delivery in real time. ➤ Unidirectional, bidirectional and multi-directional transmission.	Conferencing services
IP-M IP Multicast	➤ Messages are transmitted to a specified group. ➤ Group members must join the IP-M call to become participants. ➤ Delivery in real time. ➤ Multi-directional transmission.	Live multimedia transmissions Corporate messages to employees

4.4.8. Security in GPRS

Security is an important issue in mobile networks and has gained a special attention in the GSM world. GPRS provides a security function similar to that of GSM. It is responsible for authentication and service request validation to prevent unauthorized service usage.

User confidentiality is also protected using temporary identification during connections to the GPRS network. Finally, user data is protected using ciphering techniques.

4.4.8.1. Authentication in GPRS

Authentication in cellular systems often means the use of a PIN code as a means of identification. Such method is not very secure since it is possible in a radio environment to capture the PIN and, therefore, to break the confidentiality of the subscriber. Especially dangerous is the fact that this PIN is assigned once at subscription and, therefore, it can be captured in many ways. The GSM/GPRS approach addresses this problem by varying the access code for every connection. A secret parameter *Ki* specific to the user is used to compute a shared value using an operator-dependent one-way (trap-door) algorithm. Figure 4.28 shows the authentication computation using the parameter *Ki* which is never transmitted via the air interface. The sequence of the figure is triggered when the MS requests to attach to the GPRS network or may be requested by the SGSN if the MS is roaming.

The MS is issued a new random number RAND whenever authentication is required. The MS computes the value SRES (Signed Result) using *Ki* and RAND and forwards it back to the SGSN for comparison. The authentication procedure in GPRS is executed from the SGSN instead of the MSC/VLR as in GSM.

4.4.8.2. Ciphering in GPRS

Ciphering between the MS and SGSN starts only after the MS is authenticated. Similarly, a secret ciphering key *Kc* is used at the MS and the SGSN to encrypt the exchange of messages. The ciphering scope is from the MS to the SGSN while in GSM it is from the MS to the BSS. This is done to simplify the key management since cell selection by the MS can occur frequently and, therefore, packets may travel via different BTSs. In GSM, however, a single logical channel is used between the MS and the BTS at each given time. Also, by extending the scope of ciphering in GPRS, ciphering algorithms can be changed without updating every element of the BSS. Alternatively, radio elements can be upgraded or replaced without modifying the ciphering mechanism.

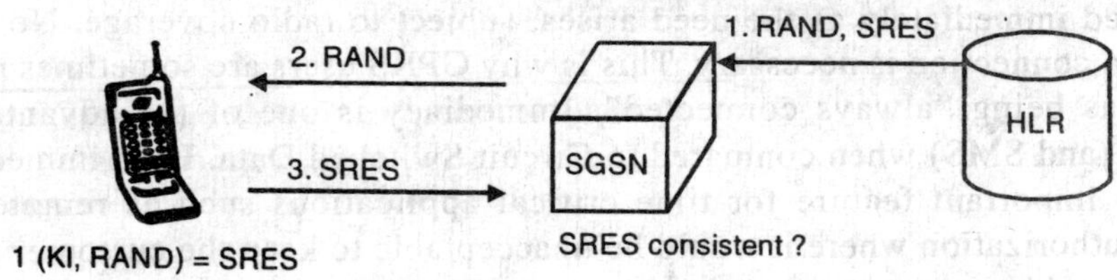

Fig. 4.12.

4.4.8.3. Security in the GPRS Backbone

Authentication of subscribers in GPRS is performed by the SGSN, and the scope of ciphering is between the MS and the SGSN. The GPRS standard has left to the implementation, the requirements for security beyond the SGSN and GGSN. For instance, an IP based GPRS backbone can ensure security using IPsec (IP Security Protocol) which groups a set of mechanisms used to protect traffic at the IP level. Secured services offered by IPsec are integrity in connectionless mode, authentication of data origin, and protection of data confidentiality. IPsec is interesting because it provides security of data at the network level rather than at the application or physical levels.

An operator can ensure security in the GPRS backbone by relying on several options such as the following :

- Installing firewalls at the *Gi* reference point (between a GGSN and an external data network).
- Using border gateways when interfacing to other GPRS networks.
- Improving the backbone network protection by using some security products such as IPsec for an IP-based backbone as mentioned above.
- Using some packet filtering mechanisms.

4.4.9. Features of GPRS

Key User Features of GPRS: The General Packet Radio Service (GPRS) is a new non voice value added service that allows information to be sent and received across a mobile telephone network. It supplements today's Circuit Switched Data and Short Message Service. GPRS is NOT related to GPS (the Global Positioning System), a similar acronym that is often used in mobile contexts. GPRS has several unique features which can be summarized as :

Speed

Theoretical maximum speeds of up to 171.2 kilobits per second (kbps) are achievable with GPRS using all eight timeslots at the same time. This is about three times as fast as the data transmission speeds possible over today's fixed telecommunications networks and ten times as fast as current Circuit Switched Data services on GSM networks. By allowing information to be transmitted more quickly, immediately and efficiently across the mobile network, GPRS may well be a relatively less costly mobile data service compared to SMS and Circuit Switched Data.

Immediacy

GPRS facilitates instant connections whereby information can be sent or received immediately as the need arises, subject to radio coverage. No dial-up modem connection is necessary. This is why GPRS users are sometimes referred to be as being "always connected". Immediacy is one of the advantages of GPRS (and SMS) when compared to Circuit Switched Data. High immediacy is a very important feature for time critical applications such as remote credit card authorization where it would be unacceptable to keep the customer waiting for even thirty extra seconds.

New Applications, Better Applications

GPRS facilitates several new applications that have not previously been available over GSM networks due to the limitations in speed of Circuit Switched Data (9.6 kbps) and message length of the Short Message Service (160 characters). GPRS will fully enable the Internet applications you are used to on your desktop from web browsing to chat over the mobile network. Other new applications for GPRS, profiled later, include file transfer and home automation—the ability to remotely access and control in-house appliances and machines.

4.4.10. Applications for GPRS

A wide range of corporate and consumer applications are enabled by nonvoice mobile services such as SMS and GPRS. This section will introduce those that are particularly suited to GPRS.

Chat

Chat can be distinguished from general information services because the source of the information is a person with chat whereas it tends to be from an Internet site for information services. The "information intensity" the amount of information transferred per message tends to be lower with chat, where people are more likely to state opinions than factual data. In the same way as Internet chat groups have proven a very popular application of the Internet, groups of likeminded people—so called communities of interest have begun to use non-voice mobile services as a means to chat and communicate and discuss.

Because of its synergy with the Internet, GPRS would allow mobile users to participate fully in existing Internet chat groups rather than needing to set up their own groups that are dedicated to mobile users. Since the number of participants is an important factor determining the value of participation in the new group, the use of GPRS here would be advantageous. GPRS will not, however, support point to multipoint services in its first phase, hindering the distribution of a single message to a group of people. As such, given the installed base of SMS capable devices, we would expect SMS to remain the primary bearer for chat applications in the foreseeable future, although experimentation with using GPRS is likely to commence sooner rather than later.

Textual and Visual Information

A wide range of content can be delivered to mobile phone users ranging from share prices, sports scores, weather, flight information, news headlines, prayer reminders, lottery results, jokes, horoscopes, traffic, location sensitive services and so on. This information need not necessarily be textual—it may be maps or graphs or other types of visual information.

The length of a short message of 160 characters suffices for delivering information when it is quantitative—such as a share price or a sports score or temperature. When the information is of a qualitative nature, however, such as a horoscope or news story, 160 characters is too short other than to tantalize or annoy the information recipient since they receive the headline or forecast but little else of substance. As such, GPRS will likely be used for qualitative information services when end users have GPRS capable devices, but SMS will continue to be used for delivering most quantitative information services. Interestingly, chat applications are a form of qualitative information that may remain delivered using SMS, in order to limit people to brevity and reduce the incidence of spurious and irrelevant posts to the mailing list that are a common occurrence on Internet chat groups.

Still Images

Still images such as photographs, pictures, postcards, greeting cards and presentations, static web pages can be sent and received over the mobile network as they are across fixed telephone networks. It will be possible with GPRS to post images from a digital camera connected to a GPRS radio device directly to an Internet site, allowing near real-time desktop publishing.

Moving Images

Over time, the nature and form of mobile communication is getting less textual and more visual. The wireless industry is moving from text messages to icons and picture messages to photographs and blueprints to video messages and movie previews being downloaded and on to full blown movie watching via data streaming on a mobile device.

Sending moving images in a mobile environment has several vertical market applications including monitoring parking lots or building sites for intruders or thieves, and sending images of patients from an ambulance to a hospital. Video conferencing applications in which teams of distributed sales people can have a regular sales meeting without having to go to a particular physical location, is another application for moving images.

Web Browsing

Using Circuit Switched Data for web browsing has never been an enduring application for mobile users. Because of the slow speed of Circuit Switched Data, it takes a long time for data to arrive from the Internet server to the browser. Alternatively, users switch off the images and just access the text on the web, and end up with difficult to read text layouts on screens that are difficult to read from. As such, mobile Internet browsing is better suited to GPRS.

Document Sharing/Collaborative Working

Mobile data facilitates document sharing and remote collaborative working. This lets different people in different places work on the same document at the same time. Multimedia applications combining voice, text, pictures and images can even be envisaged. These kinds of applications could be useful in any problem solving exercise such as fire fighting, combat to plan the route of attack, medical treatment, advertising copy setting, architecture, journalism and so on. Even comments on which resort to book a holiday at could benefit from document sharing to save everyone having to visit the travel agent to make a decision. Anywhere somebody can benefit from having and being able to comment on a visual depiction of a situation or matter, such collaborative working can be useful. By providing sufficient bandwidth, GPRS facilitates multimedia applications such as document sharing.

Audio

Despite many improvements in the quality of voice calls on mobile networks such as Enhanced Full Rate (EFR), they are still not broadcast quality. There are scenarios where journalists or undercover police officers with portable professional broadcast quality microphones and amplifiers capture interviews with people or radio reports dictated by themselves and need to send this information back to their radio or police station. Leaving a mobile phone on, or dictating to a mobile phone, would simply not give sufficient voice quality to allow that transmission to be broadcast or analyzed for the purposes of background noise analysis or voice printing, where the speech autograph is taken and matched against those in police storage. Since even short voice clips occupy large file sizes, GPRS or other high speed mobile data services are needed.

Job Despatch

Non-voice mobile services can be used to assign and communicate new jobs from office-based staff to mobile field staff. Customers typically telephone a call center whose staff take the call and categorize it. Those calls requiring a visit by field sales or service representative can then be escalated to those mobile workers. Job despatch applications can optionally be combined with vehicle positioning applications—such that the nearest available suitable personnel can be deployed to serve a customer. GSM non-voice services can be used not only to send the job out, but also as a means for the service engineer or sales person can keep the office informed of progress towards meeting the customer's requirement. The remote worker can send in a status message such as "Job 1234 complete, on my way to 1235".

The 160 characters of a short message are sufficient for communicating most delivery addresses such as those needed for a sales, service or some other job despatch application such as mobile pizza delivery and courier package delivery. However, 160 characters does require manipulation of the customer data such as the use of abbreviations such as "St" instead of "Street". Neither does 160 characters leave much space for giving the field representative any information about the problem that has been reported or the customer profile. The field representative is able to arrive at the customer premises but is not very well briefed beyond that. This is where GPRS will come in to allow more information to be sent and received more easily. With GPRS, a photograph of the customer and their premises could, for example, be sent to the field representative to assist in finding and identifying the customer. As such, we expect job despatch applications will be an early adopter of GPRS-based communications.

Corporate E-mail

With up to half of employees typically away from their desks at any one time, it is important for them to keep in touch with the office by extending the use of corporate e-mail systems beyond an employee's office PC. Corporate e-mail systems run on Local Area computer Networks (LAN) and include Microsoft Mail, Outlook, Outlook Express, Microsoft Exchange, Lotus Notes and Lotus cc:Mail.

Since, GPRS capable devices will be more widespread in corporations than amongst the general mobile phone user community, there are likely to be more corporate e-mail applications using GPRS than Internet e-mail ones whose target market is more general.

Internet E-mail

Internet e-mail services come in the form of a gateway service where the messages are not stored, or mailbox services in which messages are stored. In the case of gateway services, the wireless e-mail platform simply translates the message from SMTP, the Internet e-mail protocol, into SMS and sends to the SMS Center. In the case of mailbox e-mail services, the emails are actually stored and the user gets a notification on their mobile phone and can then retrieve the full e-mail by dialing in to collect it, forward it and so on.

Upon receiving a new e-mail, most Internet e-mail users do not currently get notified of this fact on their mobile phone. When they are out of the office, they have to dial in speculatively and periodically to check their mailbox contents. However, by linking Internet e-mail with an alert mechanism such as SMS or GPRS, users can be notified when a new e-mail is received.

Vehicle Positioning

This application integrates satellite positioning systems that tell people where they are with nonvoice mobile services that let people tell others where they are. The Global Positioning System (GPS) is a free-to-use global network of 24 satellites run by the US Department of Defense. Anyone with a GPS receiver can receive their satellite position and thereby find out where they are. Vehicle positioning applications can be used to deliver several services including remote vehicle diagnostics, ad-hoc stolen vehicle tracking and new rental car fleet tariffs.

The Short Message Service is ideal for sending Global Positioning System (GPS) position information such as longitude, latitude, bearing and altitude. GPS coordinates are typically about 60 characters in length. GPRS could alternatively be used.

Remote LAN Access

When mobile workers are away from their desks, they clearly need to connect to the Local Area Network in their office. Remote LAN application encompasses access to any applications that an employee would use when sitting at their desk, such as access to the intranet, their corporate e-mail services such as Microsoft Exchange or Lotus Notes and to database applications running on Oracle or Sybase or whatever. The mobile terminal such as handheld or laptop computer has the same software programs as the desktop on it, or cut down client versions of the applications accessible through the corporate LAN. This application area is, therefore, likely to be a conglomeration of remote access to several different information types—e-mail, intranet, databases. This information may all be accessible through web browsing tools, or require proprietary software applications on the mobile device. The ideal bearer for Remote LAN Access depends on the amount of data being transmitted, but the speed and latency of GPRS make it ideal.

File Transfer

As this generic term suggests, file transfer applications encompass any form of downloading sizeable data across the mobile network. This data could be a presentation document for a traveling salesperson, an appliance manual for a service engineer or a software application such as Adobe Acrobat Reader to read documents. The source of this information could be one of the Internet communication methods such as FTP (File Transfer Protocol), telnet, http or Java- or from a proprietary database or legacy platform. Irrespective of source and type of file being transferred, this kind of application tends to be bandwidth intensive. It, therefore, requires a high speed mobile data service such as GPRS, EDGE or 3GSM to run satisfactorily across a mobile network.

Home Automation

Home automation applications combine remote security with remote control. Basically, you can monitor your home from wherever you are on the road, on holiday, or at the office. If your burglar alarm goes off, not only do you get alerted, but you get to go live and see who are perpetrators and perhaps even lock them in. Not only can you see things at home, but you can do things too. You can program your video, switch your oven on so that the preheating is complete by the time you arrive home (traffic jams permitting) and so on. Your GPRS capable mobile phone really does become like the remote control devices we use today for our television, video, hi-fi and so on. As the Internet Protocol (IP) will soon be everywhere—not just in mobile phones because of GPRS but all manner of household appliances and in every machine—these devices can be addressed and instructed. A key enabler for home automation applications will be Bluetooth, which allows disparate devices to interwork.

4.5. EDGE TECHNOLOGY

Enhanced data for global evolution (EDGE) is the next step in the evolution of global system for mobile communication (GSM). The objective of this technology is to increase data transmission rates and spectrum efficiency and to facilitate new applications and increased capacity for mobile use. It is a cost-effective way of migration to full-blown 3G services considering GSM and general packet radio system (GPRS) as the base networks as shown in Fig. 4.13.

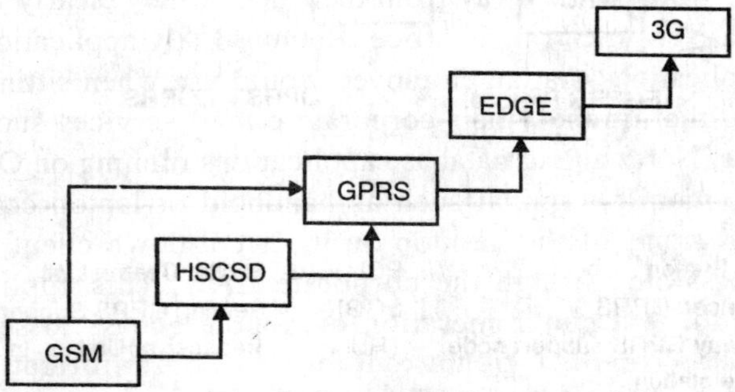

Fig. 4.13. Migration path from 2G to 3G.

GPRS allows data rates of 115 kbps (theoretically, up to 160 kbps) on the physical layer, whereas EDGE, or extended general packet radio system (EGPRS), is capable of offering data rates of 384 kbps (theoretically, up to 473.6 kbps). A new modulation technique and error-tolerant transmission method, combined with improved link-adaptation mechanisms, make these EGPRS rates possible. This is the key to increased spectrum efficiency and enhanced applications, such as wireless Internet access, e-mail and file transfers.

4.5.1. GPRS vs EDGE/EGPRS

EDGE is a method to increase the data rates on the radio link for GSM. Basically, it introduces only a new modulation technique and new channel coding that can be used to transmit both packet-switched and circuit-switched voice and data services. EDGE is, therefore, an add-on to GPRS and cannot work alone.

GPRS and EGPRS have different protocols and different behaviour on the base station system side. However, on the core network side, GPRS and EGPRS share the same packet-handling protocols and, therefore, behave in the same way. Reuse of the existing GPRS core infrastructure emphasises the fact that EGPRS is only an 'add-on' to the base station system and is, therefore, much easier to introduce than GPRS (Fig. 4.14).

In addition to enhancing the throughput for each data user, EDGE also increases capacity. With EDGE, the same time slot can support more users. EDGE makes it easier for circuit-switched and packet-switched traffic to coexist while making more efficient use of the same radio resources. It can transmit three times as many bits as GPRS during the same period of time. This is the main reason for higher EDGE bitrates.

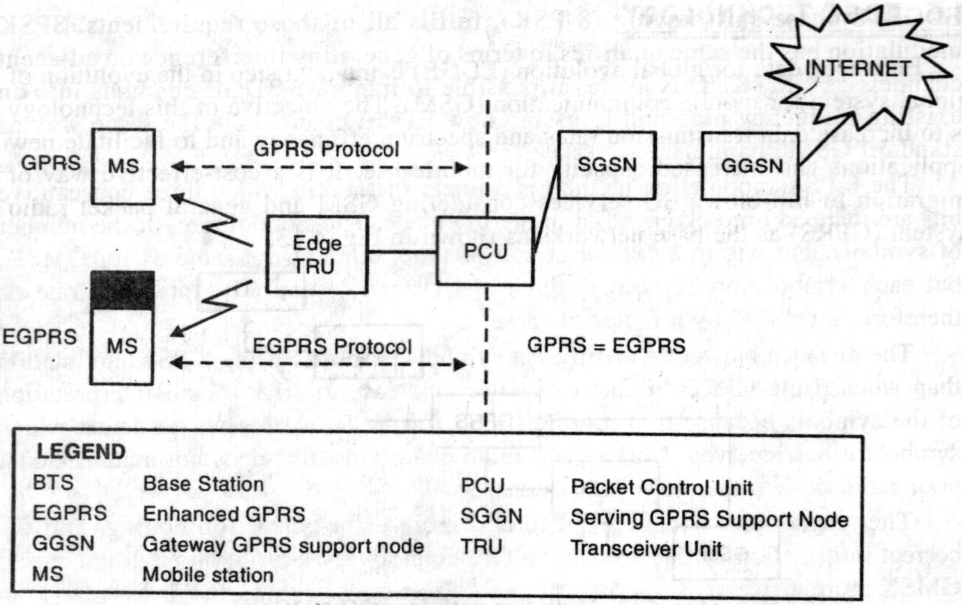

Fig. 4.14. EDGE or EGPRS on GPRS network.

4.5.2. Modulation Technique

The modulation technique used in GSM is the Gaussian minimum shift keying (GMSK), which is a kind of phase modulation. This can be visualised in an I/Q diagram that shows the real (I) and imaginary (Q) components of the transmitted signal (Fig. 4.15). Transmitting a '0' or '1' bit is then represented by changing the phase by increments of ±*p*. Every symbol that is transmitted represents '1' bit, that is, each shift in the phase represents '1' bit.

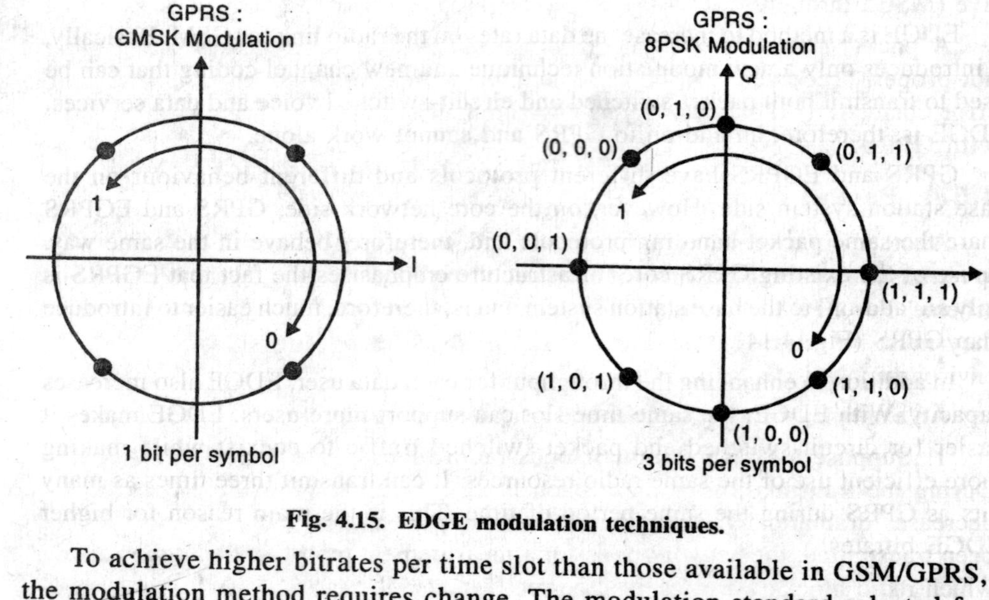

Fig. 4.15. EDGE modulation techniques.

To achieve higher bitrates per time slot than those available in GSM/GPRS, the modulation method requires change. The modulation standard selected for

EDGE, 8-phase shift keying (8 PSK), fulfils all of those requirements. 8PSK modulation has the same qualities in terms of generating interference on adjacent channels as GMSK. This makes it possible to integrate EDGE channels into an existing frequency plan and to assign new EDGE channels in the same way as standard GSM channels.

The 8 PSK modulation method is a linear method in which three consecutive bits are mapped onto one symbol in the I/Q plane. The symbol rate, or the number of symbols sent within a certain period of time, remains the same as for GMSK, but each symbol now represents three bits instead of one. The total data rate is, therefore, increased by a factor of three.

The distance between the different symbols is shorter using 8 PSK modulation than when using GMSK. Shorter distances increase the risk for misinterpretation of the symbols because it is more difficult for the radio receiver to detect which symbols it has received. Under good radio conditions, this does not matter. Under poor radio conditions, however, it does.

The 'extra' bits will be used to add more error-correction coding, and the correct information can be recovered. Only under very poor radio environments is GMSK more efficient. Therefore, the EDGE coding schemes are a mixture of both GMSK and 8 PSK.

4.5.3. Coding Schemes

For GPRS, four different coding schemes, designated CS1 through CS4, are defined. Each has different amounts of error-correction coding that is optimised for different radio environments.

For EGPRS, nine modulation coding schemes, designated MCS1 through MCS9, are introduced. These fulfil the same task as the GPRS coding schemes. The lower four EGPRS coding schemes (MSC1 through MSC4) use GMSK, whereas the upper five (MSC5 through MSC9) use 8PSK modulation.

A packet sent with a higher coding scheme (less error correction) that is not properly received, can be retransmitted with a lower coding scheme (more error correction) if the new radio environment requires it. This is not possible with GPRS.

4.5.4. Packet Handling

Another improvement that has been made to the EGPRS standard is the ability to retransmit a packet that has not been decoded properly with a more robust coding scheme. For GPRS, resegmentation is not possible. Once packets have been sent, they must be retransmitted using the original coding scheme even if the radio environment has changed.

An example of packet transfer and retransmission for GPRS is given below :

1. Suppose a GPRS terminal sends radio blocks '1' through '4' with CS3. During the transmission of these packets, the carrier-to-interference ratio (C/ I) decreases dramatically, changing the radio environment. After the packets have been transmitted, the network polls for a measurement report that tells the network which radio blocks were received correctly.

2. The GPRS handset replies with a packet downlink acknowledged/ unacknowledged message containing the information about the link quality and the bitmap. In this scenario, it is assumed that packets '2' and '3' were sent erroneously.

3. Based on the new link quality information, the GPRS link adaptation algorithm will adapt the coding scheme to the new radio environment using CS1 for the new packets '5' and '6.' However, because GPRS cannot resegment the old packets, packets 2 and 3 must be retransmitted using CS3, although there is a significant risk that these packets still may not be decoded correctly.

As a result, the link adaptation for GPRS requires careful selection of the coding scheme in order to avoid retransmissions as much as possible.

As mentioned earlier, with EGPRS, resegmentation is possible. Packets sent with little error protection can be retransmitted with more error protection, if required by the new radio environment. The rapidly changing radio environment has a much smaller effect on the problem of choosing the wrong coding scheme for the next sequence of radio blocks because resegmentation is possible. Therefore, the EGPRS link-controlling algorithm can be very aggressive when selecting the modulation coding schemes.

4.5.5. EDGE Benefits

Short-term benefits: capacity and performance. EGPRS introduces a new modulation technique, along with improvements to the radio protocol, that allows operators to use the existing frequency more effectively.

The simple improvements of the existing GSM/GPRS protocols make EDGE a cost-effective, easy-to-implement add-on. Software upgrades in the base station system enable use of the new protocol; new transceiver units in the base station enable use of the new modulation technique.

EDGE triples the capacity of GPRS. This capacity boost improves the performance of existing applications and enables new services such as multimedia services. It also enables each transceiver to carry more voice and/or data traffic.

EDGE enables new applications at higher data rates. This will attract new subscribers and increase operators' customer base. Providing the best and most attractive services will increase customer loyalty.

Mid-term benefits: complementary technology. EDGE is designed to integrate into the existing network. The installed base evolves; it is not replaced or built from scratch, making implementation seamless. Fast, easy rollout means shorter time to market, which, in turn, can lead to increased market share.

With EDGE, operators can offer more wireless data applications, including wireless multimedia, e-mail, and Web infotainment and positioning services, for both consumer and business users. Subscribers will be able to browse the Internet on their mobile phones, personal digital assistants or laptops at the same speed as on stationary personal computers.

Conclusion

EDGE is a cost-effective way of migration to full-blown 3G services. It can play an important role in the evolution toward wideband code-division multiple-access (WCDMA) leading to 3G technologies.

4.6. DROPPED CALL RATES AND THEIR EVALUATIONS

The link capacity between GSM and CDMA is always an interesting issue. A new idea has come from GSM engineers that uses a Ping-Pong handoff arrangement in one handoff region that can achieve the same selective diversity between two neighbouring cells. This section analyzes the performance of this arrangement. There is viterbi's paper published for discussion of handoff. Viterbi's paper points out that the required margin for GSM hard handoff is 10.3 dB but the required margin for CDMA soft handoff is only 6.2 dB under the same assumed conditions. The difference is 4.1 dB, which is related to the coverage area extension of 1.6 times. The 4 dB advantage is due to selective diversity between two neighboring cells used in the soft handoff area for a CDMA systems.

Using rapid Ping-Pong handoff instead of a single handoff in a handoff region on a GSM system was suggested to gain the same 4 dB selective diversity as CDMA. When the mobile unit is in a handoff region, the two log normal fading signals received by two respective neighboring cells (on the reverse link) can be used to select the strongest signal strength in real time. With this information, the rapid Ping-Pong handoff operation can take place at the switches. In the meantime, the two neighboring cell long normal fading signals received by the mobile unit can also be used for MAHO (mobile assisted handoff) to assist switches in rapid handoffs. Although the idea is very good, the dropped call rate for this Ping-Pong handoff arrangement should be analyzed.

4.6.1. Drop Call Rate for Handoff

When a call has gone through n Ping-Pong handoffs in a handoffs region, the call may have k regular handoffs while the mobile unit is passing through $k + 1$ cells, the probability of a dropped call can be expressed as

$$P = 1 - X^{n+k} \qquad \text{...(1)}$$

where

$$X = (1 - \delta)\,(1 - \mu)\,(1 - \tau\theta)\,(1 - \beta)^2 \qquad \text{...(2)}$$

where δ = the probability that the signal is below a specified reception threshold (in a noise limited system)

μ = The probablity that the signal is below a specified cochannel interference level (in an interference limited system)

τ = The probability that no traffic channel is available upon handoff when moving into a new cell

θ = The probability that the cell will return to the original cell.

β = The probability of blocking circuits between BSC and MSC.

In a noise limited system (startup system), there is no frequency re-use configuration. The call drop rate P_A is based on the signal coverage. In an interference-limited system (mature system), the frquency re-use configuration is applied, and the droppεd rate P_B is based on the interference level. The case of P_A and P_B do not occur at the same time. When capacity is bvased on frequency reuse, the interference level is high, the size of the cell is small, and coverae is not an issue. The call drop rate totally depends on interference.

In a well designed network, the value of τ, θ, β are assumed to be very small and can be neglected and

$$X \text{ becomes } = (1 - \delta)\,(1 - \mu) \quad ...(3)$$

Furthermore, in a noise limited case, $\mu \to 0$ in equation (3) and equation (1) becomes

$$P_A = 1 - (1 - \delta)^{n+k} \quad ...(4)$$

And in an interference limited system, $\delta \to 0$ in equation (3) and equation (1) becomes

$$P_B = 1 - (1 - \mu)^{n+k} \quad ...(5)$$

SUMMARY

1. A cellular system is divided into three parts containing MS (Mobile station), (Base station), MTSO (Mobile Terminal Switching Office).
2. In a cellular system whole of the geographical area is divided into Small Areas ca'led Cells.
3. Ideally these cells are of hexagonal shape, but in practice they are in circular.
4. Uplinking frequency of Mobile Phone is always lower than Downlinking frequency. Due to limited power of mobile phone.
5. Base Antennas are generally established at line of sight communication and operated at Microwave frequency.
6. Mobile phone connects with Base Antenna at HF (No need of line of sight communication).
7. Allocation of cellular spectrum done by TRAI in India.
8. Ist Generation Network, known as AMPS which was fully Analog Cellular System.
9. 2nd generation network fully Digital and Automatic.
10. In India, whole Cellular system is operated under two technologies. GSM, CDMA. These technologies offered services like SMS, GPRS, Internet Browsing etc.

REVIEW QUESTIONS

1. What will be the effect if signal propagates over water and in free space ?
2. What are factors on which cell coverage will depend ?
3. What is difference between omni and directive antenna ?
4. Explain handoff.
5. What are different types of handoff ?
6. What is role of BSC and MSC in handoff ?
7. Describe a Basic cellular system, and explain with the block diagram.
8. What are the limitations of Analog Cellular system ?
9. How digital cellular system is advantageous over Analog cellular system ?
10. Describe the GPRS, what are the features of GPRS.

11. **Why uplinking frequency is less than downlinking frequency ? Explain.**

12. **In Satellite Communication, the uplinking frequency is higher than Down linking frequency just reverse of Cellular Phone Communication, why and justify your answer with question ?**

13. **What are the different methods possible to increase the system capacity ?**

14. **What are the two functions performed by an antenna, when we design a cellular system, then what should be the characteristics of an antenna ?**

15. **What do you mean by radiation pattern and isotropic antenna ?**

16. **What are the steps involved in a A/D converter, and write any three advantages of digital transmission over analog transmission ?**

17. **What is soft handoff ? Explain the handoff performance in CDMA systems.**

18. **Distinguish between dropped calls and terminated calls.**

❑❑❑

5

CHAPTER

Modulation Methods and Coding for Error Detection and Correction

5.0. INTRODUCTION

INSIDE THIS CHAPTER

- **5.0.** Introduction
- **5.1.** Analog Modulation
- **5.2.** Digital Modulation
- **5.3.** Modulation Method in Cellular Wireless System
- **5.4.** Spread Spectrum
- **5.5.** Orthogonal Frequency Division Multiplexing (OFDM)
- **5.6.** Error Detection and Error Correction
- **5.7.** Error Correction
- **5.8.** Data Compression
- **5.9.** Block Codes
- **5.10.** Convolutional Codes
- **5.11.** Turbo Coding

A thorough knowledge of modulation and access techniques is essential for the study of wireless communication. Modulation technique is a method to superimpose base band signal over high frequency carrier signal so that message signal can be reach at destination without loss of energy and message content.

There are two types of modulation techniques—Analog modulation and digital modulation technique.

The basic difference between analog modulation and digital modulation is that analog modulation uses continuous carrier wave while digital modulation uses pulse signal as carrier wave.

Frequently used nationally and internationally standardized modulation methods are of digital modulation type.

Digital modulation technique can be classified as Binary phase shift key (BPSK), Quadrature Phase Shift Key (QPSK), $\pi/4$ DQPSK and Gaussian Minimum Shift Keying (GMSK).

Multiple access is a technique that permits the communication resources of a wireles channel to be shared by a large number of users in different locations.

There are four basic forms of multiple access applied to wireless communications depending on which particular resource is exploited.

1. Frequency division multiple access (FDMA).
2. Time division multiple access (TDMA)

3. Code division multiple access (CDMA)
4. Space division multiple access (SDMA).

FDMA operates by dividing the bandwidth of a wireless channel equally among a number of users wanting to access the channel.

While TDMA operates on same bandwidth at different times and CDMA operates on different code, SDMA operates on different space irrespective of time, bandwidth, code, space.

5.1. ANALOG MODULATION

Modulation means shift or change and it is very important process in the communication theory. Without modulation wireless communication is not possible.

Definition: Analog modulation is a process in which some specific parameter of continuous carrier signal is varied in accordance with the message signal.

Signal: Signal contains information about a variety of things and activities in our physical world.

Example: Information about the weather is contained in signals that represent the air temperature, pressure, wind, etc. So signal is a mathematical representation of any physical energy.

For example, an alternating sinusoidal current is represented as follows :

$$Y(t) = A \sin (\omega t + \phi) \qquad \text{...(5.1)}$$

Modulation can be classified in following way :

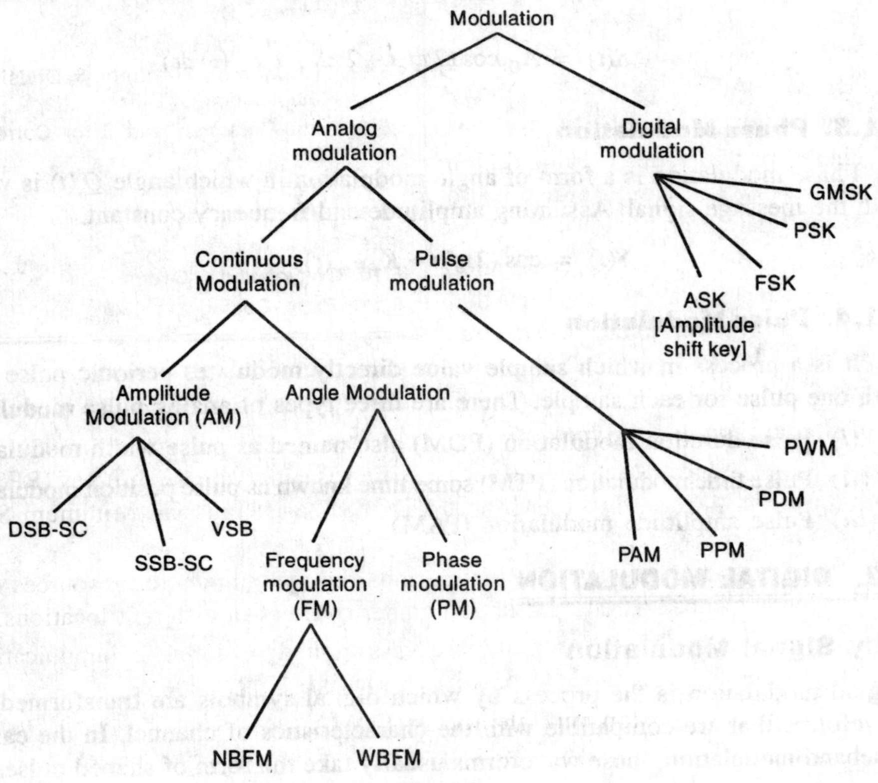

Fig. 5.1.

FSK (Frequency Shift Keying)
PSK (Phase Shift Keying)
GMSK (Gaussian Minimum Shift Keying)
PAM (Pulse Amplitude Modulation)
PPM (Pulse Position Modulation)
PDM (Pulse Duration Modulation)
PWM (Pulse Width Modulation)
NBFM (Narrowband FM)
WBFM (Wideband FM)

5.1.1. Amplitude Modulation (AM)

It is a process in which amplitude of carrier wave is varied according to mean value of message signal. Assuming frequency and phase constant mathematical equation of AM wave is

$$S(t) = A_C [1 + Ke_m (t)] \cos (2\pi f_c t) \quad ...(5.2)$$

5.1.2. Frequency Modulation (FM)

It is a process in which frequency of carrier wave varied according to message signal. Keeping amplitude and phase constant mathematical equation of FM wave is

$$S(t) = A_C \cos (2\pi f_c t + 2\pi K_f \int_0^t e_m (t)\, dt) \quad ...(5.3)$$

5.1.3. Phase Modulation

Phase modulation is a form of angle modulation in which angle $Q_i(t)$ is varied with the message signal. Assuming amplitude and frequency constant.

$$S(t) = \cos (2\pi f_c t + K_P e_m (t)) \quad ...(5.4)$$

5.1.4. Pulse Modulation

It is a process in which sample value directly modulates periodic pulse train with one pulse for each sample. There are three types of analog pulse modulation

(i) Pulse duration modulation (PDM) also named as pulse width modulation.
(ii) Pulse time modulation (PTM) some time known as pulse position modulation.
(iii) Pulse amplitude modulation (PAM).

5.2. DIGITAL MODULATION

Why Signal Modulation

Digital modulation is the process by which digital symbols are transformed into waveforms that are compatible with the characteristics of channel. In the case of baseband modulation, these waveforms usually take the form of shaped pulses, for this purpose we use PCM (Pulse code modulation).

Note that PCM is not a digital modulation technique, it is a coding technique which is a part of *A/D* converter. By using PCM we represent a baseband signal (message signal) in form of binary pulses.

While in the case of bandpass modulation the shaped pulses modulate a sinusoidal called a *carrier wave*. Here shaped pulses is message, which is obtained from the PCM, then it is digitally modulated, the modulation technique may be Amplitude, (ASK), Frequency (FSK), Phase (PSK), etc.

After modulation the modulated wave is transmitted as Electromagnetic wave for desired destination.

One may ask question, why it is necessary to use a carrier for the radio transmission of message signal.

Actually, the transmission of Electromagnetic energy is possible with the help of Antenna.

Antenna of height is inversely proportional to frequency. And for the baseband transmission (frequency range is 20 to 20 kHz) at this point height of Antenna becomes around 75 km long which is impossible to construct.

Some more advantages are also, like by modulation we can minimise the effect of interference. Modulation can also be used to place a signal in a frequency band where design requirements, such as filtering and amplification can be easily met. This is the case when radio frequency (RF) Signals are converted to an intermediate frequency (IF) in a receiver.

5.2.1. Digital Modulation and Demodulation Techniques

Modulation (Analog or Digital) is a technique by which an information signal is converted to a sinusoidal waveform. As we know the sinusoidal wave has just three characteristics—amplitude, phase, frequency. So, modulation is also of basically only of three types.

Let us consider a carrier wave

$$S_c(t) = A_c \cos \theta_c(t) \qquad ...(5.5)$$

this can be written as $\theta_c(t) = w_c t + \phi(t)$ ...(5.6)

so that $S_c(t) = A_c \cos [w_c t + \phi(t)]$ ...(5.7)

where w_c is the radian frequency of the carrier and $\phi(t)$ is the phase, *f* and *w* both are used for frequency *f* in Hz and *w* in radian per second.

The possible modulation technique will be discussed in next section. After modulation and transmission we want to receive the signal at the receiver and then recover the original signal from modulated received signal.

In digital communication, the term demodulation and detection are often used for same purpose. But demodulation emphasizes waveform recovery, and detection entails the process of symbol decision.

The demodulation/detection may be classified in two parts. When the receiver have the knowledge of the carrier's phase to detect the signals, the process is called coherent detection, when the receiver does not have such reference information of transmitting signal it is called non-coherent detection. In ideal coherent detection, there is availability at the receiver a prototype of each possible arriving signal.

These prototype waveforms attempt to duplicate the transmitted signal set in every respect, even RF phase.

The receiver is then said to be phase locked to the incoming signal.

During demodulation, the receiver multiples and integrates (correlates) the incoming signal with each of its prototype replicas.

Non-coherent demodulation refers to systems employing demodulators that are designed to operate without knowledge of the absolute value of the incoming signal's phase.

Therefore, phase estimation is not required.

Thus the advantage of non-coherent over coherent systems is reduced complexity but at the price of increased probability error.

5.2.2. Amplitude Shift Keying (ASK)

ASK is amplitude modulation technique in digital, it is parallel to AM in Analog communication.

It is one of oldest digital modulation technique used for communication. But now-a-days it is not used in any form due to higher bit error rate.

In ASK system, binary symbol 1 is represented by transmitting a sinusoidal carrier wave of fix amplitude A_c with a fixed frequency f_c for the duration T_b seconds, whereas binary symbol 0 is represented by switching of the carrier for T_b seconds.

Due to this ON-OFF characteristics, it is also known as (OOK) or ON-OFF keying.

Let the sinusoidal carrier be represented by

$$S_c(t) = A_c \cos (2\pi f_c t) \quad ...(5.8)$$

then the binary ASK is simply

$$\begin{aligned} S_c(t) &= A_c \cos (2\pi f_c t), && \text{symbol 1} \\ &= 0 && \text{symbol 2} \end{aligned} \quad ...(5.9)$$

A Fig. 5.2 illustrates the concept.

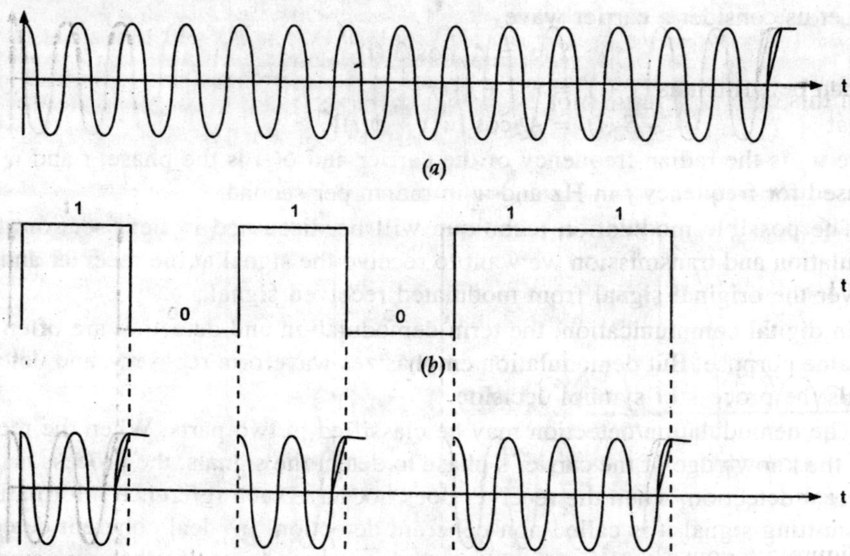

Fig. 5.2. (*a*) Carrier (*b*) In unipolar form message (*c*) ASK.

Let a binary data is (101011) transmitted with ASK modulation then after modulation it will be look like as follows.

5.2.2.1. Generation of ASK

ASK can be generated by the help of a product modulator shown in Fig. 5.3.

As we know from the equation of ASK, it is simply product of binary signal (baseband) and carrier signal.

The power spectral density of ASK is same as baseband signal (ON-OFF).

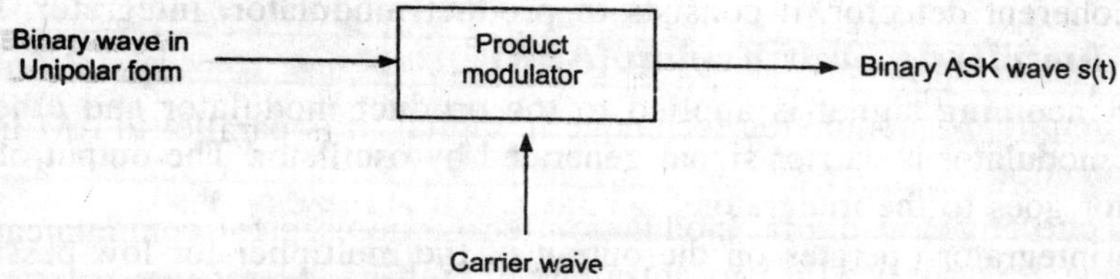

Fig. 5.3. Generation of ASK.

Note that to find the power spectral density of baseband signal or ASK, first find the Fourier transform.

It will be shifted sinc pulse in frequency domain at $\pm f_c$.

As we know that an impulse signal consists of infinite spectra of frequency mean has infinite bandwidth because

$$f[\delta(t)] = 1 \quad \text{...(5.10)}$$

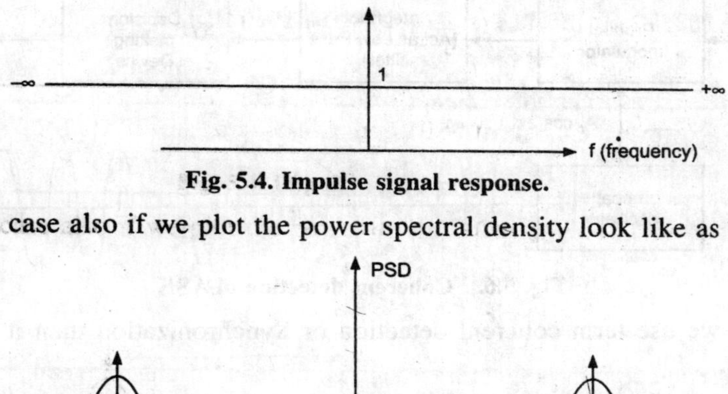

Fig. 5.4. Impulse signal response.

In this case also if we plot the power spectral density look like as in Fig. 5.5.

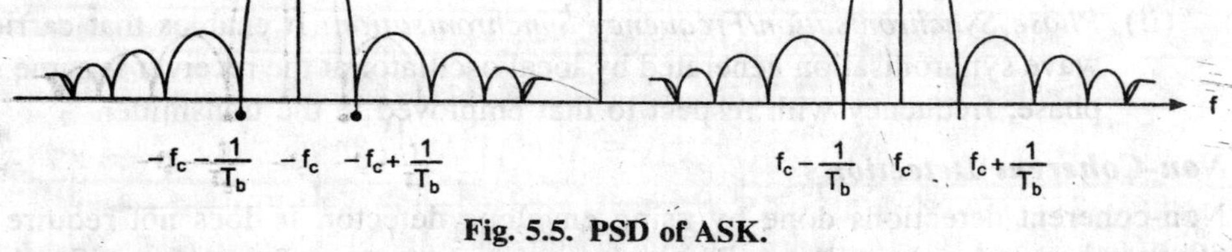

Fig. 5.5. PSD of ASK.

Figure 5.5 consists of two impulses at $\pm f_c$ which indicates that bandwidth of ASK is infinite, but from the definition of bandwidth, bandwidth is defined as bandwidth of an ideal bandpass filter centered at f_c whose output contains around 96% of the total average power content of the ASK signal

According to this criteria the bandwidth of an ASK signal is around $3/T_b$ Hz.

or Bandwidth of ASK = $3f_b$ ...(5.11)

5.2.2.2. Demodulation/Detection of ASK

In demodulation section we have discussed, that this is of two types—coherent and non-coherent.

In coherent detector, it consists of product modulator, Integrator, Decision making device.

The incoming signal is applied to the product modulator and other *I/P* of product modulator is carrier signal generated by oscillator. The output of product modulator goes to the integrator.

The integrator operates on the output of the multiplier for low-pass filtering action.

This *O/P* goes to the decision making device. And decision making device compares the *O/P* of integrator with a preset threshold value.

If integrator output of signal level crosses the threshold level of decision device the *O/P* comes, if not crosses the *O/P* then it comes 0 (zero).

The coherent detection involves the use of linear operation. It is assumed that the carrier signal generated by local oscillator have same carrier frequency and phase as those of transmitted carriers.

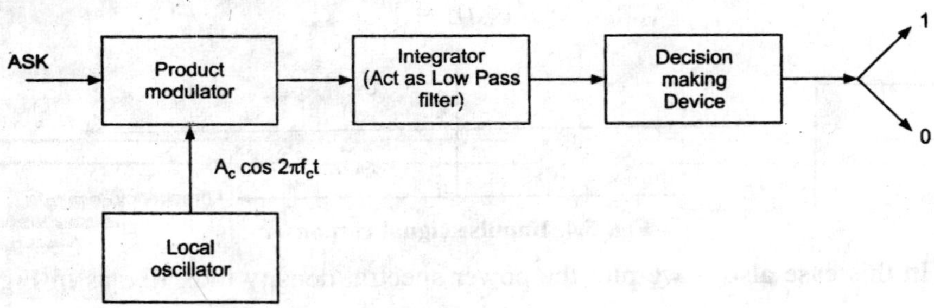

Fig. 5.6. Coherent detection of ASK.

When we use term coherent detection or Synchronization then it must be of two forms.

(*i*) *Timing Synchronisation:* It enables the proper timing of the decision in the receiver with respect to switching instant for bit 1 and 0.

(*ii*) *Phase Synchronisation/Frequency Synchronisation:* It enables that carrier wave synhronisation generated by local oscillator at the receiver is same in phase, frequency with respect to that employed in the transmitter.

Non-Coherent Detection

Non-coherent detections done by using envelope detector. It does not require a phase coherent local oscillator. This scheme involves some form of rectification (like Amplification) and low-pass filtering at the receiver.

Figure 5.7 illustrates operation of non-coherent detection.

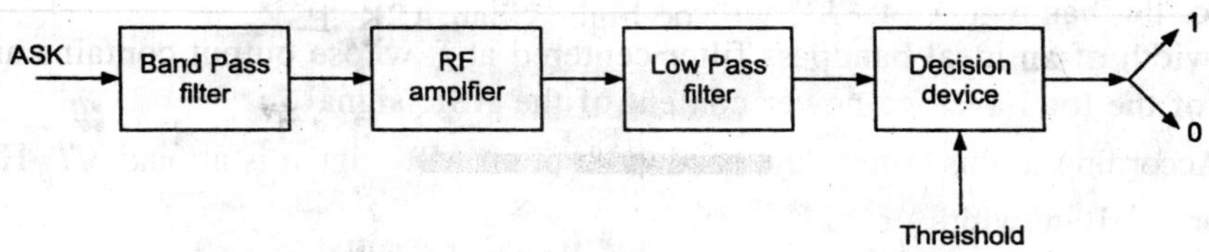

Fig. 5.7. Non-coherent detection.

5.2.3. Frequency Shift Keying (FSK)

In FSK, two sinusoidal carrier waves of same amplitude with different carrier frequencies are used to represent 1 and 0 as follows:

$$S(t) = A_c \cos 2\pi f_{c_1} t \text{ for symbol 1}$$
$$= A_c \cos 2\pi f_{c_2} t \text{ for symbol 2} \quad \text{...(5.12)}$$

FSK work is same as of FM in Analog communication for bit 1 frequency increased and for bit 0 frequency is decreased as shown in Fig. 5.8 for bit stream [10110].

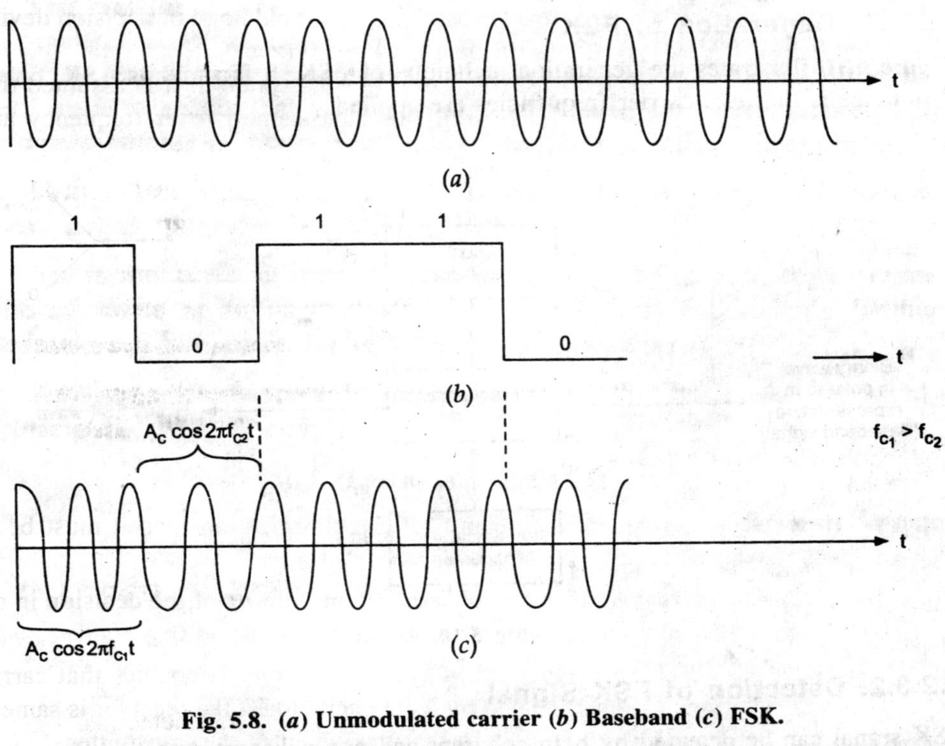

Fig. 5.8. (*a*) Unmodulated carrier (*b*) Baseband (*c*) FSK.

It can be easily understood that FSK is a combination of two ASK waveforms. For bit '1' one type of ASK and for bit '0' another type of ASK. Both ASK having different carrier frequencies f_{c_1} and f_{c_2} and $f_{c_1} > f_{c_2}$. Therefore, the power spectral density of FSK is the sum of two ASK spectra at frequencies f_{c_1} and f_{c_2}. And in the spectrum of FSK there will not be any discrete components.

So, the bandwidth of FSK will be higher than ASK, PSK.

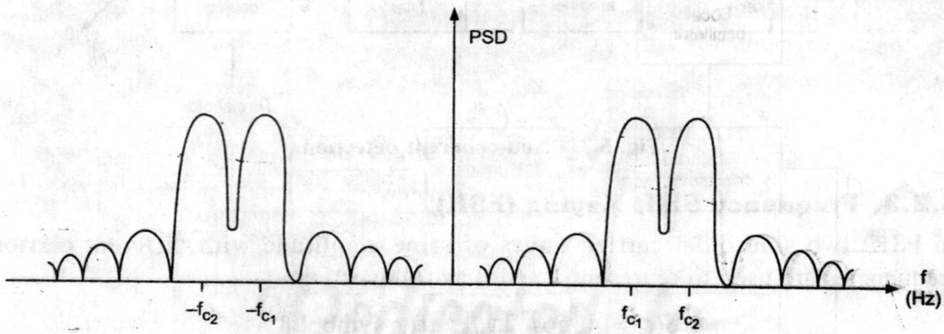

Fig. 5.9. PSD of FSK.

FSK scheme found variety of applications in digital transmission like wireless communication, specially in cellphone communication. As compared to ASK, PSK, the hardware implementation is very easy.

Power spectral density of FSK is shown in Fig. 5.9.

5.2.3.1. Generation of FSK

Figure 5.10 illustrates the generation technique of FSK. It is same as ASK, only difference is that two carrier frequencies are applied.

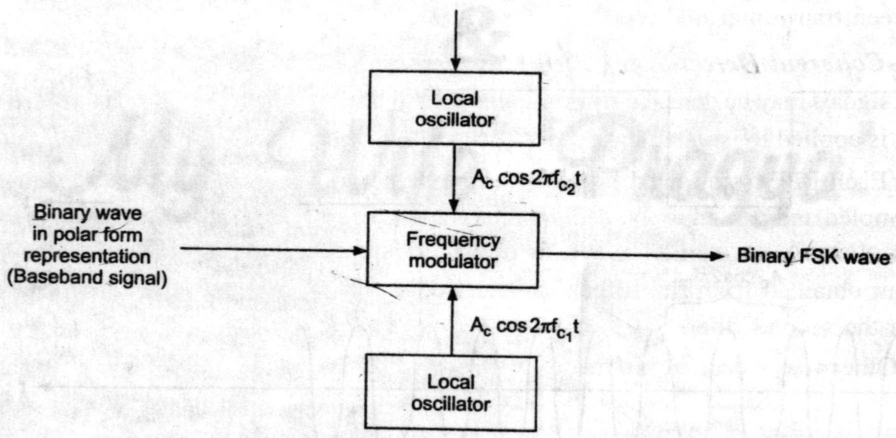

Fig. 5.10.

5.2.3.2. Detection of FSK Signal

FSK signal can be detected by both coherent and non-coherent detection. Firstly, we are discussing the coherent detection-technique.

Like modulator in detector there are two local oscillators as in transmitter for proper synchronization. A detector circuit consists of two product modulators followed by two integrators then output of both integrators is passed to the decision making device (it is a type of comparator).

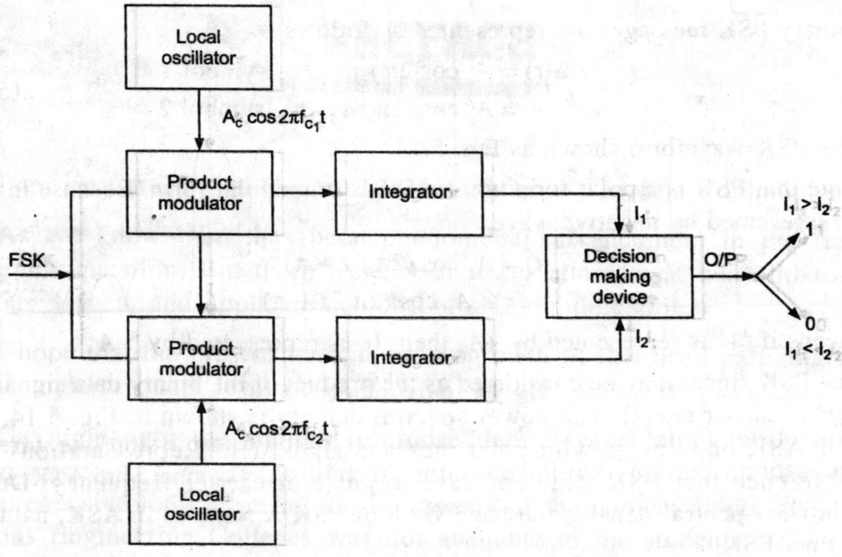

Fig. 5.11. FSK demodulator (coherent detection).

If I_1 is greater than I_2 then output is one and if I_2 is greater than I_1 then output is zero.

But keep in mind, to get error free output there must be proper synchronization between transmitter and receiver.

Non-Coherent Detection of FSK (Frequency Shift Keying)

FSK signals may be detected by non-coherently using envelope detector. The received FSK is applied to isolated bandpass flter tuned to two different frequencies f_{c_1} and f_{c_2}. Each BPF is followed by an envelope detector. The envelope detector output is sampled and passed to the decision making device (Comparator) to compare with each other. A decision is made in favour of symbol 1 if the envelope detector output obtained from the filter tuned to frequency f_{c_1} is larger than that derived from the second filter.

Otherwise a decision is made in favour of bit 0 (zero).

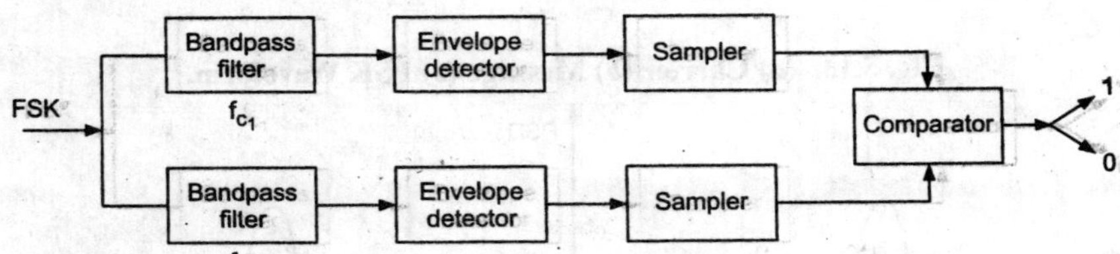

Fig. 5.12. Non-coherent detection of FSK binary signals.

5.2.4. Phase Shift Keying (PSK)

In a PSK system, only one carrier signal of a fixed frequency is used to modulate the bit-1 and 0. Both are modulated by a phase difference of π as follows :

$$S_c(t) = A_c \cos 2\pi f_c t \quad \text{...(5.13)}$$

then binary PSK messages are represented as follows

$$S(t) = A_c \cos 2\pi f_c t \quad \text{symbol 1}$$
$$= A_c \cos (2\pi f_c t + \pi) \quad \text{symbol 2} \quad \text{...(5.14)}$$

The PSK waveform shown as Fig. 5.13.

Note that PSK is a polar form while ASK is in unipolar form. Because in PSK '0' is represented by negative wave.

$$S_c(t) = A_c \cos (2\pi f_c t + \pi)$$
$$= - A_c \cos (2\pi f_c t + \pi) \quad \text{...(5.15)}$$

Means if '1' is represented by $+A_c$ then '0' is represented by '$-A_c$'.

The PSK signal may be considered as the product of the binary data signal $b(t)$ and cosine carrier signal. The power spectral density is shown in Fig. 5.14, it is same as ASK or same as with polar message signal (Bit Representation) with one difference that PSK does not have impulse at carrier frequency. Due to same power spectral density the bandwidth of PSK is same as in ASK, naturally lower than FSK.

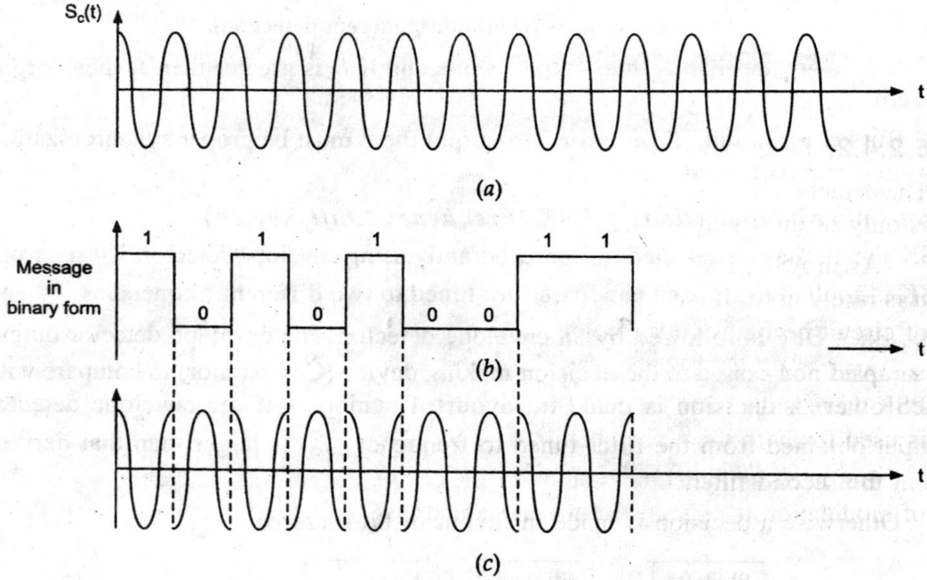

Fig. 5.13. ***(a)* Carrier *(b)* Message *(c)* PSK Waveform.**

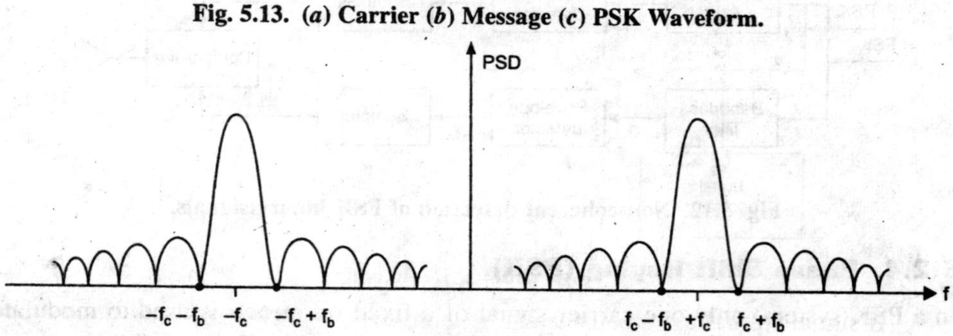

Fig. 5.14.

One more difference is that ASK is a linear modulation technique, while PSK is a nonlinear modulation technique. And PSK has better noise performance than ASK for same peak power. FSK has more better performance with respect to noise than ASK, PSK for same input peak power.

The reason is that FSK has better noise perfomance due to higher bandwidth of FSK.

5.2.4.1. Generation of Phase Shift Keying

The generation technique of PSK is same as that of ASK, the difference is only, the incoming message signal must be in polar digital form.

If we compare PSK with analog modulation technique then PSK can be think similar to doublc sideband suppressed carrier (DSB-SC) modulated wave.

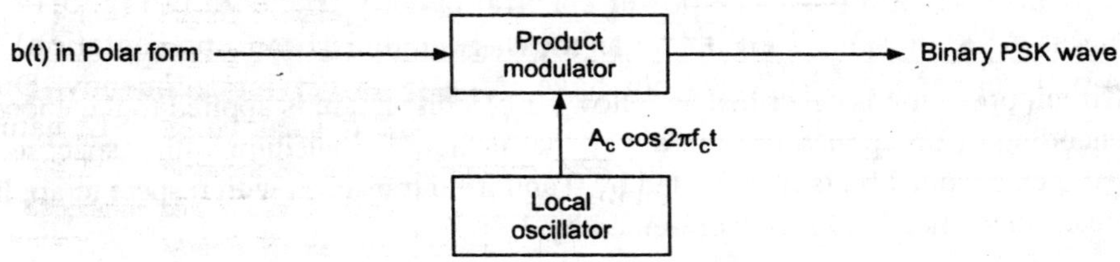

Fig. 5.15. PSK Modulator.

5.2.4.2. Detection of PSK

The detection technique of PSK is same as for ASK, the difference in PSK detection is only of threshold level.

As in ASK, PSK also suffers with problem of synchronisation and due to that it is rarely used. If we want to reduce this problem of synchronisation, then design of circuit becomes very complex which increases cost excessively.

And non-coherent detection of PSK is not possible because the envelope of a PSK wave is the same for both the symbols '1' and '0' and a single carrier frequency is used for both the representations.

But need for synchronisation of PSK can be improved by suitable modification of modulation technique and this is named DPSK. Differential phase shift keying.

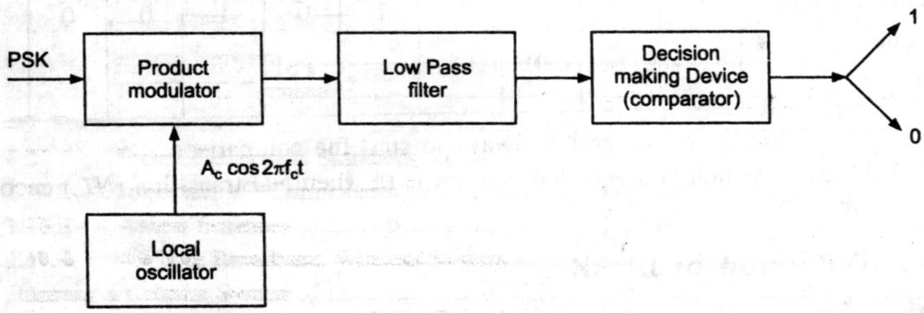

Fig. 5.16. PSK detection.

If signal strength is greater than threshold value set by comparator then the bit-1 and if less than the threshold value then bit-0.

5.2.5. Differential Phase Shift Keying

This is an improvment method of PSK modulation, to reduce the synchronisation problem. We have seen in previous modulation technique that bit is represented by rectangular pulses (Polar or Unipolar). But here bits are represented by transition. For example if there is transition then bit is '0', if no transition then bit is '1'. A circuit in Fig. 5.17 is shown to generate the DPSK bit.

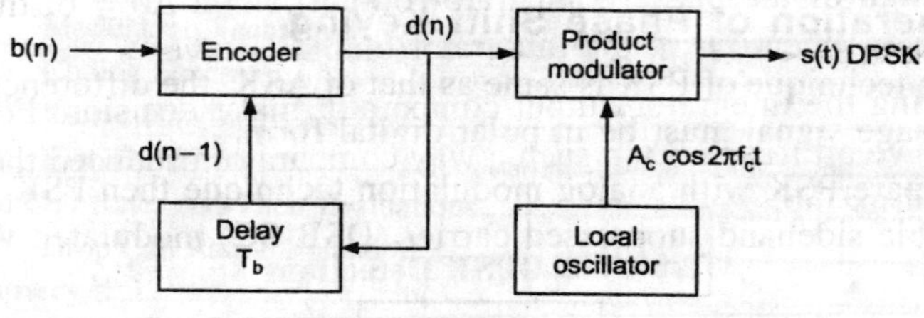

Fig. 5.17. DPSK Generation.

Circuit operation is described as follows. Let a bit stream is applied to the encoder and encoding of bit applied in such a way that if there is transition with respect to bit applied then encoded bit is represented by 0 and if no transition with respect to applied bit stream then encoded bit is represented by 1.

Table 5.1. DPSK Coding.

Binary Data		Tr → **0**	NT → **0**	NT → **1**	NT → **1**	Tr → **0**	NT → **1**	NT → **0**	Tr → **0**	Tr → **1**
Differential encoded bit $d(n)$ (DPSK)	*1	0	1	1	1	0	0	1	0	0
Phase DPSK	0	π	0	0	0	π	π	0	π	π
Shifted differentially encoded data $d(n-1)$		1	0	1	1	1	0	0	1	0
Phase of shifted DPSK		0	π	0	0	0	π	π	0	π
Detected binary sequence at Receiver		0	0	1	1	0	1	0	0	1

Notes: 1. 1* is taken as reference bit always to start the comparison.
2. When transition (T_r) encoded then bit is 0, when NO transition (NT_r) encoded then bit is 1.

5.2.5.1. Detection of DPSK

Detection circuit is shown in Fig. 5.18. DPSK bit is first applied to the product modulator with delayed version of DPSK. Then it is passed to the integrator which is working as a low-pass filter, then output is passed to the comparator which has a fix threshold value.

On the basis of this threshold value original bit is generated.

The DPSK signal is proportional to cos (ϕ), where ϕ is the difference between the carrier phase angle of the received DPSK signal and its delayed version measured in the same time interval T_b.

The phase of DPSK and its delayed version is shown in Table below.

The phase difference between the two sequences for each bit interval is used to determine the sign of the phase comparator output, when $\Delta\phi = 0$, the integrator output is positive. When $\phi = \pi$, the integrator output is – ve.

By comparing the integrator output comparator make decision. For + ve bit it is one and for – ve bit it is 0. So in such a way comparator produced the bit stream. $\Delta\phi$ is illustrated in Table 5.2.

Table 5.2. DPSK Detection.

Original bit		0	0	1	1	0	1	0	0	1
Phase of DPSK	0	π	0	0	0	π	π	0	π	π
Phase of Delayed		0	π	0	0	0	π	π	0	π
DPSK $d(n-1)\ \Delta\phi$		–	–	+	+	–	+	–	–	+
Output bit stream based on $\Delta\phi$		0	0	1	1	0	1	0	0	1

From Table 5.2, it is understood that original bit transmitted and bit detected at the receiver is exactly same.

Which indicates that by using DPSK we can reconstruct original data without synchronisation in the absence of noise.

Here we are not using synchronisation, so we can say that DPSK is non-coherent version of PSK.

It is important to note that reconstruction of bit does not depend upon selection of reference bit, you can choose reference bit 1 or 0 as your choice.

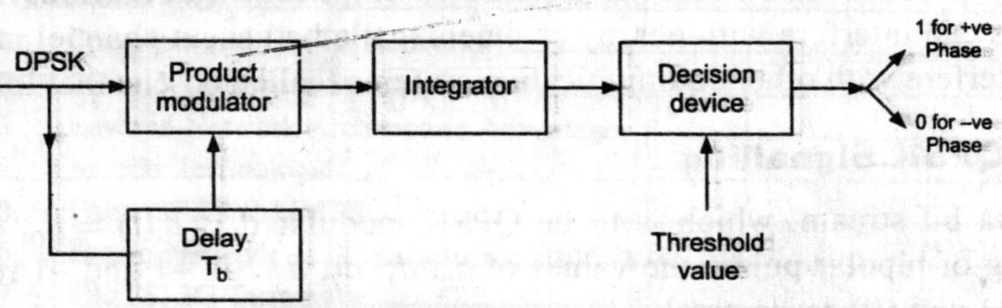

Fig. 5.18. DPSK Detector.

5.3. MODULATION METHOD IN CELLULAR WIRELESS SYSTEM

There are so many modulation techniques used in cellular system but mainly used modulation techniques are :

Quadrature Phase Shift Keying (QPSK)

Minimum Shift Keying (MSK)

Gaussian Minimum Shift Keying (GMSK)

Spread Spectrum Modulation (SSM).

5.3.1. Quadrature Phase Shift Keying (QPSK)

Till now we have discussed many of digital modulation techniques but they are insufficient to use the full channel bandwidth. Here we are going to discuss another two techniques, in which channel efficiency is improved as compared to earlier used digital modulation techniques, those techniques are QPSK, a improved version of PSK and MSK (Minimum shift keying).

In simple binary data transmission, we transmit bit '1' or '0' with each duration of T_b, while if we are using M-arry data transmission then we can increase the data transmission in duration T_b.

In digital data transmission, the number of possible signals is

$$[M = 2^n] \qquad ...(5.16)$$

where n is the integer. And signalling interval for n bit is equal to nT_b.

On the basis of this we can say that every binary data transmission system is a case of M-arry. When we transmit only '1' or '0' then $m = 2$ and $n = 1$. But if $m = 4$, $n = 2$ then bits are 00, 01, 10, 11, and QPSK is example of $m = 4$. In QPSK any one of four data is transmitted in each signalling interval, So if we compare the PSK, then in this case, it requires only half of the bandwidth, because in PSK we are transmitting a single bit 1 or 0 in duration T_b.

While in QPSK in T_b duration we transmit two bits that may be 00 or 01 or 10 or 11.

In PSK to transmit the 1 and 0 (each bit separately), we require $2T_b$ time duration while in QPSK, in $2T_b$ duration we transmit 8 bits in the combination of 2 bits.

The advantages of QPSK, MSK are that, they are bandwidth efficient, give maximum efficiency for non-linear system also.

For example, satellite communication is very non-linear. Transponders of satellite systems are very non-linear in nature, therefore, for better performance they require a constant envelope modulation. If modulated envelope is not constant then they produce extra sidebands. Due to amplitude fluctuations these extra sidebands can interfere with nearby channels (called adjacent channel interference) or can interfere with other communication systems (called co-channel Interference).

5.3.2. QPSK Signalling

Let a data bit stream, which is to be QPSK modulated is $d_i(t) = d_0, d_1, d_2, d_3...$ consisting of bipolar pulses the values of $d_0, d_1, d_2, d_3...$ is +1 and –1 representing the bit (1) and (0) respectively.

From $d_i(t)$, we can separate even and odd bit

$$d_i(t) = d_0, d_2, d_4, d_6 \quad \text{...even bit} \qquad ... (5.17)$$

$$d_q(t) = d_1, d_3, d_5, d_7 \quad \text{...odd bit}$$

where $d_i(t)$ is called in-phase bit

$d_q(t)$ is called quadrature phase bit.

To modulate the $d_i(t)$ and $d_q(t)$, we require two carrier waves of same amplitude, frequency with a phase of 90°.

And the carrier waves are:

$$S_I(t) = A_c \cos(2\pi f_o t + \pi/4) \qquad ...(5.18)$$
$$S_q(t) = A_c \sin(2\pi f_o t + \pi/4)$$

and complete QPSK modulated wave will be

$$\Rightarrow \quad s(t) = A_c\, d_I(t) \cos(2\pi f_o t + \pi/4) + A_c\, d_q(t) \sin(2\pi f_o t + \pi/4) \quad ...(5.19)$$

From $s(t)$ it is understood that $d_I(t)$ modulate with $\cos(2\pi f_o t)$ which is simple BPSK and same with $d_q(t)$.

So, we can say that QPSK is sum of the BPSK modulated wave. In QPSK, the carrier phase changes only once in $2T_b$ sec. The carrier phase during any $2T_b$ interval can be any one of the 4 phases. (0, ± 90°, 180°) depending upon the bit value of $d_I(t)$ and $d_q(t)$.

During $2T$ interval if no bit stream changes the sign then carrier phase remain the same.

If any one bit changes the sign then a phase shift of ±90° occurs. While if change in both the streams occur then phase of 180° occurs.

The following waveforms show the QPSK waveform.

5.3.2.1. OQPSK (Offset QPSK)

Sometimes OQPSK also called staggered QPSK, in QPSK $d_I(t)$ and $d_q(t)$ started from same reference so there is proper alignment in both while the base band waveform in OQPSK it is shifted by T_b.

Let $\quad d_i(t) = 10\;1\;0\;1\;1\;0\;0\;1\;0$

or $\quad d_i(t) = d_0, d_1, d_2, d_3, d_4, d_5, d_6, d_7, d_8, d_9$

and $\quad d_I(t) = d_0, d_2, d_4, d_6, d_8 \Rightarrow \{1\;1\;1\;0\;1\}$

$\quad d_q(t) = d_1, d_3, d_5, d_7, d_9 \Rightarrow \{0\;0\;1\;0\;0\}$

OQPSK has same representation as QPSK, and the change of phase also occur as in QPSK. The difference is only that in OQPSK, the data bit $d_I(t)$ and $d_q(t)$ is shifted with phase difference of T_b.

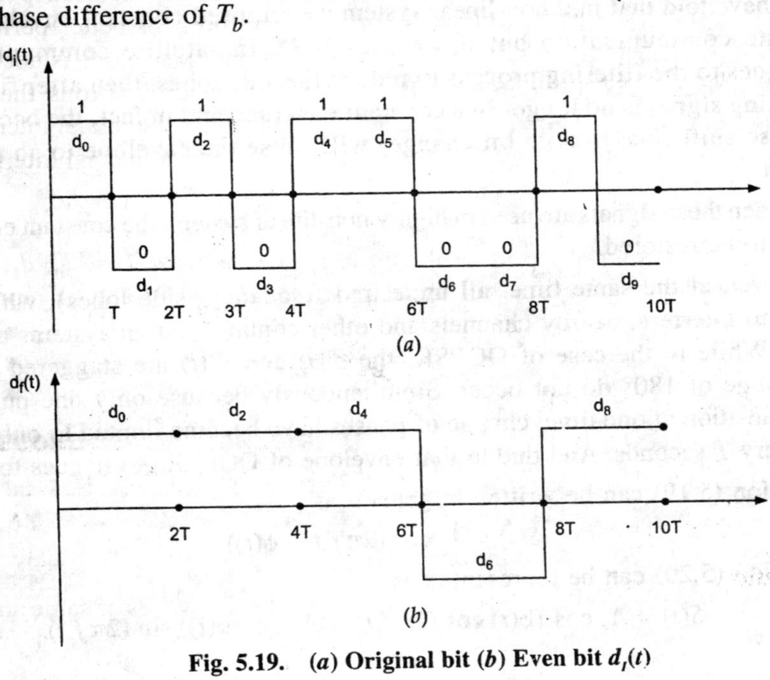

Fig. 5.19. (a) Original bit (b) Even bit $d_I(t)$

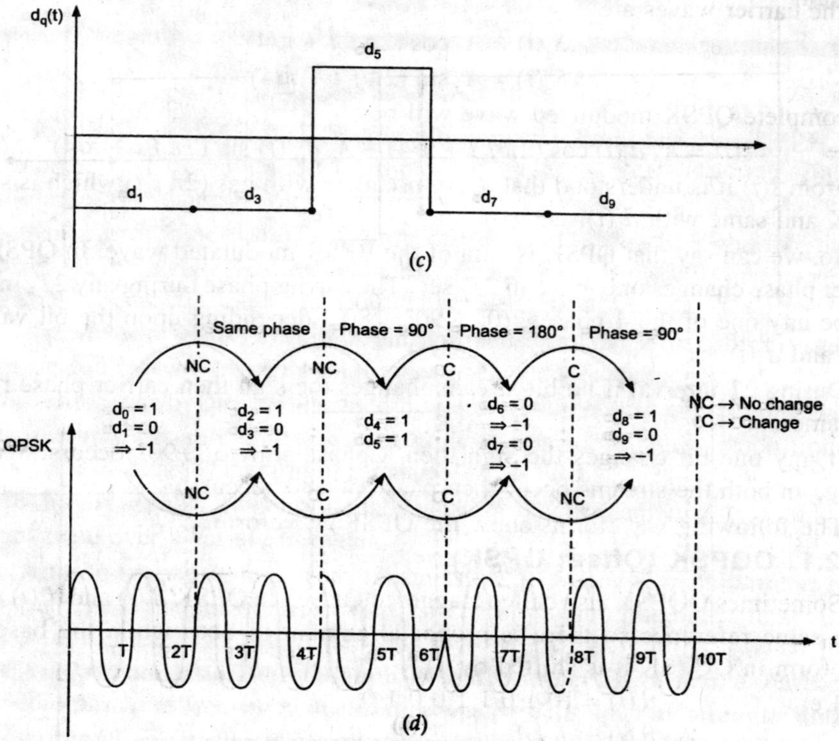

Fig. 5.19. (c) Odd bit $d_q(t)$ (d) QPSK waveform

5.3.2.2. OQPSK Waveform

In the case of OQPSK, waveform will be of 1 T duration. In discussion of QPSK, I have told that in a non-linear system we required constant envelope like in satellite communication but if we use QPSK in satellite communication, when it goes to the filtering process to reduce the side lobes, then after filtering, the resulting signal is no longer in a constant envelope and in fact, the occasional 180° phase shift (due to both bit change) will cause the envelope to go zero for a moment.

So, when these signals are used in highly non-linear system, the constant envelope will tend to be restored.

However, at the same time. all undesired frequency (side lobes), which has tendency to interfere, nearby channels and other communication systems are also restored. While in the case of OQPSK, the $d_I(t)$ and $d_q(t)$ are staggered so that phase change of 180° do not occur. Simultaneously because only one pulse can make a transition at one time, change of phases have become limited to only 0 and 90° in every T seconds. And due to that envelope of OQPSK never goes to zero.

Equation (5.19) can be written in general as

$$S(t) = A_c \cos (2\pi f_c t + \phi(t)) \qquad ...(5.20)$$

Equation (5.20) can be represented as

$$S(t) = A_c \cos (\phi(t)) \cos (2\pi f_c t) - A_c \sin (\phi(t)) \sin (2\pi f_c t) \qquad ...(5.21)$$

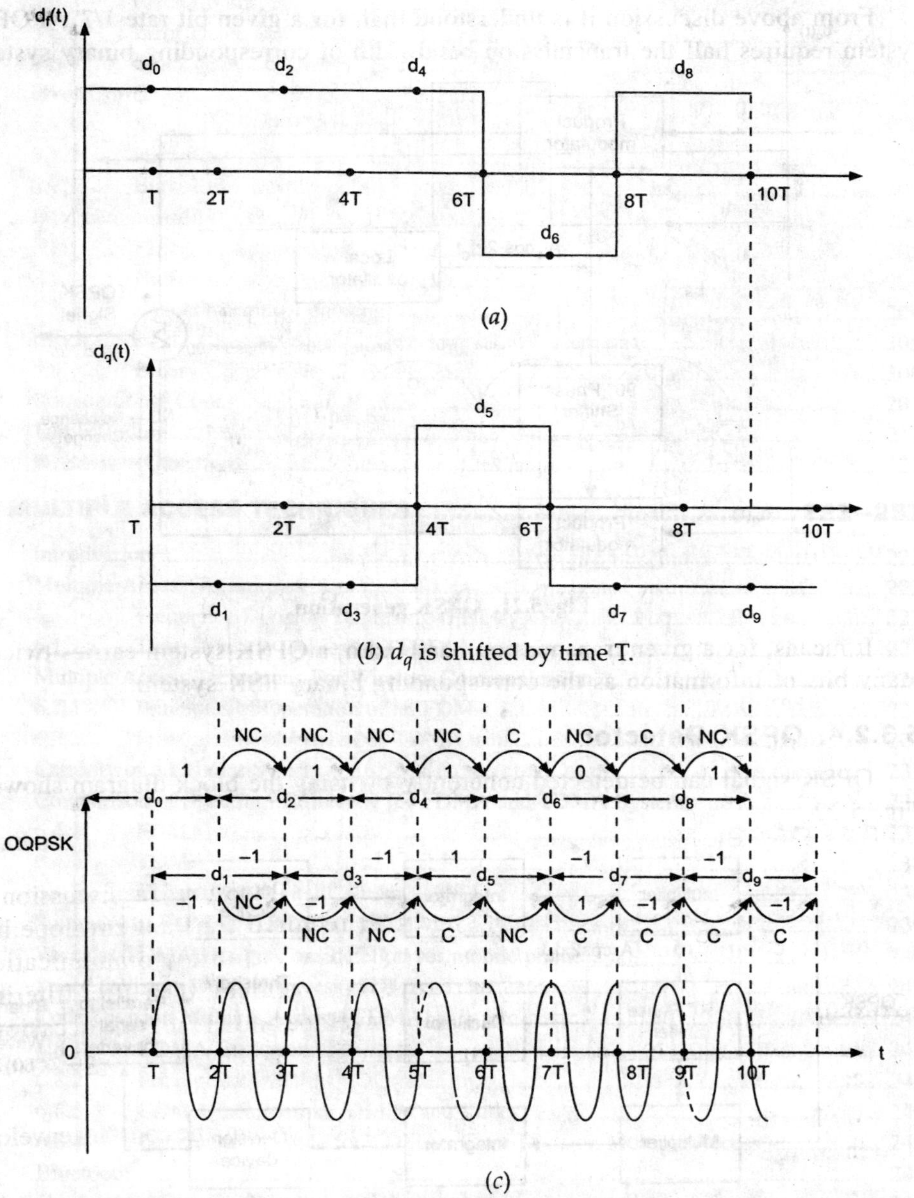

Fig. 5.20. *(a)* $d_I(t)$ *(b)* **Staggered** $d_q(t)$ *(c)* **OQPSK**

Cos $2\pi f_c t$ is said to be inphase component and sin $2\pi f_c t$ is said to be the quadrature component.

5.3.2.3. QPSK Generation

Figure 5.21 shows the arrangement for the generation of QPSK.

One bit of data bit is applied to the in-phase channel of the transmitter and the other bit applied to the quadrature channel.

Finally, the output of two product modulator is subtracted from each other.

From above discussion it is understood that, for a given bit rate $1/T_b$, a QPSK system requires half the transmission bandwidth of corresponding binary system.

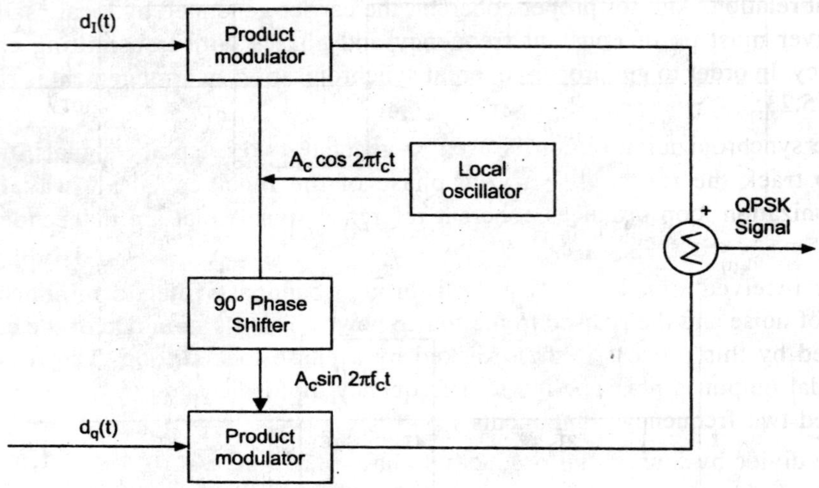

Fig. 5.21. QPSK generation.

It means, for a given transmission bandwidth, a QPSK system carries twice as many bits of information as the corresponding binary PSK system.

5.3.2.4. QPSK Detector

QPSK signal can be detected coherently by using the block diagram shown in Fig. 5.22.

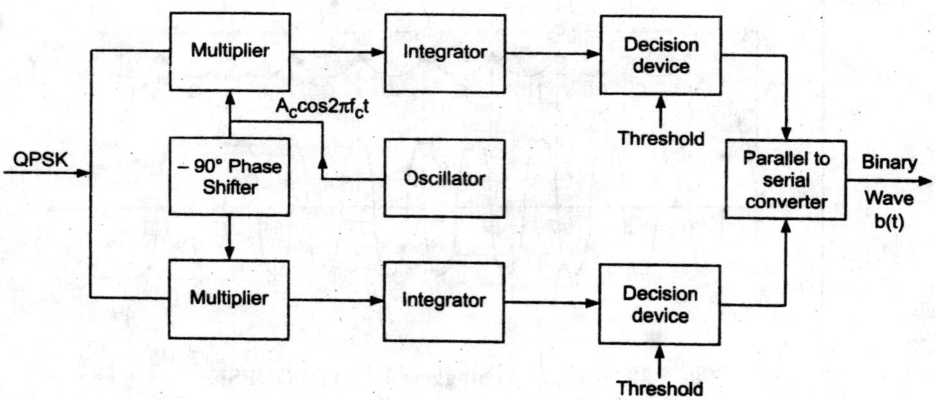

Fig. 5.22. QPSK detector.

In the Fig. 5.22, parallel to serial converter, interleaves the decision made by in-phase and quadrature part of the detector and then reconstructed the binary data stream. In the absence of noise, the detected binary stream is same as transmitted.

We may say, QPSK scheme is a combination of two PSK schemes that work in parallel and employed two carrier in phase quadrature.

So, PSK is a quadrature multiplexing scheme that gives method to utilize the full bandwidth in more efficient way.

5.3.2.5. Carrier Synchronization

In order to detect transmitted bit correctly, it is necessary that there must be proper coherent relation. And for proper coherent, the carrier generated by local oscillator at receiver must be of constant frequency and phases with transmitting carrier frequency. In order to get proper coherent synchronisation an arrangement is shown in Fig. 5.23.

The synchronization circuit is used to generate two reference signals, that is used to track the randomly varying phase of the received QPSK wave. The synchronization loop produces coherent reference signals that are independent of the I^4.

The received signal is first passed through bandpass filter to minimise the effects of noise and then raised to the fourth power. The 4th harmonic of the carrier produced by this process is then tracked by a phase locked loop. The resulting sinusoidal output is next applied to a frequency divided by four($f/4$), circuit that produced two frequency components.

The divide by four circuit produces 4 phase in the interval (0, 2π). Which take care of differentially encoding incoming binary wave at the transmitter input and decoding the same at the receiver output.

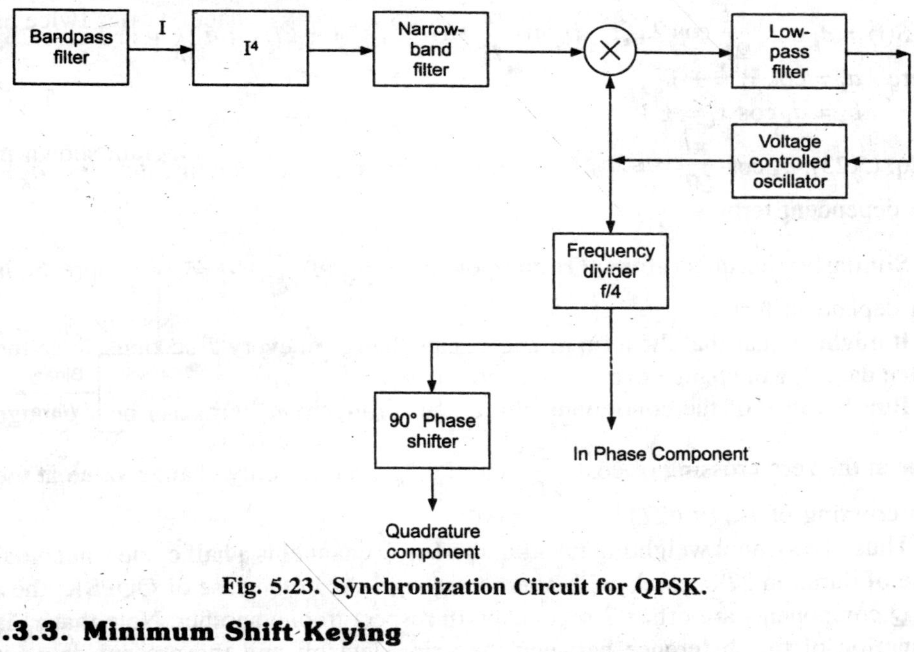

Fig. 5.23. Synchronization Circuit for QPSK.

5.3.3. Minimum Shift Keying

Minimum shift keying can be viewed as either a special case of continuous phase frequency shift keying (CPFSK) or a special case of OQPSK (Offset Quadrature Phase Shift Keying). Actually the main advantage of OQPSK over QPSK, that of suppressing out of band interference, suggests that further improvement is possible if the OQPSK format is modified to avoid discontinuous phase transition.

The MSK waveform equation can be viewed as

$$S(t) = \cos\left[2\pi\left(f_c + \frac{d_k}{4T}\right)t + x_k\right],\ kT < t < (k+1)\,T \qquad \text{...(5.21}a\text{)}$$

where amplitude is set to be 1. $d_k = \pm 1$ represents the bipolar data being transmitted at a rate $R = \frac{1}{T}$, and x_k is a phase constant which is valid over the kth binary data interval.

Note that for $d_k = 1$, frequency transmitted is $f_c + \frac{1}{4T}$ and for $d_k = -1$, frequency transmitted is $f_c - \frac{1}{4T}$.

The tone spacing in MSK is thus one half that employed for non-coherently demodulated orthogonal FSK, $\left[f_c + \frac{1}{4T} - f_c + \frac{1}{4T} = \frac{1}{2T} \right]$ giving rise to the name minimum shift keying.

During each T-second data interval, the value of x_k is constant, that is $x_k = 0$ or π, determined by the requirement that the phase of the waveform be continuous at $t = kT$.

This requirement results in the following recursive phase constraint for x_k.

$$x_k = \left[x_{k-1} + \frac{\pi k}{2}(d_{k-1} - d_k) \right] \quad ...(5.22)$$

Equation (5.21a) can be expressed as following by using trigonometric identities.

$$S(t) = a_k \cos \frac{\pi t}{2T} \cos 2\pi f_c t - b_k \sin \frac{\pi t}{2T} \sin 2\pi f_c t \quad \text{for } kT < t < (k+1)T \quad ...(5.23)$$

where $a_k = \cos x_k = \pm 1$

$b_k = d_k \cos x_k = \pm 1$ ...(5.24)

in Eq. (5.23), $a_k \cos \frac{\pi t}{2T} \cos 2\pi f_c t$ is said to be the in-phase component, where a_k is data dependent term.

Similarly, the quadrature (Q) component is $b_k \sin \frac{\pi t}{2T} \sin 2\pi f_c t$ where b_k is data dependent term.

It might appear that the term a_k and b_k can change in every T seconds since the source data d_k can change every T seconds.

But, because of the continuous phase constraint, the a_k term can only change value at the zero crossing of $\cos\left(\frac{\pi t}{2T}\right)$ and the b_k term can only change value at the zero crossing of $\sin(\pi t/2T)$.

Thus, the symbol weighting in either the I or Q channel is a half cycle sinusoidal pulse of duration $2T$ seconds with alternating sign. As in the case of OQPSK, the I and Q components are offset T seconds with respect to one another. Note that x_k is a function of the difference between the prior data bit and the present data bit (differential encoding). Hence the a_k and b_k can be considered as differentially encoded components of the d_k source data.

However, for bit to bit independent data d_k, the sign of successive I or Q channel pulses are also random from one $2T$ second pulse interval to the next.

Thus as a special case of OQPSK, Eq. (5.23) can be seen rewritten with more straight forward (non-differential) data encoding as follows :

$$S(t) = d_I(t) \cos \frac{\pi t}{2T} \cos 2\pi f_c t + d_Q(t) \sin \frac{\pi t}{2T} \sin 2\pi f_c t \quad ...(5.25)$$

where
$$d_I(t) = d_0, d_2, d_4 \quad \text{... even bits}$$
$$d_q(t) = d_1, d_3, d_5 \quad \text{... odd bits} \qquad ...(5.26)$$

and (*d*), (*f*) represent the sinusoidal weighting of the *I* and *Q* channel pulses.

These sequences represent the same data sequences as $d_I(t)$ and $d_q(t)$, but here multiplication by a sinusodial result to in more gradual phase transitions compared to those of the original data representation.

Figure 5.24 (*f*), (*g*) represents the modulation of the orthogonal components cos $2\pi f_c t$ and sin $2\pi f_c t$, respectively, by the sinusoidal data streams.

While Fig. 5.24 (*h*) illustrates the summation of the (*f*) and (*g*) and that is required MSK waveform.

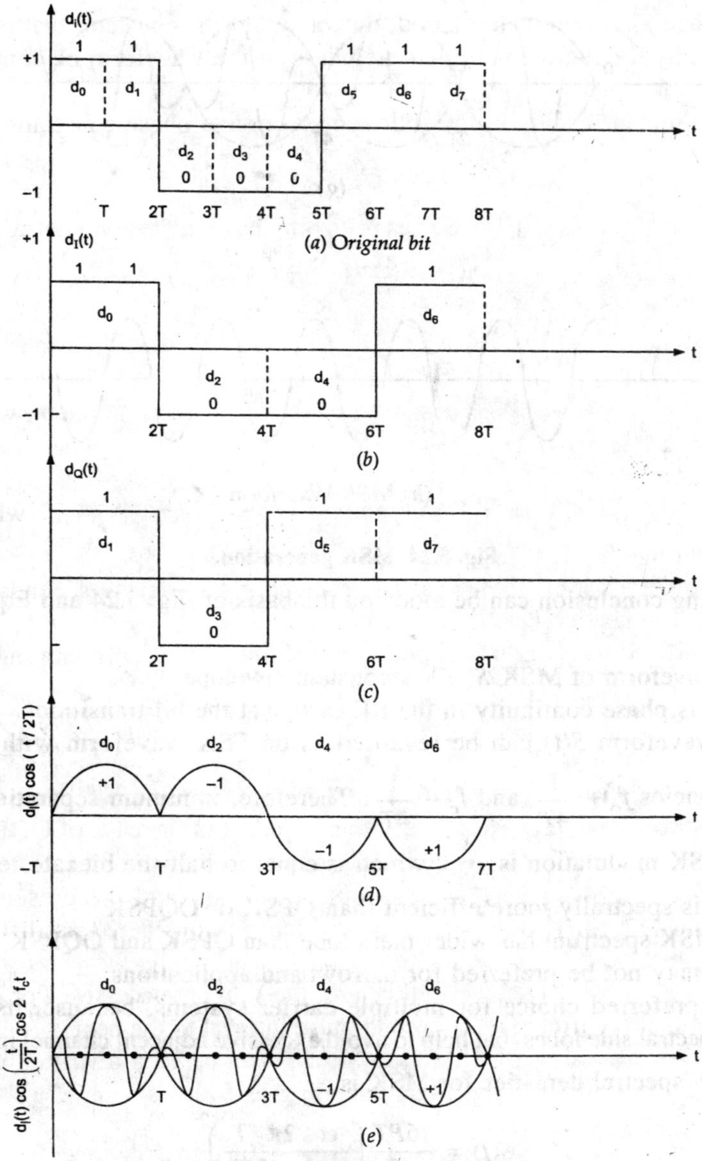

Fig. 5.24. (*b*) Only even bit of $d_i(t)$ (*c*) only odd bit of $d_i(t)$

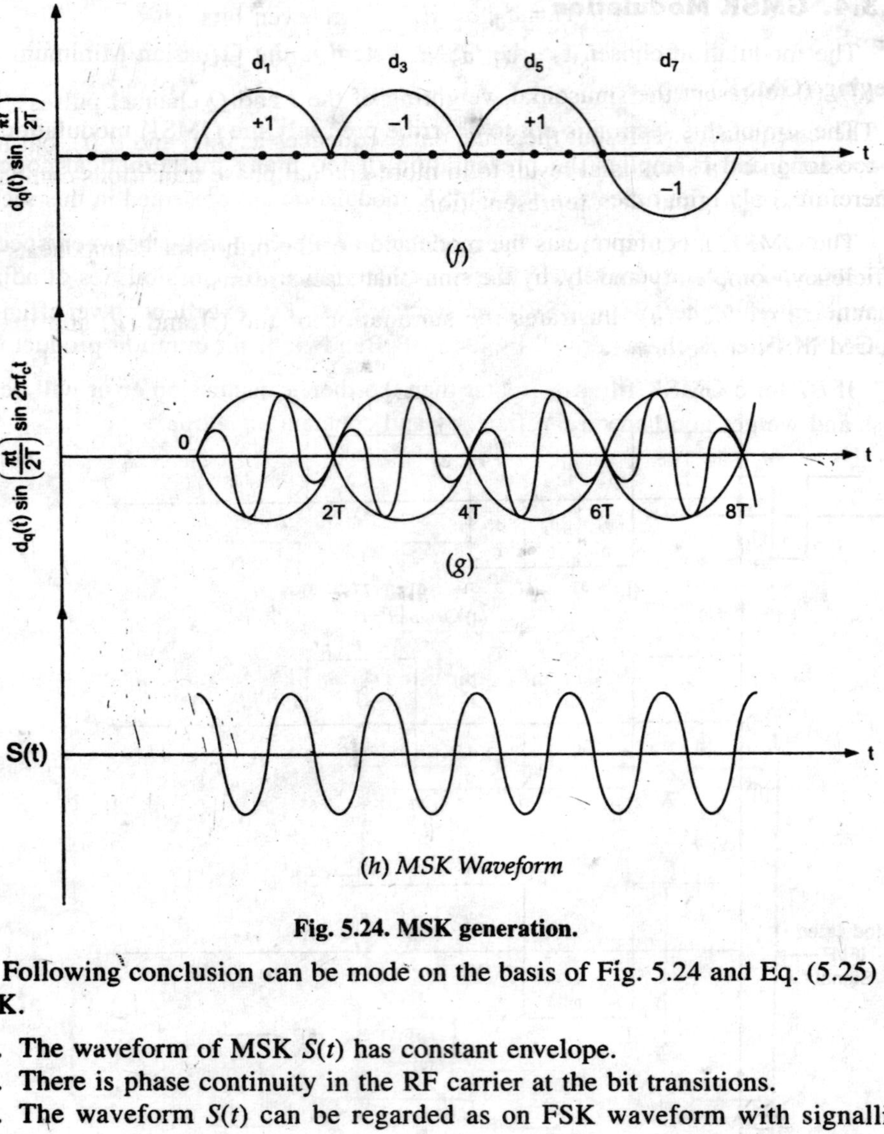

(h) *MSK Waveform*

Fig. 5.24. MSK generation.

Following conclusion can be mode on the basis of Fig. 5.24 and Eq. (5.25) for MSK.

1. **The waveform of MSK $S(t)$ has constant envelope.**
2. **There is phase continuity in the RF carrier at the bit transitions.**
3. **The waveform $S(t)$ can be regarded as on FSK waveform with signalling frequencies $f_c + \frac{1}{4T}$ and $f_c - \frac{1}{4T}$. Therefore, minimum separation required for MSK modulation is $\frac{1}{2T}$, which is equal to half the bit rate.**
4. **MSK is spectrally more efficient than QPSK or OQPSK.**
5. **The MSK spectrum has wider main lobe than QPSK and OQPSK. Therefore, MSK may not be preferred for narrowband applications.**
6. **MSK preferred choice for multiple carrier systems, because, its relatively low spectral side lobes that help to avoid excessive adjacent channel interference.**

The low spectral densities for MSK is

$$S(f) = \frac{16PT}{\pi^2}\left(\frac{\cos 2\pi f T}{1 - 16 f^2 T^2}\right)^2 \qquad ...(5.27)$$

5.3.4. GMSK Modulation

The modulation chosen for the GSM system is the Gaussian Minimum Shift Keying (GMSK).

The aim of this section is not to describe precisely the GMSK modulation as it is too long and it implies the presentation of too many mathematical concepts. Therefore, only brief aspects of the GMSK modulation are presented in this section.

The GMSK modulation has been selected as a compromise between spectrum efficiency, complexity and low radiations (that reduces the possibilities of adjacent channel interference). But GMSK is most attractive for its excellent power efficiency. A GMSK filter is characterized by its 3 dB Bandwidth-bit duration product (BT).

If *BT* for a GMSK filter is greater than 0.5 then transmission error will be very less and we get good spectral efficiency and constant envelope.

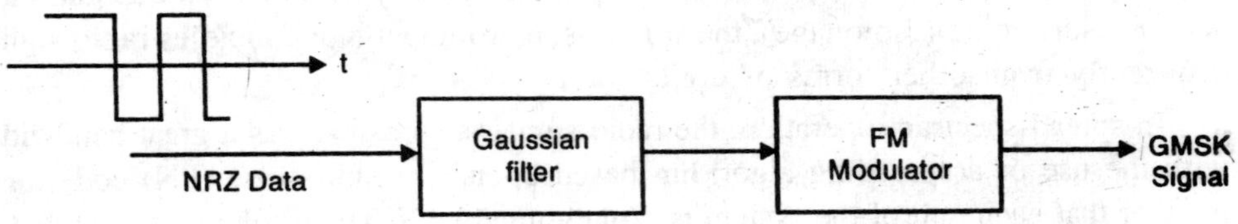

***(a)* GMSK transmitter using direct FM generation.**

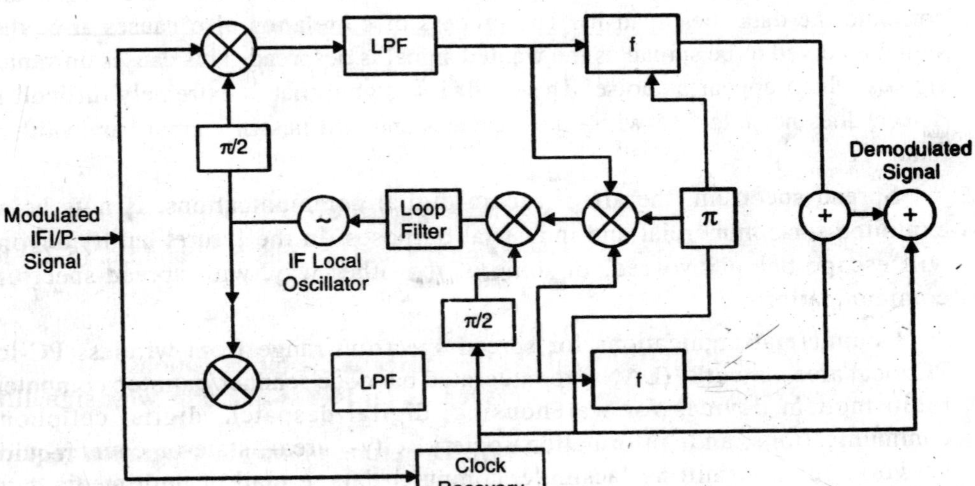

***(b)* GMSK receiver.**

Fig. 5.25.

Impulse response of a GMSK filter is given by

$$h_G(t) = \frac{\sqrt{\pi}}{\beta} \exp\left(-\frac{\pi^2}{\beta^2} t^2\right) \qquad \text{...(5.28)}$$

and transfer function $H_G(f) = \exp(-\beta^2 f^2)$ ...(5.29) where $\left\{\beta = \frac{\sqrt{\ln 2}}{\sqrt{2}\, B} \Rightarrow \frac{0.58}{B}\right\}$

Easiest method to get a GMSK signal is to pass a Non Return Zero (NRZ) data through a Gaussian low pass filter (also known as Gaussian Base band filter) having an impulse response $h_G(t)$, followed by an FM modulator. Finally we get GMSK signal at Radio frequency. This modulation technique is used in variety of analog and digital communications. For example, used in US Cellular Digital Packet Data (US CDPD) system as well as in GSM system. Basic block diagram of GMSK transmitter and receiver is shown in Fig. 5.25(*a*) and (*b*) respectively.

5.4. SPREAD SPECTRUM

Advances in technology have brought to us a new form of digital radio service called 'spread spectrum.' First developed by the military as a deterrent to jamming and eavesdropping (espionage), the spread-spectrum technique handles radio signal differently from other forms of digital radio.

In spread-spectrum operation, the radio signal is spread across a great bandwidth with the use of a spreading algorithm based upon a pseudo-noise (PN) code, or a number that each unit of the system is programmed with. The result is a signal that is essentially 'buried' in the noise floor of the radio band.

The receiver is programmed to examine the bandwidth of the spread signal and correlate the data (despread it). The process of correlation also causes any other signal received to be spread as the wanted signal is despread. This causes unwanted signals which appear as noise. The result is a signal that is extremely difficult to detect, does not interfere with other services and still passes a great bandwidth of data.

Spread spectrum, the art of secure digital communications, is now being exploited for commercial and industrial purposes. In the future, hardly anyone will escape being involved, in some or the other way, with spread-spectrum communications.

Commercial applications for spread spectrum range from wireless PC-to-PC local area networks (LANs) to integrated bar code scanner/palmtop computer/radio modem devices for warehousing, digital despatch, digital cellphone communications, and 'information society' city-, area-, state-or country-wide networks for transmitting facsimile, computer data, e-mail or multimedia data.

5.4.1. History of Spread Spectrum

Early spark-gap wireless transmitters actually used spread spectrum, since their RF bandwidths were much wider than their information bandwidth. The first intentional use of spread spectrum was probably by Armstrong in the late 1920s or early 1930s with wide-band frequency modulation (FM). However, the real impetus for spread spectrum came with World War II.

Both the allies and the axis powers experimented with simple spread-spectrum systems. Much of what was done is still shrouded in secrecy, however.

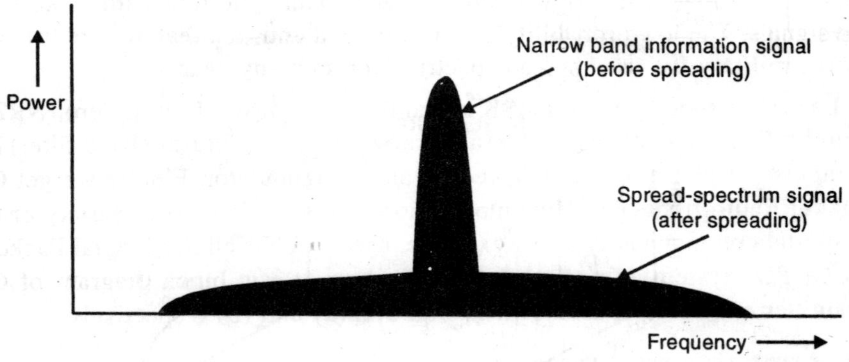

Fig. 5.26. Spread-spectrum signals are hard to detect on narrowband equipment because the signal's energy is spread over a bandwidth of may be 100 times the information bandwidth.

The first publicly available patent on spread spectrum came from Hedy Lamarr, the Hollywood movie actress, and George Antheil, an *avant-garde* music composer. This patent was granted in 1942, but the details were a military secret for many years. The inventors never realised a time for their invention; they simply turned it over to the US government for use in the war effort, and commercial use was delayed until the patent had expired.

Most of the work done in spread spectrum throughout the 1950s, 1960s and 1970s was heavily backed by the military and drowned in secrecy. The global positioning system (GPS) is now the world's single largest spread-spectrum system. Most of the details on GPS are now public information.

Spread spectrum was first used for commercial purposes in the 1980s when Equatorial Communications of Mountain View, CA, used direct sequence for multiple-access communications over synchronous satellite transponders. In the late 1980s, the US Federal Communications Commission (FCC) opened up the industrial, scientific and medicine (ISM) frequency bands for unlicenced spread-spectrum communications.

5.4.2. How Spread Spectrum Works

The term 'spread spectrum' describes a modulation technique that makes the sacrifice of bandwidth in order to gain signal-to-noise (S/N) performance. Basically, in a spread-spectrum system, the transmitted signal is spread over a frequency much wider than the minimum bandwidth required to send the signal.

The fundamental premise is that in channels with narrow-band noise, increasing the transmitted signal bandwidth results in an increased probability that the received information will be correct. If total signal power is interpreted as the area under the spectral density curve, signals with equivalent total power may have either a large signal power concentrated in a small bandwidth or a small signal power spread over a large bandwidth.

Spread signals are intentionally made to be much widerband than the information they are carrying to make them more noise-like. Because spread-spectrum signals are noise-like, they are hard to detect. They are also hard to intercept or demodulate.

Further, spread-spectrum signals are harder to jam (interfere with) than narrow-band signals. The low probability of intercept and anti-jam features are the reasons why the military has used spread spectrum for so many years.

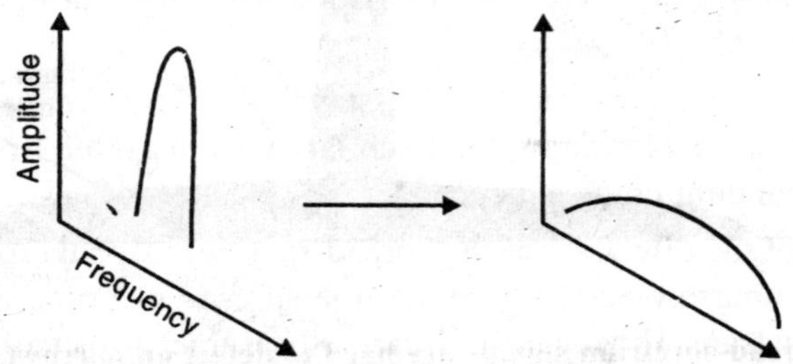

Fig. 5.27. Generation of the direct-sequence spread-spectrum signal (spreading).

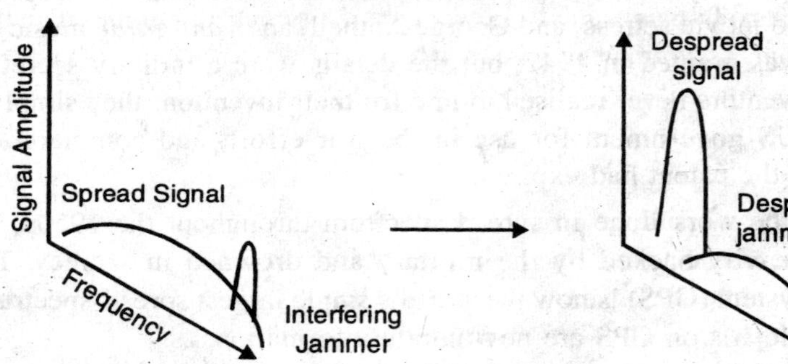

Fig. 5.28. At the receiving end, the spread-spectrum signal is despread to generate the original narrowband signal.

Spread-spectrum signals use fast codes that run many times the information bandwidth or data rate. These special 'spreading' codes are called 'pseudo random' or 'pseudo noise' codes. They are called 'pseudo' because they are not real Gaussian noise.

The use of special pseudo-noise codes in spread-spectrum communications makes signals appear wide-band and noise-like. It is this very characteristic that makes spread-spectrum signals possess the quality of low probability of intercept.

Spread-spectrum transmitters use the same transmit power levels as narrow-band transmitters. Because spread-spectrum signals are very wide, they transmit at a much lower spectral power density, measured in watts per hertz, than narrow-band transmitters. The lower transmitted power density characteristic gives spread signals a big plus. Spread and narrow-band signals can occupy the same band, with little or no interference. This capability is the main reason for all the interest in spread spectrum today.

Since the total integrated signal density or signal-to-noise ratio (SNR) at the correlator's input determines whether there will be interference or not, all spread-spectrum systems have a threshold or tolerance level of interference beyond which useful communication ceases. This tolerance or threshold is related to the spread-spectrum processing gain. Processing gain is essentially the ratio of the RF bandwidth to the information bandwidth.

Direct sequence and frequency hopping are the most commonly used methods for the spread-spectrum technology. Although the basic idea is the same, these two methods have many distinctive characteristics that result in completely different radio performances.

The carrier of direct-sequence radio stays at a fixed frequency. The narrow-band information is spread out into a much larger (at least ten times) bandwidth by using a pseudo-random chip sequence.

Generation of the direct-sequence spread-spectrum signal (spreading) is shown in Fig. 5.28. The narrow-band signal and the spread-spectrum signal both use the same amount of transmit power and carry the same information. However, the power density of the spread-spectrum signal is much lower than the narrow-band signal. (The power density is the amount of power over a certain frequency.) As a result, it is more difficult to detect the presence of the spread-spectrum signal. In this case, the narrow-band signal's power density is ten times higher than the spread-spectrum signal, assuming the spread ratio as '10.'

At the receiving end, the spread-spectrum signal is despread to generate the original narrow-band signal as shown in Fig. 5.29.

If there is an interference jammer in the same band, it will be spread out during despreading. As a result, its impact is greatly reduced. This is the way the direct-sequence, spread-spectrum radio fights the interference. It spreads out the offending jammer by the spreading factor, which is at least '10.' In other words, the offending jammer's amplitude is greatly reduced by at least 90 per cent.

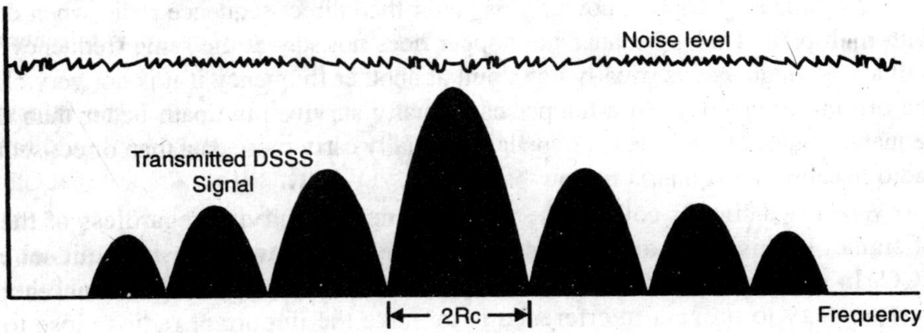

Fig. 5.29. Direct-sequence spread-spectrum signal.

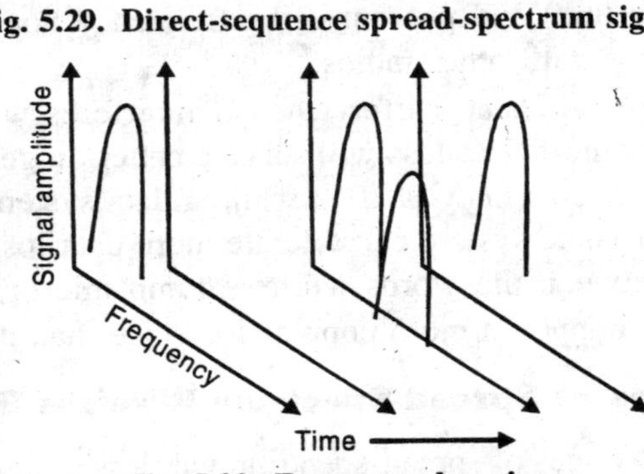

Fig. 5.30. Frequency hopping.

Frequency hopping achieves the same result by using a different carrier frequency at a different time. Its carrier will hop around within the band, so it will avoid the jammer at some frequencies. The frequency hopper is more popular and the only way to survive in the 2.45 GHz band because the leakages from the microwave oven (from 2.4 to 2.5 GHz) sometimes exceed 10 W. Frequency hopper is not needed in the 915 MHz band because there is no legal big jammer in this frequency.

The frequency-hopping technique is shown in Fig. 5.30. It does not spread the signal, so there is no processing gain. The processing gain is the increase in power density when the signal is despread and it improves the received signal's SNR. In other words, the frequency hopper needs to put out more power in order to have the same SNR as a direct-sequence radio.

The frequency hopper is also more difficult to synchronise the receiver to the transmitter because both the time and frequency need to be in tune. Whereas, in a direct-sequence radio, only the timing of the chips needs to be synchronised. The frequency hopper will need to spend more time to search the signal and lock to it. As a result, the latency time is usually longer. Whereas, a direct-sequence radio can lock in the chip sequence in just a few bits.

Usually, to make the initial synchronisation possible, the frequency hopper will park at a fixed frequency before hopping or communication begins. If the jammer happens to locate at the same frequency as the parking frequency, the hopper will not be able to hop at all. And once it hops, it will be very difficult to re-synchronise if the receiver ever lost the sync.

The hopper usually costs more and is more complicated than direct-sequence radio because it needs extra hopping and synchronising circuits to implement the synchronisation algorithm.

The frequency hopper, however, is better than direct-sequence radio when dealing with multipath. This is because the hopper does not stay at the same frequency and a null at one frequency is usually not a null at another frequency if it is not very close to the original frequency. So a hopper can usually survive multipath better than direct-sequence radio. The frequency hopper can usually carry more data than direct-sequence radio because the signal is narrow-band.

When two signals collide, the stronger one may survive regardless of the kind of signal. In this band, all radio must not exceed the power density limit set by the FCC. In other words, all radios are equal when interfering with one another. The best strategy to prevent interference is to make the important radios close to each other (strengthen the link) and prevent using frequency hopper because they are guaranteed to interfere with other radios.

The hopper itself will also suffer when it interferes with other radio. Any system that can suffer more data loss will survive better. In general, a voice system can survive an error rate as high as 10^{-2}, while a data system must have an error rate lower than 10^{-4}. Voice system can tolerate more data loss because the human brain can 'guess' between the words while a dumb microprocessor can't. As a result, the frequency hopper is more popular for voice than data communications.

5.4.3. Advantages of Spread-Spectrum Wireless Systems

Some of the advantages of spread-spectrum wireless systems over conventional systems are :

1. No crosstalk interference. Conventional cordless phones frequently suffer from crosstalk interference, especially when used in densely populated residential areas (such as apartment complexes). This problem disappears in spread-spectrum cordless phone systems because :

(*i*) Crosstalk interference is greatly attenuated due to the processing gain of the spread-spectrum system as described earlier.

(*ii*) The effect of the suppressed crosstalk interference can be essentially removed with digital processing where noise below certain threshold results in negligible bit errors. These negligible bit errors will have little effect on voice transmissions.

2. Better voice quality/data integrity and less static noise. Due to the processing gain and digital processing nature of spread-spectrum technology, a spread-spectrum based system is more immune to interference and noise. This greatly reduces the static noise induced by consumer electronics devices that is commonly experienced by conventional analogue wireless system users.

3. Lowered susceptibility to multipath fading. Because of its inherent frequency diversity properties (thanks to wide spectrum spread), a spread-spectrum system is much less susceptible to multipath fading. This makes reception of a spread-spectrum based cordless phone much less sensitive to the location and pointing direction of the handset than a conventional analogue wireless system.

4. Inherent security. In a spread-spectrum system, a PN sequence is used to either modulate the signal in the time domain (direct sequencing) or select the carrier frequency (frequency hopping). Due to the pseudorandom nature of the PN sequence, the signal in the air is 'randomised.' Only a receiver that has exactly the same pseudo-random sequence and synchronous timing can despread and retrieve the original signal. Consequently, a spread-spectrum system provides signal security that is not available to conventional analogue wireless systems.

5. Co-existence. A spread-spectrum system is less susceptible to interference than other non-spread-spectrum systems. In addition, with proper designing of pseudo-random sequences, multiple spread-spectrum systems can coexist without causing severe interference to other systems. This further increases the system capacity for spread-spectrum systems or devices.

6. Longer operating distances. A spread-spectrum device operated in the ISM band is allowed to have higher transmit power due to its non-interfering nature. Because of the higher transmit power, the operating distance of such a device can be significantly longer than for a traditional analogue wireless communication device.

7. Hard to detect. Spread-spectrum signals are transmitted over a much wider bandwidth than conventional narrowband transmissions—20 to 254 times the bandwidth of narrow-band transmissions. Since the communication band is spread, these can be transmitted at a low power without suffering interference from background noise. This is because when despreading takes place, the noise at one frequency is rejected, leaving the desired signal.

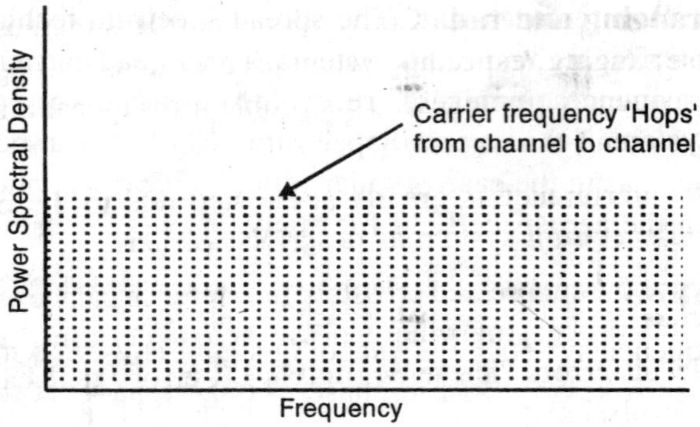

Fig. 5.31. Frequency hopping.

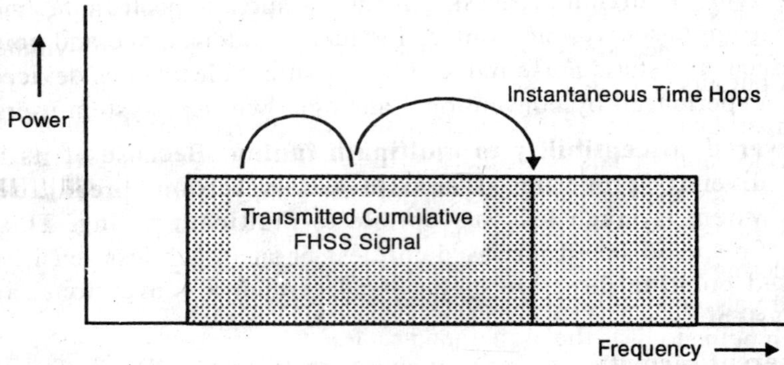

Fig. 5.32. FHSS spectrum.

8. Hard to intercept or demodulate. The very foundation of the spreading technique is the code used to spread the signal. Without knowing the code, it is impossible to decipher the transmission. Also, because the codes are so long (and quick), simply viewing the code would still be next to impossible to solve the code, hence interception is very hard.

9. Harder to jam than narrow bands. The most important feature of the spread-spectrum technique is its ability to reject interference. At first glance, it may be considered that spread-spectrum transmission would be most affected by interference. However, any signal is spread in the bandwidth, and after it passes through the correlator, the bandwidth signal is equal to its original bandwidth plus the bandwidth of local interference.

An interference signal with 2 MHz bandwidth being input into a direct-sequence receiver whose signal is 10 MHz wide gives 12 MHz output from the correlator. The wider the interference bandwidth, the wider the output signal. Thus the wider the input signal, the less the effect on the system because the power density of the signal after processing is lower, and less power falls in the band-pass filter.

Conversely, it may be guessed that the most effective interference to a direct-sequence receiver is one with the narrowest bandwidth (a continuous-wave carrier). This is the most effective because power density in the correlator output due to narrow-band signals is higher than due to wide-band signals.

10. Use for ranging and radar. The spread-spectrum technique can be used to construct precise ranging and radar systems. The spread carrier, modulated with the pseudo noise sequence, permits the receiver to measure very precisely the time when the signal was sent; thus, spread-spectrum technique can be used to time the distance to an object, as in the case of radar reflection. Both applications have been commonly used in the aerospace field for many years.

5.4.4. Modulation

For direct-sequence systems, the encoding signal is used to modulate a carrier, usually by phase-shift keying (*e.g.*, biphase or quadriphase) at the code rate.

Frequency-hopping systems generate their wide band by transmitting at different frequencies, hopping from one frequency to another according to the code sequence. Typically, such a system may have a few thousand frequencies to choose from, and unlike direct-sequence signal, it has only one output rather than symmetrically distributed outputs.

The important thing to note is that both direct sequencing and frequency hopping generate wide-band signals controlled by the code-sequence generator. For direct-sequence systems the code is direct-carrier modulation, while frequency hopping commands the carrier frequency.

Considering direct sequencing, balance modulation is an important tool in any suppressed carrier system, used to generate the transmitted signal. Balanced modulation helps to hide the signal, and there is no power wasted in transmitting a carrier that would contribute to interference rejection or information transfer. When a signal has poor balance in either code or carrier, spikes are seen in its spectrum. With these spikes, or spurs, the signal is easily detectable, since once these spikes are noticed above the noise, it is obvious to look for the hidden signal.

In CDMA spread-spectrum transmission, user channels are created by assigning different codes to different users. This type of system provides privacy by controlling distribution of user-unique code sequences.

Information modulation in spread-spectrum systems is possible in most of the conventional ways; both amplitude modulation (AM) and angle modulation are satisfactory. Normally, AM is not used for spread-spectrum signals because it tends to be detectable when examining the spectrum. Frequency modulation (FM) is more useful because it is a constant envelope signal, but information is still readily observed. In both AM and FM, no knowledge of the code is needed to receive the transmitted information.

Clock modulation is actually frequency modulation of the code clock. It is usually avoided (*e.g.*, for frequency hopping) because the loss in correlation due to phase slippage between received and local clocks can degrade the performance. For direct sequence, an FM demodulator tuned to radio frequency (RF) carrier plus/minus the clock could recover the data.

Another technique is code modification, where the code is changed such that the information is embedded in it, then modulated by phase transitions on an RF carrier.

5.4.5. Demodulation

Once the signal is coded, modulated and then sent, the receiver must demodulate the signal. This is usually done in two steps :

1. Spectrum-spreading (*e.g.*, direct-sequence or frequency-hopping) modulation is removed.
2. The remaining information bearing signal is demodulated by multiplying with a local reference identical in structure and synchronised with the received signal.

5.4.6. Spread Spectrum (SS) Communication Concepts

In Code Division Multiple Access (CDMA) systems all users transmit in the same bandwidth simultaneously. Communication systems following this concept are "spread spectrum systems". In this transmission technique, the frequency spectrum of a data-signal is spread using a code uncorrelated with that signal. As a result the bandwidth occupancy is much higher than required.

The codes used for spreading have low cross-correlation values and are unique to every user. This is the reason that a receiver which has knowledge about the code of the intended transmitter is capable of selecting the desired signal.

The main parameter in spread spectrum systems is the processing gain : the ratio of transmission and information bandwidth :

$$G_P = \frac{BW_t}{BW_t} = \frac{W}{R} \qquad ...(5.30)$$

where W = Transmission bandwidth; R = Original messages bandwidth (or the signal rate R) which is basically the "spreading factor". The processing gain determines the number of users that can be allowed in a system, the amount of multi-path effect reduction, the difficulty to jam or detect a signal etc. For spread spectrum systems it is advantageous to have a processing gain as high as possible.

There exist different techniques to spread a signal: Direct-Sequence (DS), Frequency-Hopping (FH), Time-Hopping (TH) and Multi-Carrier CDMA. It is also possible to make use of combinations of them. We will now concentrate on the two most popular techniques: Direct-Sequence and Frequency-Hopping.

Direct Sequence (DS)

Direct Sequence is the best known Spread Spectrum Technique. The data signal is multiplied by a Pseudo Random Noise Code (PNcode). A PNcode is a sequence of chips valued –1 and 1 (polar) or 0 and 1 (non-polar) and has noise-like properties. This results in low cross-correlation values among the codes. Several families of binary PN codes exist; they are addressed a little later. A usual way to create a PN code is by means of at least one shift-register. When the length of such a shift-register is n, the following can be said about the period N_{DS} of the above mentioned code-families :

$$N_{DS} = 2^n - 1 \qquad ...(5.31)$$

In direct-sequence systems the length of the code is the same as the spreading-factor with the consequence that :

$$G_P(DS) = N_{DS} \qquad ...(5.32)$$

Each bit is represented by multiple bits using spreading code which spreads signal across wider frequency band. In proportion to number of bits used, *e.g.*, 10 bit spreading code spreads signal across 10 times bandwidth of 1 bit code

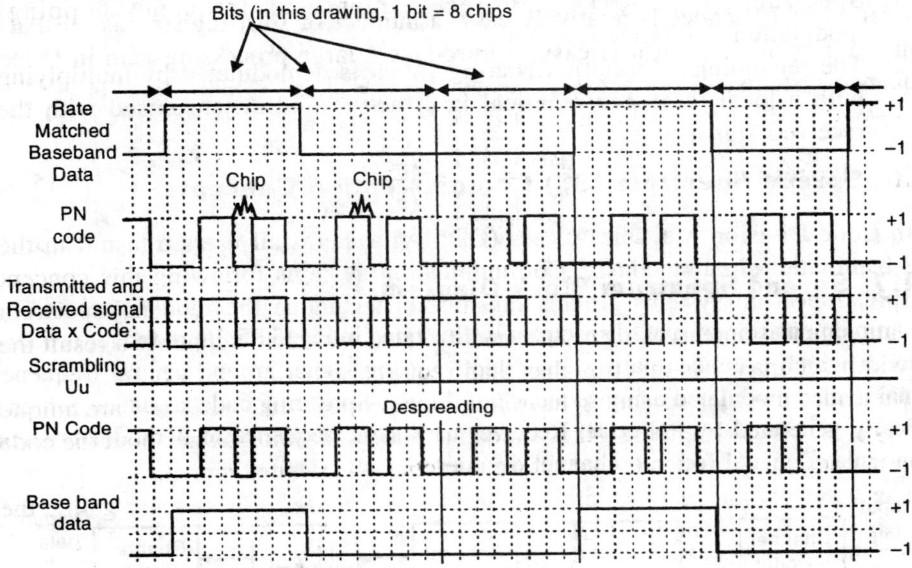

Fig. 5.33. Direct-sequence spreading and despreading.

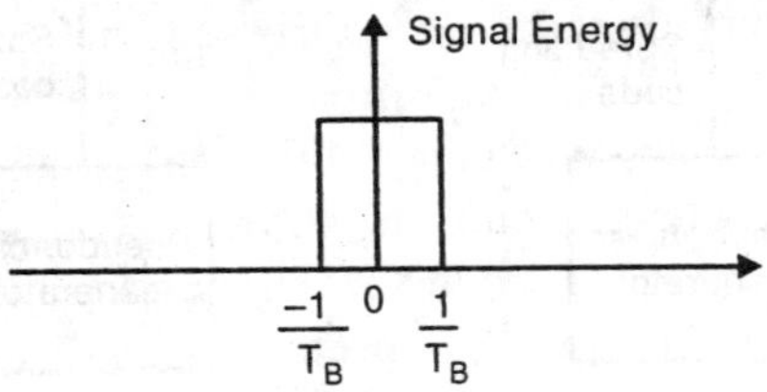

(a) Spectrum of Original Signal

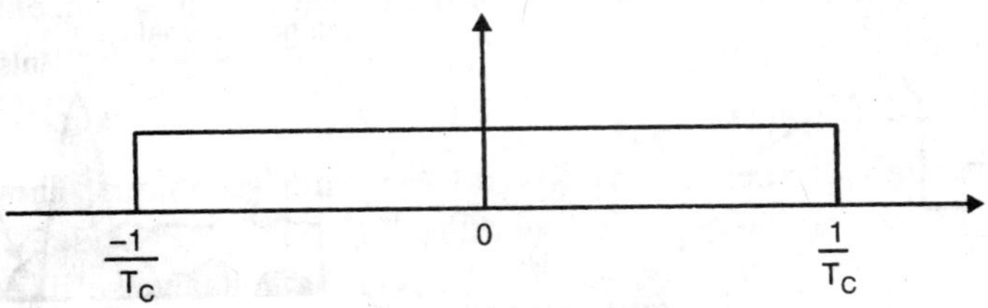

(b) Spectrum of Pseudocode signal

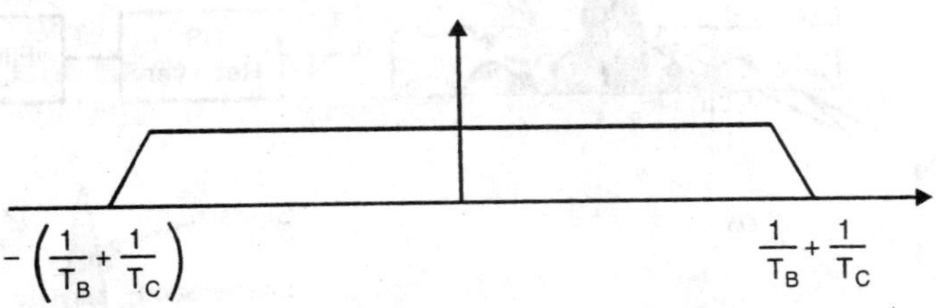

(c) Spectrum of combined signal

Fig. 5.34. Spread spectrum concept.

This can also be seen from Fig. 5.33, where we show how the PN code is combined with the data-signal, in this Example $N_{DS} = 10$. The bandwidth of the data signal is now multiplied by a factor N_{DS} (Fig. 5.33). The power contents, however, stays the same, with the result that the power spectral density lowers. The generation of PN codes is relatively easy; a number of shift-registers is all that is required. For this reason it is easy to introduce a large processing-gain in Direct-Sequence systems.

$$\text{Processing Gain } G_P = \frac{W}{R} = \frac{\left(\frac{1}{T_C}\right)}{\left(\frac{1}{T_B}\right)} = \frac{T_B}{T_C} \qquad \text{...(5.33)}$$

5.4.7. Spread Spectrum Block Diagram

Implementation of spread spectrum is illustrated in Fig. 5.35. Input fed into channel encoder which produces narrow bandwidth analog signal around central frequency. Signal is then modulated using sequence of digits—Spreading code/sequence, which is typically generated by PN code. Receiver uses same sequence to demodulate signal Demodulated signal fed into channel decoder.

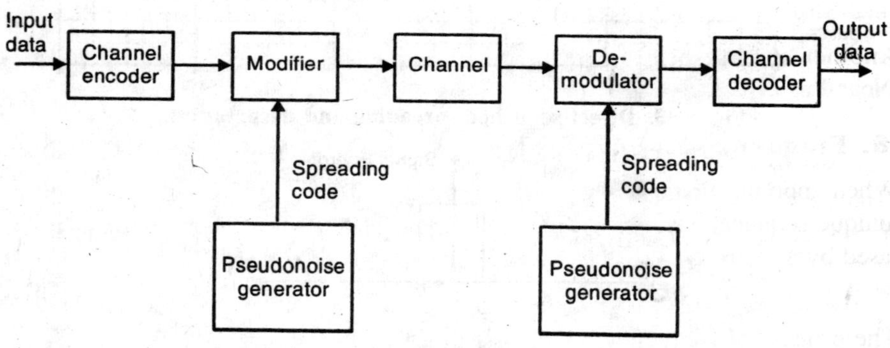

Fig. 5.35. Block diagram of spread spectrum.

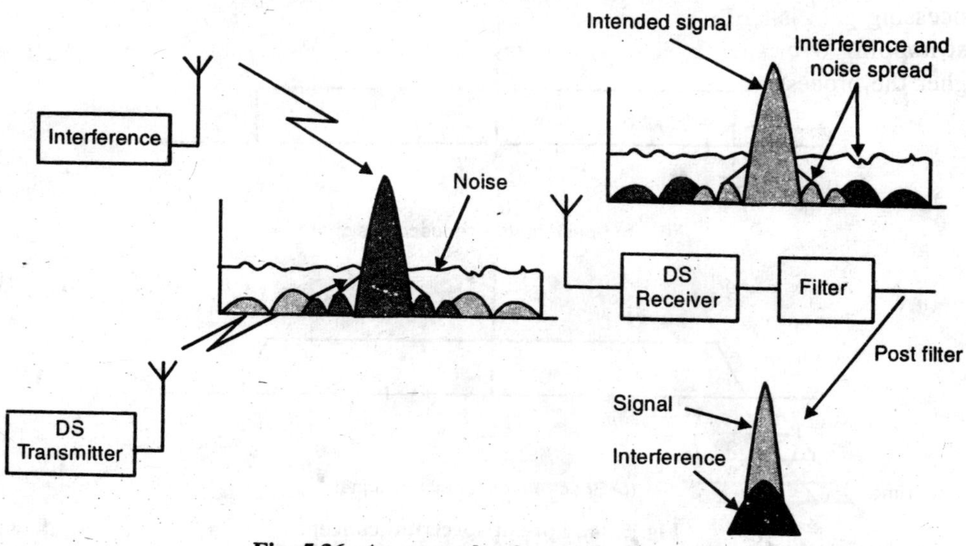

Fig. 5.36. An example of spread spectrum.

Figure 5.36 gives an example of spread spectrum application. Signal is recovered from interference.

The main problem with applying Direct Sequence spreading is the so-called Near-Far effect which is illustrated in Fig. 5.37. This effect is present when an interfering transmitter is much closer to the receiver than the intended transmitter. Although the cross-correlation between codes A and B is low, the correlation between the received signal from the interfering transmitter and code A can be higher than the correlation between the received signal from the intended transmitter and code A. The result is that proper data detection is not possible.

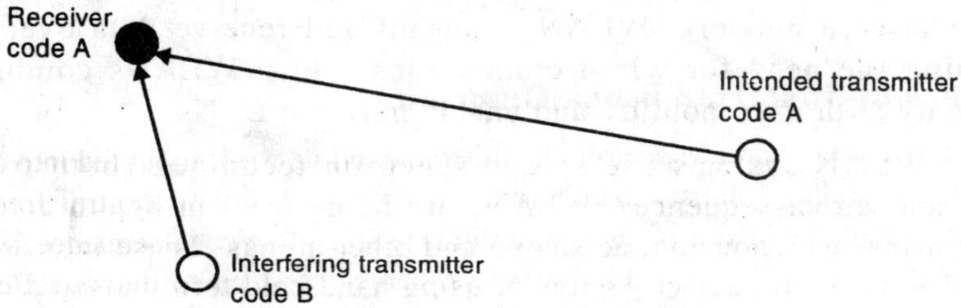

Fig. 5.37. Near-far effect illustrated.

Another spread spectrum technique: Frequency-Hopping is less affected by this Near-Far effect.

5.4.8. Frequency Hopping (FH)

When applying Frequency Hopping, the carrier frequency is 'hopping' according to a unique sequence (an FH-sequence of length N_{FH}). In this way the bandwidth is increased by a factor N_{FH} (if the channels are non-overlapping):

$$G_P\ (FH) = N_{FH} \qquad \text{...(5.34)}$$

The process of frequency hopping is illustrated in Fig. 5.38. A disadvantage of Frequency-Hopping as opposed to Direct-Sequence is that obtaining a high processing-gain is hard. There is need for a frequency-synthesizer able to perform fast-hopping over the carrier-frequencies. The faster the "hopping-rate" is, the higher the processing gain.

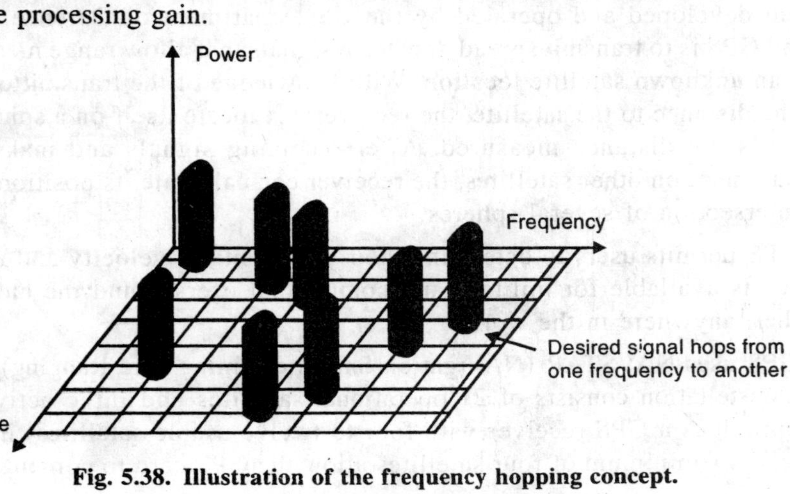

Fig. 5.38. Illustration of the frequency hopping concept.

On the other hand, Frequency-Hopping is less affected by the Near-Far effect than Direct-Sequence. Frequency-Hopping sequences have only a limited number of "hits" with each other. This means that if a near-interferer is present, only a number of "frequency-hops" will be blocked instead of the whole signal. From the "hops" that are not blocked it should be possible to recover the original data-message.

5.4.9. Applications of Spread Spectrum

Wireless local area network (WLAN). A WLAN is a flexible data communication system implemented as an extension to or an alternative for a wired local area network. WLANs transmit and receive data over the air, minimising the need for wired connections. Thus, WLANs combine data connectivity with user mobility and enable movable LANs.

Most WLAN systems use spread-spectrum technique (both frequency hopping and direct sequence). WLANs are being used in health care, retail, manufacturing, warehousing, academic and other arenas. These industries have profited from the productivity gains of using hand held terminals and notebook computers to transmit real-time information to centralised hosts for processing. WLANs offer productivity, convenience and cost advantages over wired networks.

Space systems: In space stations, which are continuously accessible to interference, spread-spectrum methods have proved effective. This is especially true for communication satellites.

In general, satellites do not employ processing on-board as it adds to the complexity and would limit the number of satellite users. A simple repeating satellite is used, so all the spread-spectrum modulation and demodulation must be done on the ground. With no on-board processing, the satellite is forced to transmit an uplink interference signal, which reduces the spacecraft transmitter power to send the desired signal. Another disadvantage of no on-board processing is that every receiver would have to acquire a spread-spectrum demodulator.

Global Positioning System (GPS): GPS is a satellite-based navigation system developed and operated by the US Department of Defence. The idea behind GPS is to transmit spread-spectrum signals that allow range measurement from an unknown satellite location. With knowledge of the transmitter location and the distance to the satellite, the receiver can locate itself on a sphere whose radius is the distance measured. After receiving signals and making range measurement on other satellites, the receiver can calculate its position based on the intersection of several spheres.

GPS permits users to determine their 3-D position, velocity and time. This service is available for military and commercial users round the clock, in all weather, anywhere in the world.

GPS uses NAVSTAR (NAVigation Satellite Timing And Ranging) satellites. The constellation consists of 21 operational satellites and three active spares. This provides a GPS receiver with four to twelve usable satellites 'in view' at any time. A minimum of four satellites allow the GPS card to compute latitude, longitude, altitude and GPS system time. The NAVSTAR satellites orbit the earth at an altitude of 10,898 Nautical miles in six 55-degree orbital planes, with four satellites in each plane. The orbital period of each satellite is approximately 12 hours.

The GPS satellite signal contains information to identify the satellite, as also positioning, timing, ranging data and satellite status. The satellites are identified by the space vehicle number or the pseudo-random code number. They transmit on two L-band frequencies : 1.57542 GHz (LI) and 1.22760 GHz (L2). The LI signal has a sequence encoded on the carrier frequency by a modulation technique that contains two codes, a precision (P) code and a course/acquisition (C/A) code. The L2 code contains only P code, which is encrypted for military and authorised commercial users.

5.5. ORTHOGONAL FREQUENCY DIVISION MULTIPLEXING (OFDM)

OFDM is a modulation technique based on Minimum frequency shift keying (MFSK).

OFDM is more useful modulation technique for high speed data transmission based on protocol IEEE 802.11a to provide 54 Mbps WLAN connections. As well as for high speed line of sight and non line of sight microwave connections.

In MFSK, the transmitted signals are defined by

$$S_i(t) = \sqrt{\frac{2E_S}{T_s}} \sin\left[\frac{\pi}{T_s}(n_c + i)t\right] \quad 0 \le t \le T_s$$

$$\text{and } i = 1, 2, \dots m \quad \dots(5.35)$$

where $f_c = \dfrac{n_c}{2T_s}$ for some fixed integer n_c.

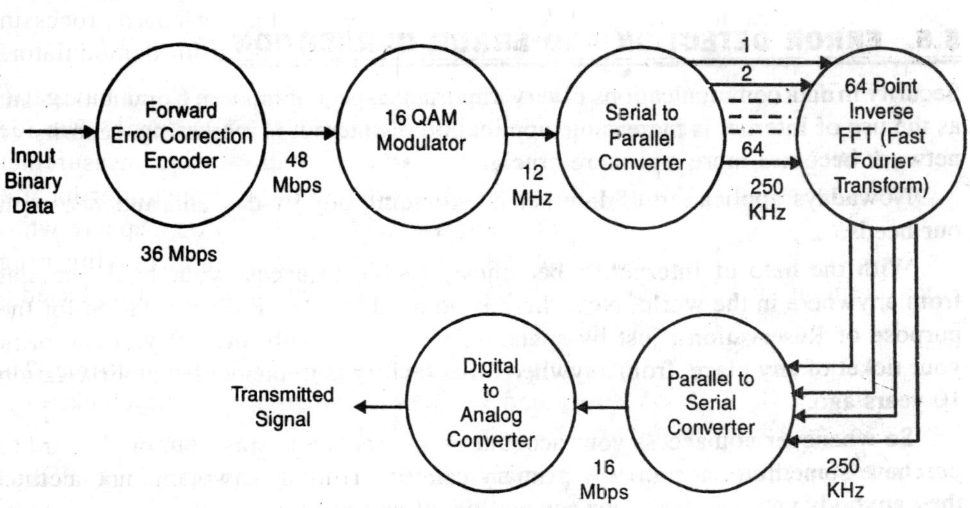

Fig. 5.39. OFDM transmitter.

The M transmitted signals are of equal energy and equal duration and the signal frequency are separated by $\frac{1}{2T_s}$ Hz, making the signals orthogonal to one another.

This orthogonal characteristic of MFSK has motivated the engineers to explore OFDM as a means of providing power efficient signal for a large number of users on the same channel. Each frequency in Eqn. (5.35) is modulated with binary data ON/OFF to provide a number of parallel carriers each containing a portion of user data.

OFDM transmitter and receiver are shown in following Fig. 5.39.

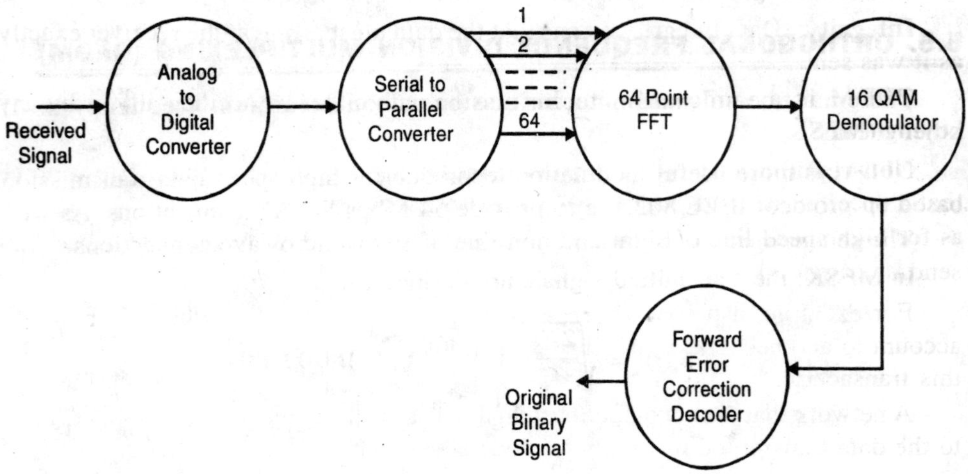

Fig. 5.40. OFDM Receiver.

5.6. ERROR DETECTION AND ERROR CORRECTION

Security in data communications is very important aspect of modern Communication, as the use of Internet is increasing, application of Internet is increasing, security of network becomes more and more crucial.

Nowadays application of Internet is extending day-by-day and attached with our needs.

With the help of Internet, it becomes possible to access your bank account from anywhere in the world. Now there is no need to go to Railway station for the purpose of Reservations, just by spending some time with internet you can print your ticket of any place, from anywhere, this facility is implemented in air aviation 10 years ago.

So whenever you access your account on Internet and using your credit card to purchase something, accuracy is primary concern, if your network is not secured they anybody can get steal your account detail and misuse.

Therefore, when people use Internet they have certain expectations, they expect confidentiality and data integrity. They want to be able to identify the sender of message. They want to be able to prove that a message has in fact been sent by a certain sender even if the sender denies it.

There are four aspects of security, those are as follows:

Privacy: Privacy means confidential, when a person sends some information to a person then message should make sense to that person only and it must be unintelligible, or hide to other person.

Authentication: Network must have authentication capability, simply verification capability before allowing to use the network application.

Suppose, you want to access your Bank account, then Bank network must verify to you before allowing to use particular application like to see the detail of account or to allow for any transaction.

Verification is done by providing password, used in etc.

Integrity: Data Integrity means that the data must arrive at the receiver exactly as it was sent.

There must be no change in transmission path either by accidentally or by any other means.

Non-repudiation: Non-repudiation means that a receiver must be able to prove that a specific message comes from right source.

The sender must not be able to deny sending a message that he, in fact, did not send.

For example, a person requests to his bank to transfer the money from one account to another. The bank must have proof that the customer actually requested this transaction.

A network that cannot guarantee that the data received by one device are identical to the data transmitted by another device is essentially a useless network.

5.6.1. Transmission Errors

Errors may be of any type and can be generated at any position of transmissions.

Whenever a signal is transmitted from one point to another, then in between transmission path, it is very difficult to predict temperature, weather, environmental condition. So it is not possible to control the noise and to control the error generated in signal due to that external noise. If signal is transmitted in digital format then by change of a single bit meaning of message signal will be changed.

Therefore, transmission error is that error, which occurs in between the transmission path. This may be classified as follows.

One Bit Error

One bit error or some time called single bit error, means only one bit of a given data unit has been changed. This happened very rarely, mostly more than one bit is changed.

Figure 5.41 shows the impact of change of one bit.

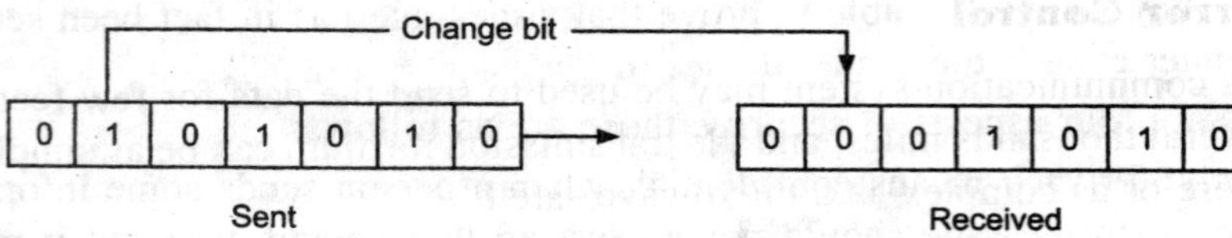

Fig. 5.41.

If we consider the data in simple binary code and take decimal equivalent of transmitting data, then it was 42 while receiving data is 10.

Single bit error occurs mostly in parallel transmission than in serial transmission.

If 8 wire is used to transmit a byte in parallel transmission and due to any reason one line becomes corrupted then in each byte one bit will be corrupted.

While in serial communication let a data is transmitted at the rate of 2 Mbps, it means that each bit lasts only in 1/2 Mbps = 0.5 μ second. A noise can corrupt the data bit if noise duration is of 0.5 μ second.

Which is very rare case, so we can say that in serial transmission, bit error depends upon rate of transmission, while in parallel transmission bit error depends upon number of transmission line used to transmit the data bit.

Multiple Bit Error

Multiple bit errors some time called Burs error, and it is a type of transmission error in which more than one bit is damaged. This type of error mostly occurs in data transmission. It is important to note that corruption of two bit is not necessary at consecutive positions, it may be may not be.

This type of error generally occurs in serial data transmission. Actually the noise duration is more than bit duration hence whenever noise corrupted data bit, it is set of bit not a single bit.

The number of affected bits depends upon data rate and duration of each bit.

If we are transmitting the data bit at the rate of 1 Mbps then noise of duration 1/1000 can affect 10000 bit simultaneously and a noise of 1/100 sec can affect the 10000 bit of data.

Example illustrated is shown in Fig. 5.42.

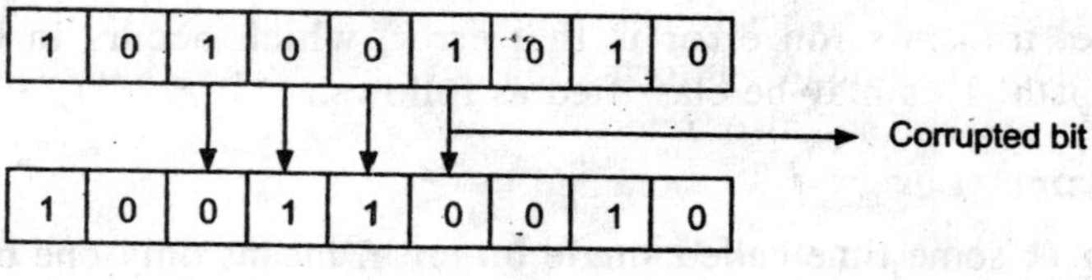

Fig. 5.42. Multiple error of bit.

There is possibility to more than two bits damaged consecutively but not necessary.

5.6.2. Error Control

A data communication system may be used to send the data for few feet or as long as several thousands miles, and the transmission medium can be as simple as a piece of wire or as complex as a microwave satellite.

Therefore, due to non-ideal characteristics of equipment, devices, some noise in physical channel will also occur and there is a chance to get the distorted message at the receiver. Hence it becomes necessary to develop and implement procedures for error control.

Error control can be divide in two parts as follows.

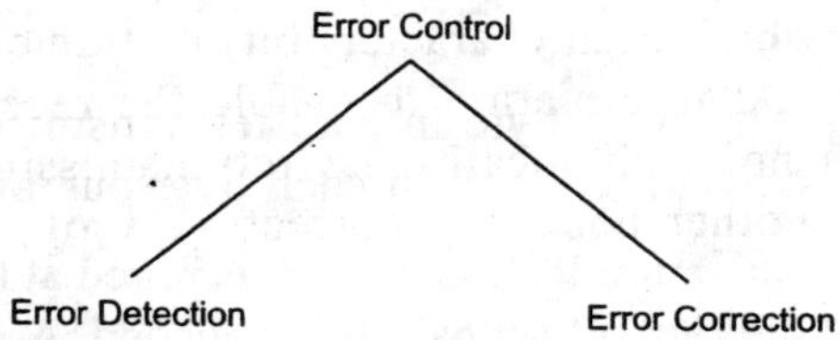

Fig. 5.43.

5.6.3. Error Detection

Error detection is a process of monitoring the transmission and reception. Error detection only determines the received data is corrupted but at which position data bit is corrupted and which data bit is corrupted, it cannot be notified by error detection technique.

The purpose of error detection is not to prevent errors from occurring but to prevent undetected errors from occurring.

The performance of a system when received data bit is corrupted it varies system to system.

To ensure that information received by a destination has high probability of being same as that transmitted by the transmitter, there must be some way for the receiver to deduce, to a high probability, when received information contains errors.

If errors are detected, a mechanism needed to obtain a copy of the correct information. To achieve this following approaches are used.

5.6.3.1. Forward Error Control

Forward error control is technique in which each transmitted character or frame contains additional (redundant) information so that receiver cannot only detect when errors are present but also determine wherein the received bit stream the errors are, the correct data is then obtained by inverting these bits.

In practice, the number of additional bit required to achieve reliable forward error control increases rapidly as the number of information bits increases with an automatic repeat request (ARQ) error control scheme, additional check digits are added to each transmitted message to enable the receiver to detect when an error is present in a received message, assuming certain types of error.

If an error is detected, additional control procedures are used to request another copy of the message to be transmitted.

With forward error control, sufficient additional check digits are added to each transmitted message to enable the receiver not only to detect the presence of one or more errors in a received message but also to locate the position of errors.

Practically, the number of additional check digits required for error correction is much larger than that needed for just error detection.

There are various techniques used in forward error control approach.

5.6.3.2. Feedback Error Control

It is technique in which each character (bit) or frame of message includes only sufficient additional information to enable the receiver to detect, when errors are present but not their location. A retransmission control scheme is used to request that another hopefully correct copy of information is to be sent.

Feedback error control is predominant method over forward error control. If number of data bits are large, feedback error control can be divided in two parts.

1. The techniques that are used to achieve reliable error detection, and
2. To control algorithms that are available to perform the associated retransmission control schemes.

There are two factors that determine the type of error detection scheme—the bit error rate (BER) of the line or circuit and the type of errors, that is, whether the errors occur is single bit error or as group of bit errors. A group of bit error is some time called Burst error.

The BER is the probability *P* of a single bit being corrupted in a defined time interval.

Thus a BER of 10^{-3} means that, on average, 1 bit in 10^3 will be corrupted during a defined time period.

Let us take example, if we are transmitting blocks of characters using synchronous transmission (say 125 characters per block, each of 8 bits), then the probability of a block (frame) containing an error is approximately 1. This means that an average every block will contain an error and must be retransmitted.

Clearly, this length of frame is too long for this type of line and must be reduced to obtain an acceptable throughput. Throughput is defined as number of bits crossed to channel per second.

The type of errors scheme detect different types of errors.

Also, the number of bits used in some schemes determines the bursts length that are detected. Some of error detection techniques commonly used are as follows.

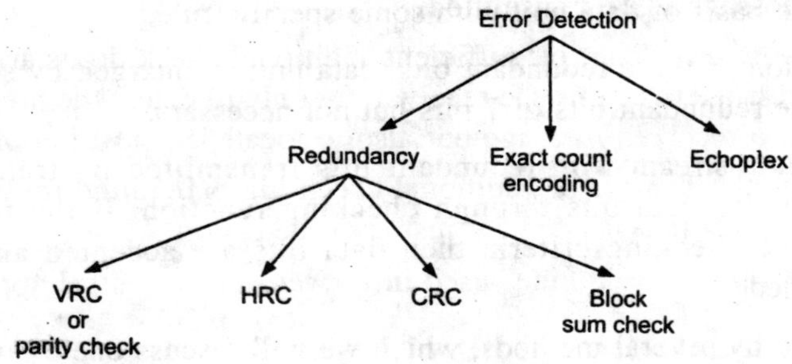

VRC — Vertical Redundancy check or Parity check
HRC — Horizontal Redundancy check
CRC — Cyclic Redundancy check

Fig. 5.44. Error Detection.

5.6.3.3. Redundancy

Redundancy involves transmitting every data unit twice, and then at receiver, receiving device after receiving both data starts comparison of both data received. Any change in both data indicates an error and an appropriate correction mecha. ism could be set in place. In other words, if at the receiver, same data bit is not received twice in succession, a transmission error has occurred.

This technique is very appropriate up to some extent but it takes more time so that communication systems performance become very slow.

Because in this case transmission time is doubled and the time taken by receiving device to compare the each bit also affects the speed.

So we use the concept of including some extra bit with data. These extra bits are calls redundancy bits, a process is illustrated in Fig. 5.45.

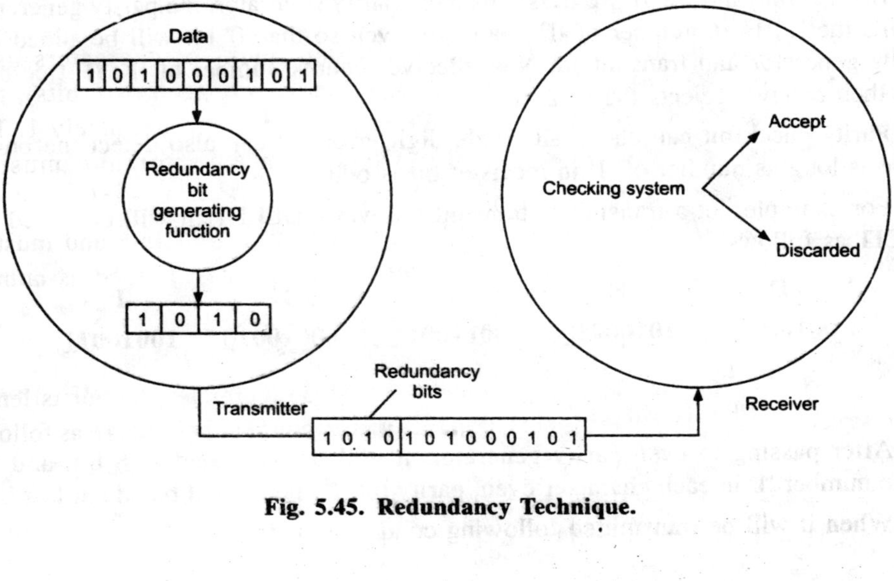

Fig. 5.45. Redundancy Technique.

In Fig. 5.45, some redundant bit is coded with data unit. This redundant is generates on the basis of data unit with some specific rule.

After addition of these redundant bits, data unit is enlarged by several bits in this example the redundant bits of 7 bits but not necessary.

Now this data stream with redundant bits transmitted by transmitter. The receiver puts entire data bits through checking function. If the received data bits satisfied the checking criteria then data bits are accepted and redundant bits are discarded.

This is done by several methods, which we will discuss one by one.

5.6.3.3.1. Parity Check (Vertical Redundancy Check)

It is most common and economical method for the purpose of detection. In this technique we added a parity bit as redundant bit to original data bit so that number of '1s' in data bit become even or odd.

Here we are using even parity bit checker, we may use odd parity bit checker, in which we will add '1' to data bit so that number of '1' become odd. Procedure will be same but generating function of parity will be different.

A example is illustrated in Fig. 5.46. Let us consider, we want to transmit a letter 'B' in ASCII (B = 0100001).

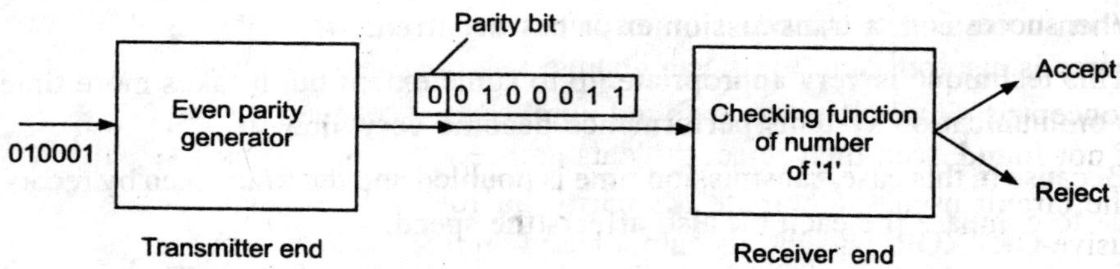

Fig. 5.46. Parity Concept.

Before transmitting 'B' we pass it through parity generator, the parity generator counts the '1'. Here number of '1' is already even so that '0' bit will be added by parity generator and transmitted. Now receiver counts number of '1', if it comes odd then receiver rejects the data bit.

Parity check bit can check all single digit errors, it can also detect the burst error as long as number of '1' in received bit is odd.

For example, let a transmitter transmit the word 'DELHI', it will be coded in ASCIT as follows

D	E	L	H	I
0010001<u>0</u>	1010001<u>0</u>	0011001<u>1</u>	0001001<u>0</u>	1001001<u>1</u>
↓ Parity Bit				

After passing to even parity generator, it will be converted to 8 bit, and to make number '1' in each character even, parity bit '1' and '0' will be add at last bit.

When it will be transmitted following conditions will arise.

Case 1: No bit converted from 1 to 0 or 0 to 1 and all the characters received as transmitted, therefore, receiver will accept the data.

Case 2: Transmitted bit corrupted in the path as follows and receiver found odd number of '1' in some character. Therefore, receiver discard, the whole data and requests to retransmit.

D	E	L	H	I
10100010	10100011	01110011	00010010	10010011

In 'D' and 'L' number of 1 is odd.

Case 3: Error is introduced in 'DELHI' but number of '1' at receiver found even. Let error is introduced in letter E and L.

D	E	L	H	I
00100010	10100011	00110011	00010010	10010011
	↓	↓		
	10111011	01110111		

In E and L due to noise error is introduced but in such a way that number of '1' is found to be even, so receiver will accept the data, not request to retransmit, This is biggest disadvantage of Parity method, it cannot detect burst error (group error) if number of '1' is not odd.

When noise is introduced, it is not necessary that always 1 is converted to 0, vice versa is also possible. Both can change at a time, may not change at a time.

Concept is simple if receiver found number of '1' is even then accept the data and if not found even then reject the data and request for retransmission.

The circuit used to compute the parity bit for each character is a XOR gate (Exclusive-OR) XOR gate and its output based on input is shown in Fig. 5.47.

A_0 Bit 1	A_1 Bit 2	XOR
0	0	0
0	1	1
1	0	1
1	1	0

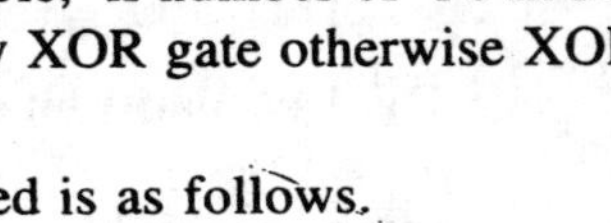

Fig. 5.47. Parity generator

It is understood from Fig. 5.47 truth table, if number of '1's in data bit is odd, only then Parity bit '1' will be generated by XOR gate otherwise XOR output will be zero.

Data bit format, that is to be transmitted is as follows.

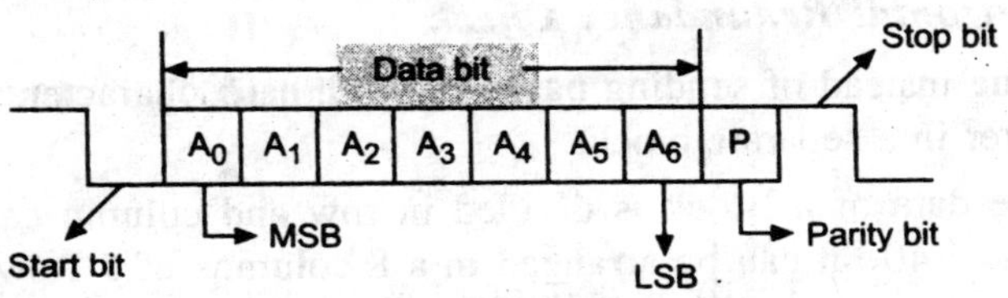

Fig. 5.48.

Parity generator of a data bit is shown in Fig. 5.49.

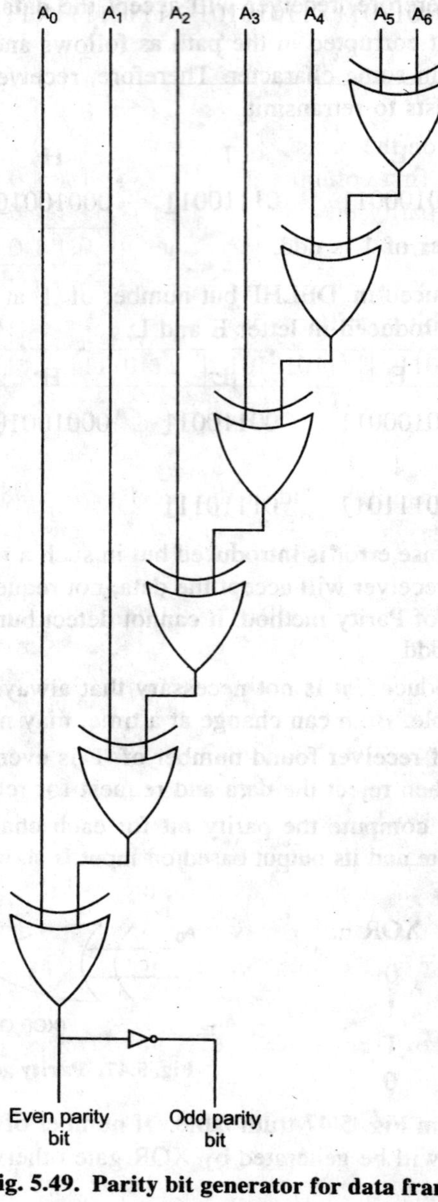

Fig. 5.49. Parity bit generator for data frame.

5.6.3.3.2. Horizontal Redundancy Check

In this technique instead of sending parity bit with each character, we send parity of each character in a separate block.

And all the data of a Block is divided in row and column called *table*. For example a data of 40 bit can be arranged in a 8 columns of 5 Rows and parity is generated for each column and created a new row of 8 bits. Which are the parity bits for whole block of Parity data? Procedure is as follows.

Let data bit is

11101011 01011011 11010101 11101111 11101101

Now arranged in 8 × 5 Table. →

1 1 1 0 1 0 1 1
0 1 0 1 1 0 1 1
1 1 0 1 0 1 0 1
1 1 1 0 1 1 1 1
1 1 1 0 1 1 0 1
0 1 1 0 0 1 1 1 → Parity bit (LRC)

Parity is generated on the basis of each column. In first column number of '1' is 5, parity bit is 1 to make even parity

and complete data bit with Parity is

11101011	01011011	11010101	11101111	11101101	01100111

Parity bit (LRC)

Let this data is transmitted and at the receiver second and third blocks are damaged as follows.

Transmitted Data.

11101011 01011011 11010101 11101111 11101101 01100111

Receive data ↓↓ ↓ ↓↓→ changed bit due to noise.

11101011 01101011 01011001 11101111 11101101 01100111 (LRC)

When Receiver checks the LRC corresponding to second and third block.

It is found that third Parity does not follow the even parity rule hence whole the data is discarded and requested to retransmit the data. It is important to note that in second block data is also affected but one zero convert to 1 and one zero convert to 1 hence Parity matches. This is limitation of this method.

It is also assumed that when Parity bit is transmitted with data bits it is not affected due to noise, which is Impractical. Hence it is second limitation of this technique.

5.6.3.3.3. Cyclic Redundancy Check (CRC) or Frame Check Sequence (FCS)

In parity check method, we have seen that if more than one bit is corrupted in data then it becomes very difficult to detect the error.

CRC is somehow, one of the best methods to detect the burst error in data.

Instead of addition of the bits as in Parity, CRC is based on binary division method.

So in CRC technique to get the desired parity bits, a sequence of redundant bits called CRC or the CRC remainder is added to the end of a data unit so that resulting data unit becomes exactly divisible by a second, predetermined binary number.

At receiver end, received data is divided by same divisor and if remainder is zero, it means data is correct and accepted but if remainder is not zero it means data is corrupted and then data is discarded.

Let us take an example to illustrate the concept.

Let a data 11100110 is to be transmitted, to find the CRC (Parity bit) for this data it is divided by a divisor 11001 as follows.

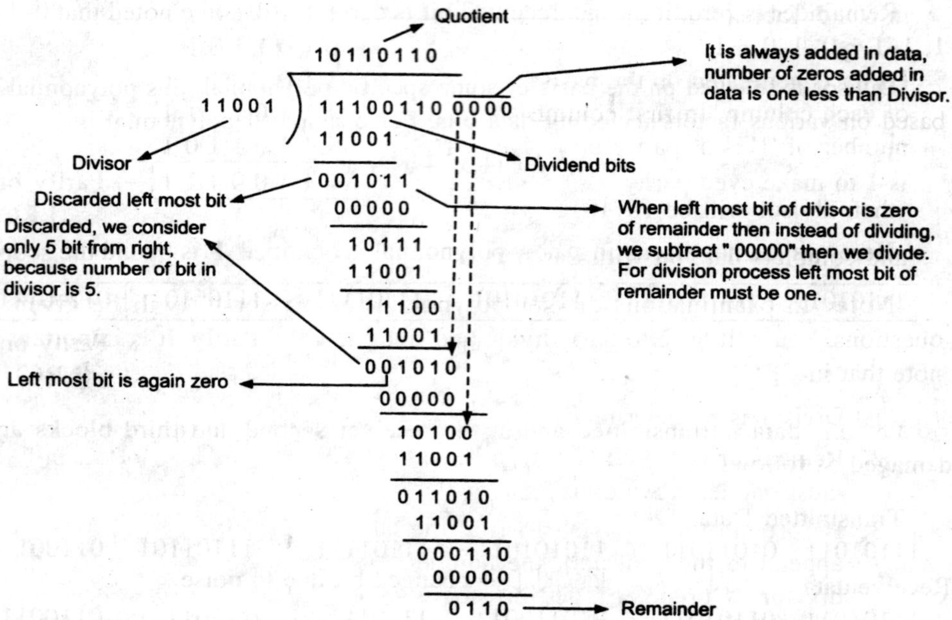

0110 cannot be divided further, because there is no more zero in dividend. Note that Quotient will be ignored, only remainder of use.

Now the data bit to be transmitted with CRC bit is

111001100110

CRC One bit less than divisor bits

At Receiver

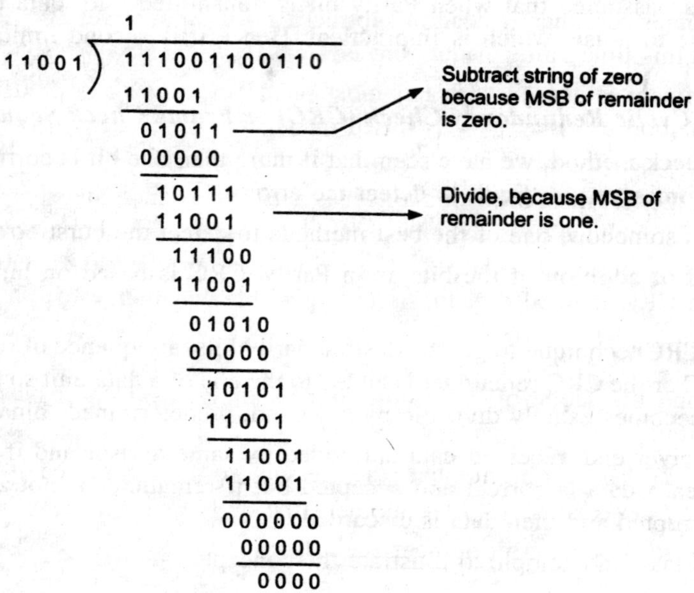

At the receiver end this is again divided by same divisor, if remainder comes zero, then received data is correct and accepted. At receiver we will not add zero bits in the divided data bits.

Remainder is zero, it means, received bit is correct. It is to be noted that 0–1 = 1, 1–1 = 0, 1–0 = 1.

Divisor is selected on the basis of some specific polynomial, this polynomial is based on various factors including data bits. For example if polynomial is

$$x^6 + x^4 + x^3 + x^1 + 1$$

then divisor will be 1011011.

Here, we are not discussing, how polynomial is obtained. It is byond the scope.

Note: In examinations divisor polynomial or divisor bit will be given in questions. You will have to only divide and find the CRC Parity. It is important to note that in CRC.

1. Divisor is predetermined.
2. Remainder is the CRC and to be valid a CRC the number of bit in CRC must one less than divisor.
3. Before starting division to find the CRC parity bit, a string of n 0s it append to the data unit, the number of zero is one less than the bits in divisor. If predetermined divisor is of 4 bit 1101 then number of zero appended to data string is "000".
4. The data to be transmitted is data string followed by CRC.
5. At the receiver end data string followed by CRC is divided by same divisor, if remainders came zero, data is correct and accepted. If remainder is not zero data is rejected.
6. In division process, divisor is substracted from corresponding bits of dividend bits, but if in a intermediate step the remainder left most bit comes zero then instead of subtracting divisor subtract a string of zero and at this time value of quotient is '0'.
7. Divisor is always with left most bit '1' and divisor is predetermined, the method to get divisor is based on algebraic polynomial functions that is beyond our scope.
8. If the left most bit of remainder is zero in intermediate steps of division, it is dropped. But if next bit is also zero then it is retained and subtract a string of zero by pulling a higher bit from dividend to remainder. Because in CRC method division is only possible when left most bit is '1'.

Polynomials for Divisor

It is beyond the scope of this book. Here I am going to give you only a brief introduction.

Left $T(x)$ be a n bit number (the message to be transmitted)

and $D(x)$ is an $(k + 1)$ bit number, (the divisor or generator)

$R(x)$ be an k bit number such that $n > k$ (the remainder)

Then if.

$$\frac{T(x)\times 2^n}{D(x)} = Q(x) + \frac{R(x)}{D(x)},\ \text{where } Q(x) \text{ is Quotient.}$$

Polynomial $D(x)$ should be selected to have at least following property:

1. It should be divisible by $(x + 1)$
2. Not divisible by x.

The first condition assures that all burst errors affecting an odd number of bits are detected while second condition is assurance for detecting all burst errors of a length equal to the degree of the polynomial are detected.

From above point it is understood that, we cannot select divisor polynomial as shown in following example.

1. $D(x) = x$

   binary 10

   Because it is divisible by x.

2. $D(x) = x^2 + x \Rightarrow$ Binary (110), but is also divisible by x.
3. $D(x) = x^3 + x^2 \Rightarrow$ (1100), it is also divisible by x.

But if polynomial is in following form then it can be selected.

Example

1. $D(x) = x + 1$ Divisible by $x + 1$ but not by x Binary (11)
2. $D(x) = x^2 + x + 1$ Divisible by $x + 1$ but not by x Binary (111).

If polynomial is $x^{12} + x^{11} + x^3 + x^2 + x + 1$, it indicates that in CRC number of bit is 12 and divisor of 13 bit. Here the divisor is

[1100000001111] of 13 bit.

and it indicates, it can detect all burst error of length less than 12 or equal to 12.

The number of check digits per frame or the degree of polynomial is selected to suit the type of transmission errors. The compute check digits are referred to as the *Frame check sequence* or *cyclic redundancy check digits.*

Conclusion

CRC is most effective and powerful method to detect the errors if divisor is selected based on above rule.

1. CRC can detect all burst errors of length less than or equal to polynomial degree.
2. CRC can detect all burst errors that affect the odd number of bits.
3. All single bit error.
4. All double bit errors.

Example 5.1. A series of 8 bits data is transmitted using CRC bits. A generator polynomial 1101 is to be used. By considering an example perform the generation process and checking process.

Let the message bit is11010101

Divisor is given 1101

```
          10000110
1101 ) 11010101000   --> Three zero to be appended in string
       1101
       00000
        0000
        00001
         0000
          0010
          0000
           0101
           0000
            1010
            1101
             1110
             1101
              0110
              0000
               110 --> CRC
```

Now data bit is to be transmitted

[11010101110]

Checking Process at the Receiver.

```
          1000011
1101 ) 11010101110
       1101
        0000
        0000
         0001
         000C
          0010
          0000
           0101
           0000
            1011
            1101
            01101
             1101
             0000  --> Remainder is zero
```

So data is not corrupted

Let Received data is corrupted at 2, 3, 4 position then,

11010101110 $\rightarrow$ Transmitted

10100101110 $\rightarrow$ Received

```
            1101011
       ___________
 1101 ) 10100101110
        1101
        ----
        01110
         1101
         ----
          0111
          0000
          ----
           1110
           1101
           ----
            0111
            0000
            ----
             1111
             1101
             ----
              0101
              0000
              ----
               1010
               1101
               ----
                111 ——→ Remainder is not zero
```

It means data is corrupted.

5.6.4. Block Sum Check (BSC)

Block sum check is also an efficient and powerful technique for error detection. It is generally used in higher layer of OSI model.

When blocks of characters (Bytes) are being transmitted, there is an increased probability that a character will contain a bit error. So there is need to increase in error detection capabilities obtained from a single parity bit per character (byte) by using an additional set of parity bits computed from the complete block of characters (bytes) in the frame.

BSC is a simple method of error detection. A check sum is simply the least significant byte of the arithmetic sum of the binary data being transmitted. While data are being transmitted, each character is add (summed) with the accumulated sum of previously transmitted characters.

When the end of the message is reached, the adder has accumulated the sum of all the characters included in the message just sent.

The least significant byte of the sum is appended to the end of the message and transmitted.

The receive terminal replicates the summing operation and determines its own sum and check sum character. The least significant byte of the receiver's sum is corporated with the check sum appended to the end of the message. If they are the same, most likely transmission error has occurred.

If they are different, a transmission error has definitely occurred, when an error is detected, a retransmission of the entire message is called *for*. Procedure to generate the check sum is divided in following parts.

In the sender (Transmitted end), there is a device called check sum generator, which generators the check sum parities as follows.

1. Check sum generator subdivides the data unit into equal segments of n (usually 16).
2. These subdivided segments are added using 1st complement method in such a way that total bit is also of n bit.
3. Then total sum is complemented which is required check sum parity.

4. Finally it is appended with data unit and transmitted. Note that if sum of data segment is N then its check sum will be $-N$.

Let us take an example, data bit to be transmitted is

110111111010110

It is of 16 bit, divided in two parts of 8 bit and add them.

```
1 1 0 1 0 1 1 1      → 8 bit
1 1 0 1 0 1 1 0      → 8 bit
---------------
1 0 1 0 1 1 0 1
            + 1
---------------
1 0 1 0 1 1 1 0      → Sum
```

Now complement the sum.

```
1 0 1 0 1 1 1 0
0 1 0 1 0 0 0 1    → 1s Complement.
```

appended with data bit and transmitted.

<u>1 1 0 1 0 1 1 1</u>   <u>1 1 0 1 1 0 1 1 0</u>   <u>0 1 0 1 0 0 0 1</u>

At receiver end the data is again added

```
1 1 0 1 0 1 1 1
1 1 0 1 0 1 1 0
0 1 0 1 0 0 0 1
---------------
1 1 1 1 1 1 1 0
              1
---------------
1 1 1 1 1 1 1 1
```

Again take one complement of sum, that comes 00000000 which means there is no error, if complement is not zero then received data is corrupted and it will be discarded by receiver.

At the receiver end there is sum checker which follows these steps. If we conclude whole the process then,

At the transmitter following steps must be followed :

1. Data sequence is divided into n section to equal bits.
2. All the sections are added by using 1s complement method.
3. Take the complement of sum, called as check sum and appended with data bit then transmitted.

At the receiver end, again whole the data is divided in the section of equal bits including check sum bit.

Find the sum using 1s complement method then take complement. If all the bits are zero, it means data bits are not corrupted.

There are some limitations of this technique also, which are as follows.

If one or more bits of a section are damaged and the corresponding bit or bits of opposite value in a second segment are also damaged, the sum of those columns not change and error will not be detected by receiver.

5.7. ERROR CORRECTION

Error correction operates in a fashion similar to error detection but is capable of correcting certain errors in a transmitted bit stream. When a digital message is transmitted due to noise in transmission path signal may be corrupt. Error correction can be handled in two ways, firstly if error occurred then receiver requests to transmit the message again but there is no guarantee that retransmitted signal will not be corrupted, so this is limitation. Second method receiver uses error correcting code and corrects the error itself. But error correction code is very complex in nature they require more redundant bits for error correction.

Therefore, this method is also limited upto 1, 2, 3 bit correction.

When a message signal is transmitted to the receiver, there possibility of following outcomes.

1. If there are no bit errors, the input to the forward error control (FEC) decoder is identical to the original codeword, and the decoder produces the original block as output.
2. For certain error patterns, it is possible for the decoder to detect and correct those errors.
3. For certain error patterns the decoder can detect but not correct the errors, in this case, the decoder simply reports and uncorrectable error.
4. For certain, typically rare error patterns, the decoder does not detect that any error has occurred.

How is it possible for the decoder to correct bit errors ? In essence, error correction works by adding redundancy to the transmitted message. The redundancy makes it possible for the receiver to deduce what the original message was, even in the face of a certain level of error rate. Here we deal error correction in following parts. First we take single bit error correction.

5.7.1. Single Bit Error Correction

In previous discussion, we have seen that single bit error detection is very simple, just by using a parity bit. But most important thing is to detect at which position this bit is corrupted. ASCII data bit there Eight possibilities occur for single bit error correction.

There is no error, error is at first bit, 2nd bit, 3rd bit . . . . 7th bit. An error occurs when the receiver reads a 1 bit as a 0 or a 0 bit as a 1. To correct the error, the receiver simply reverses the value of the altered bit. But to do so, it is must for receiver to know at which position bit error occur.

As we know from digital electronics if we take three redundancy bits then it will be able to detect all 8 possibilities of error detection in ASCII code because if 3 bit in redundancy code then 8 possibilities (000 to 111) will be there, but if an error occurs in the redundancy itself so let us first calculate the relation between data bit (n) and redundancy bit required (r) for data bit.

If in a transmitted data unit total number of bit is $n + r$ then r must be able to show $n + r + 1$ condition, where (+ 1) indicates there is no error. Hence, 2^r must be equal to r greater than.

Hence, $$2^r \geq n + r + 1 \qquad ...(5.36)$$

Value of r can be determine by putting different values of r in Eq. (5.36).

For example if $n = 7$, then if

$r = 12^1 \geq 7 + 1 + 1$ not satisfied
$r = 22^2 \geq 7 + 2 + 1$ not satisfied
$r = 32^3 \geq 7 + 3 + 1$ not satisfied
$r = 42^4 \geq 7 + 4 + 1$ satisfied

Hence $r = 4$.

So for a data bit of 7, number of redundancy bit required is 4.

Table 5.3 shows some value of r depending on different data bits.

Table 5.3

Number of Data Bits (n)	Redundancy bits	Total bits ($n + r$)
1	2	3
2	3	5
3	3	6
4	3	7
5	4	9
6	4	10
7	4	11

5.7.2. Hamming Code

In previous discussion, we have found number of redundancy bits required to cover all the possible states. But now question arises that how we will discover which state has occurred. A scientist Hamming provided concept to search the position bit, where error has occurred. According to Hamming, we place the redundancy bit at some specified position of Data unit instead of placing redundancy bit with tail of Data unit.

For example, if data unit is of 7 bit.

Then redundancy bit is (r) = 4. So these redundancy bits will be placed at position $1(2^0)$, $2(2^1)$, $4(2^2)$, $8(2^3)$

means the positions are 1, 2, 4, 8 power of 2.

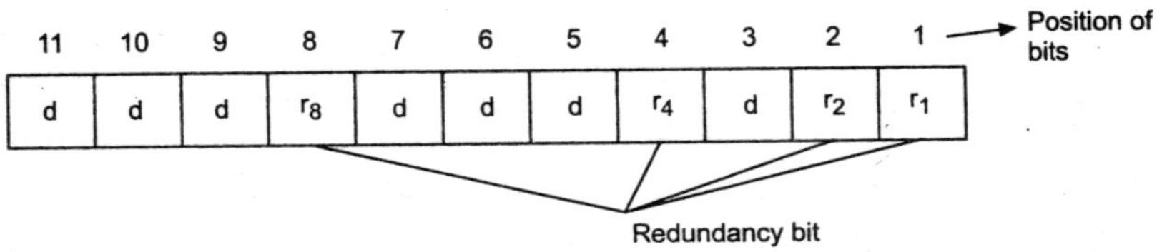

Fig. 5.50

Each redundancy bit will take care of some specified bits as discussed below. But each redundancy bit will take care of itself also, with other data bit.

r_1 will take care of those data bits in which binary equivalent, 1 at right most position LSB, (Least Significant bit) see the Table 5.4.

Table 5.4

Binary Equivalent				Decimal Number
MSB			LSB	
0	0	0	0	0
0	0	0	1	1
0	0	1	0	2
0	0	1	1	3
0	1	0	0	4
0	1	0	1	5
0	1	1	0	6
0	1	1	1	7
1	0	0	0	8
1	0	0	1	9
1	0	1	0	10
1	0	1	1	11
1	1	0	0	12
1	1	0	1	13
1	1	1	0	14
1	1	1	1	15

r_1 will take care of bits 1 3 5 7 9 11

000$\underline{1}$ 001$\underline{1}$ 010$\underline{1}$ 011$\underline{1}$ 100$\underline{1}$ 101$\underline{1}$

Each binary equivalent of its decimal number containing '1' at right position.

r_2 will take care of bits, in which binary equivalent of decimal number at second position from right is 1.

r_2 will take care of 2 3 6 7 10 11

00$\underline{1}$0 00$\underline{1}$1 01$\underline{1}$0 01$\underline{1}$1 10$\underline{1}$0 10$\underline{1}$1

r_4 will take care of those bits in which binary equivalent of corresponding number at 3rd position from right is '1'.

r_4 will take care of 4 3 6 7

0$\underline{1}$00 0$\underline{1}$01 0$\underline{1}$10 0$\underline{1}$11

r_8 will take care of those bits, in which at 4th position from right is '1'

r_8 8 9 10 11

$\underline{1}$000 $\underline{1}$001 $\underline{1}$010 $\underline{1}$011

It can be summarized as follows

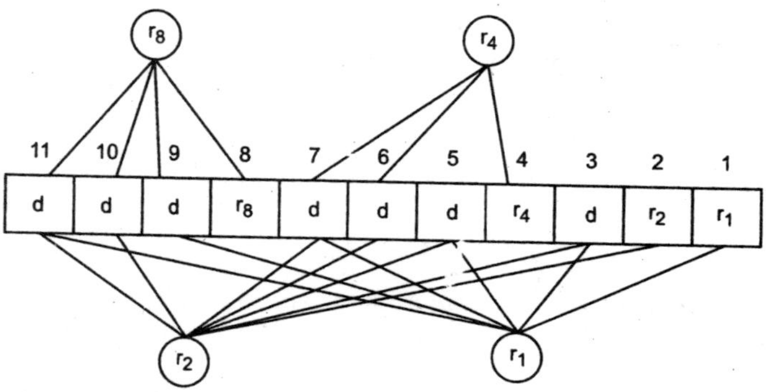

Fig. 5.51.

r_1 → 1, 3, 5, 7, 9, 11
r_2 → 2, 3, 6, 7, 10, 11
r_4 → 4, 5, 6, 7
r_8 → 8, 9, 10, 11

It is understood from Fig. 5.53, that every redundancy bit included every data bit more than one times.

Now we calculate the value of r as follows.

1. In first step we place the data bits at appropriate positions in 11 data format [For this example only 7 data bit + 4 redundancy bit].
2. Calculate the value of r_1 for even parity by considering 1, 3, 5, 7, 9, 11 data bits. It means that consider value of above positions data, if number of one for above position is even then r_1 is 0 if number of '1's is odd for above position then r_1 is 1.
3. Calculate the value of r_4 by considering the value of 2, 3, 6, 7, 10, 11 bit positions.
4. Calculate the value of r_2 in same manner.

Let data bits in ASCTI model is 1101101.

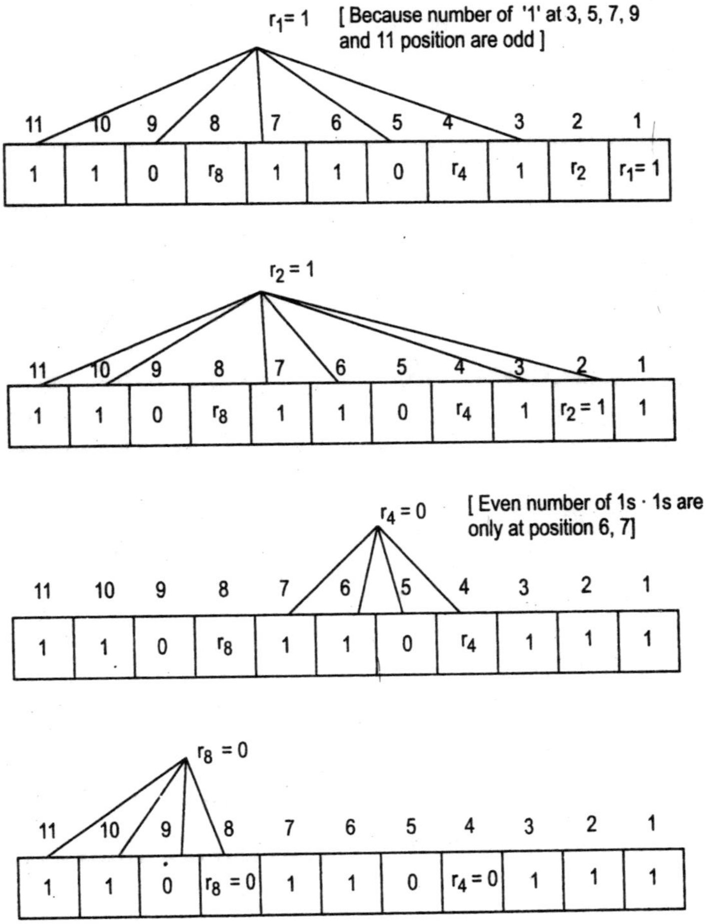

Fig. 5.52.

So, data bit that is to be transmitted is [1 0 0 1 1 0 0 1 1 1].

Now receiver, receives this data bit and again calculate the redundancy bits as above rules specified and the decimal equivalent of redundancy binary bits shows the error.

Let sixth bit is corrupted, it is inverted from 1 to 0.

11	10	9	8	7	6	5	4	3	2	1
1	1	0	0	1	0	0	0	1	1	1

[Received data bit at receiver]

Now calculate the value of even parity for r_1 by considering 1, 3, 5, 7, 9, 11

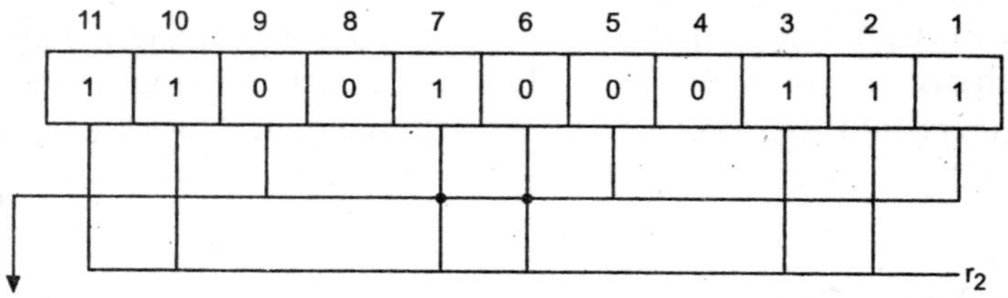

r_1 = 0 by considering at 1, 3, 7, 11 positions are 1

r_2 = by considering 2, 3, 6, 7, 10, 11 is 0

r_4 = by considering 4, 5, 6, 7 is 1

r_8 = by considering 8, 9, 10, 11 is 0

So, redundancy bit is

r_8	r_4	r_2	r_1
0	1	0	1

Therefore $(0101)_2 = (6)_{10}$

Which shows 6th position bit is corrupted and it will be inverted by receiver?

If there is no error then redundancy bit will be (0000). This method is correct to correct only one bit error.

5.7.3. Burst Error Correction

Burst error correction is very complex, it requires higher number of redundancy bits and it is beyond the scope of this level.

5.8. DATA COMPRESSION

In data communication, data compression is one of the important technique. Data compression is the technique in which source data is compressed prior to transmission. This is very common to use particularly in Public Communication System for example PSTN.

(Public Switched telephone network). Because in PSTN call charges are based on duration and distance, hence, for a particular call, if the time to transmit each block of data can be reduced, it will automatically reduce the call cost and also data transmission rate increases. For example, let us assume we are transmitting data at 3600 bps over the PSTN and time needed to transmit the data is 18 minutes then if it is compressed automatically time will be reduced, which reduce the call charges.

There are various methods for data compression, each suited to a particular type of data. Some modems named as intelligent modems offer an adaptive compression feature which selects a compression algorithm to suit type of data being transmitted. Some of the techniques are as follows.

5.8.1. Character Suppression

When data is transmitted in form of frame, there are often sequences in the frame when the same character repeats, for example the space character. The control device at the transmitter scans the frame contents prior to transmission and if a contiguous string of three or more characters is located, replaces these with the three character sequence. This character sequence is known as control character indicating that a compressed string follows, the character type, and a count of the number of characters in the string.

The receiver on detecting the compresson control character, simply reads the following character type and count values and inserts the appropriate number of these characters into the received frames at this point. This is example of more general encoding named run length encoding which we will discuss in coming section.

5.8.2. Huffman Coding

Huffman coding is a data compression technique which is based on statistical concept and uses the property that in a frame sequence some character occurs more frequently and some character very rarely. So instead of using a fixed number of bits per character, we use a different encoding scheme in which the most common characters are encoded using fewer bits than less frequent character.

In this technique, firstly the character string to be transmitted is analyzed and the character types and their relative frequency determined.

Actually transmitter transmits the data on the basis of probability.

Let us suppose we transmit a sentence.

"I am an Indian". Now it is not necessary that transmitter transmits the above sentence in given sequence. There is possibility transmitter first transmits the word 'Indian' then 'am' then 'I' then 'an'.

It depends upon probability of each letter. The letter which occurs more frequently have less probability so given more preference and which rarely occurs have higher probability given lowest preference.

Because lower probability of character occurring have containing higher information. Information is related mathematically with probability as follows.

$$I = -\log P(x) \text{ or } = \log \frac{1}{p(x)} \qquad ...(5.37)$$

To understand this, let us take one example, A Dog bytes a man, this has very high probability hence very less information to transmit, it is very common and occurs frequently. But if a man bytes dog. It has very less probability and hence more information it will spread around very fast but rarely occurs. Now question arises how receiver distinguishes and arranges in required sequence if character is transmitted randomly.

At the transmitted end each character is numbered in a specific manner and this number is transmitted with each character so that receiver can form desired sequence after receiving each character.

Let us take example.

Example 5.2. Find out the Huffman encoding of following message.

Message	x_1	x_2	x_3	x_4	x_5	x_6
Probability	0.4	0.32	0.08	0.08	0.08	0.04

Solution. In example 5.2, $x_1, x_2 \ldots . x_6$ are the message characters and probabilities of each message show how much information they have. To find the Huffman encoding bits follow the following steps:

1. Add all the Probability, and sum must be 1, if it is less than 1, then subtract from 1 and the remaining probability will be for next character.
2. Arrange all the Probability in descending order.
3. Combine last two Probability and again make another column to arrange all the Probability in descending order.
4. If sum of combining two blocks is same with any message probability then give higher position to sum.
5. We will make column until last column element will become 1.
6. If not 1 then check the Process, something is wrong.
7. Always we combine last two probability and upper probability is given to 1 and lower to zero. This is your choice it may be 0 and 1 but follow only one method.
8. To find the codeword for messages start from left to right.
9. Let us find the message code for x_1, follow the dashed line in left to right direction.
10. For x_1

I	II	III	IV	V	VI
0.4	0.4	0.4	0.4	0.4	1
no code	no code	no code	no code	code <u>0</u>	code <u>1</u>

Upto IV column no binary bit is given, in it is 0 and in VI it is 1. Hence code in left to right direction is 01.

11. Reverse it, that is required code for x_1, is '<u>1</u>0'.
12. Follow this process for all the message characters.

Message

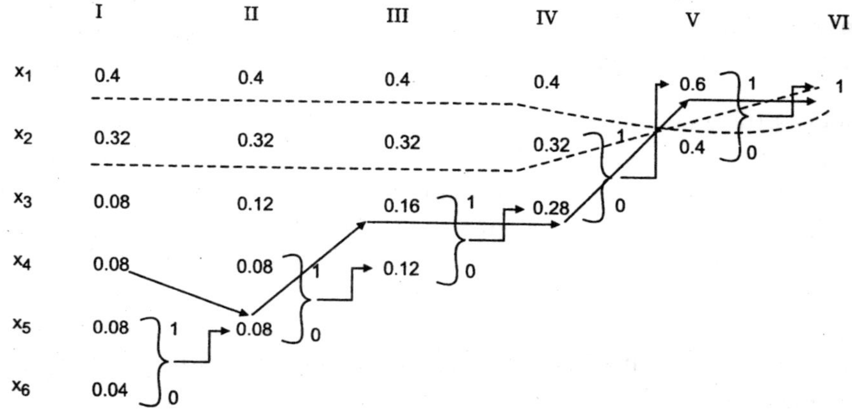

For x_2	0.32	0.32	0.32	0.32	0.6	1
	no code	no code	no code	1	1	1

So code is 111

After reversing it is '111'

For x_3	0.08	0.08	0.16	0.28	0.6	1
	×	1	1	0	1	1

for x_3 '11011'

For x_4	0.08	0.08	0.16	0.28	0.6	1
	×	0	1	0	1	1

code for x_4 "11010"

For x_5	0.08	0.12	0.12	0.28	0.6	1
	1	×	0	0	1	1

code for x_5 "11001"

For x_6	0.04	0.12	0.12	0.28	0.6	1
	0	×	0	0	1	1

Code for x_6 "11000"

It can be summarized as follows.

Message	Probability	Code	Length
x_1	0.4	01	2 bit
x_2	0.32	111	3
x_3	0.08	11011	5
x_4	0.08	11010	5
x_5	0.08	11001	5
x_6	0.04	11000	5

Notes

1. No message will have same code.
2. A coding will be said to be efficient

- If the efficiency of coding will be high as close to as 100 per cent.

Conclusion: Huffcoding technique suggests that there is no need to code to each character in equal number of bits it must depend upon information or probability.

Efficiency: Efficiency $(\eta) = \dfrac{H(x)}{\bar{L} \log_2 m}$

where $m = 2$ if bits are taken as 1 and 0

$m = 3$ if bits are $-1, 0, 1$

for $m = 2 \log_2^2 = 1$

$\bar{L} = \Sigma p_k \, n_k = P_1 \, n_2 + P_2 \, n_2 + P_3 \, n_3 + \ldots..$

$P_1, P_2, P_3 \ldots P_n$ are the possibility of corresponding messages.

$H(x)$ is named entropy of messages.

$$H(x) = -\sum_{i=1}^{K} P(x_i) \log (P(x_i))$$

$$= [P(x_1) \log P(x_1) + P(x_2) \log P(x_2) + P(x_3) \log P(x_3) + ...]$$

Note that base of log is 2 to convert in base 10 use the formula.

$$\log_2 P(x) = \frac{\log_{10} P(x)}{\log_{10}^2} = \frac{1}{0.3010} \log_{10} P(x)$$

5.8.3. Shannon-Fanno Encoding

It is another method of data compressing technique. It is also based on Probability concept. To get message code in Binary form we use following steps.

1. Arrange the Probability in descending order.
2. Divide whole of the Probability in two parts, so that sum of both parts is nearly equal.
3. Give the code '1' to every message of upper part and '0' to lower part messages.
4. Again divide upper part in two nearly equal parts.
5. Also divide lower part in two nearly equal parts and give code '1' to upper part and '0' to lower part.
6. Before start division, ensure that sum of probability of messages is 1.

Example 5.3. Find out the codes for following message. Using Shannon-Fanno coding technique.

x_1	x_2	x_3	x_4	x_5	x_6
0.4	**0.32**	**0.08**	**0.08**	**0.08**	**0.04**

Solution. It is already given in descending order. Sum of probabilities

0.4 + 0.32 + 0.08 + 0.08 + 0.08 + 0.08 = 1

Message	Probability		Codes	n
x_1	0.4		1	1
		1 division		
x_2	0.32		01	2
		2 division		
x_3	0.08		0011	4
		4 division		
x_4	0.08		0010	4
		3 division		
x_5	0.08		0001	4
		5 division		
x_6	0.04		0000	4

1. Firstly divide x_1 and x_2, x_3, x_4, x_5, x_6 and give the code x_1 to '1' and lower part to '0', we **may give upper part 0 and lower part to 1.**
2. In second step we divide x_2 and x_3, x_4, x_5, x_6 because x_2 is 0.32 and $x_3 + x_4 + x_5 + x_6$ is 0.28 so, it is best combination and because 0.32 and 0.28 are nearly equal. Given '1' to upper part and lower part to '0', these bit will be placed at second position.

3. In third step, we divide, $x_3 + x_4$ and $x_5 + x_6$.
4. In 4th step we divide x_3 and x_4.
5. In 5th step we divide x_5 and x_6.

Comparison between Both Encoding: Both the encoding is good, but technically that encoding is good which gives higher efficiency. And it is possible in that coding which gives lower ($\bar{L}$) average bit length. And if $\bar{L}$ is lower when number of bit is lesser in corresponding messages.

One problem always arises is Shannon-Fanno coding, that is equal division. And code length depends upon number of division.

5.9. BLOCK CODES

It is used to control the error, so we can say that it is an error control coding technique.

In previous discussion, we have seen that to correct the error and to detect the error we use redundancy bits that is added by channel encoder systematically to the transmitted message digits. These additional digits convey no new information, they make it possible for the channel decoder to detect and correct errors in the information learning digits. The overall probability of error is reduced due to error detection and correction.

In block codes, a group of r check bits is derived from a block of K information bits.

The structure of block codes in such a way that information bits are followed by check (redundancy) bits.

A block of n digits generated by the encoder in a particular time unit depends only on the block of K input message digits within that time unit.

A block codes some times also known as arithmetic codes or group codes. Each block of K message bits is encoded into a block of n bits ($n > K$).

The check bit or redundancy bits are derived from the messages bits and added to them.

The n-bit block of a channel encoder output is called a codeword and the codes in which message bits appear at the beginning of a codeword are called systematic codes.

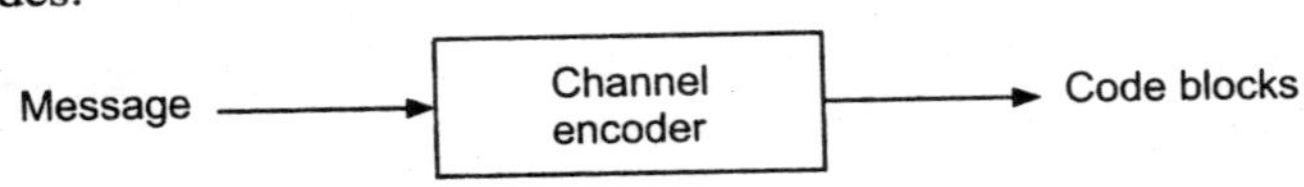

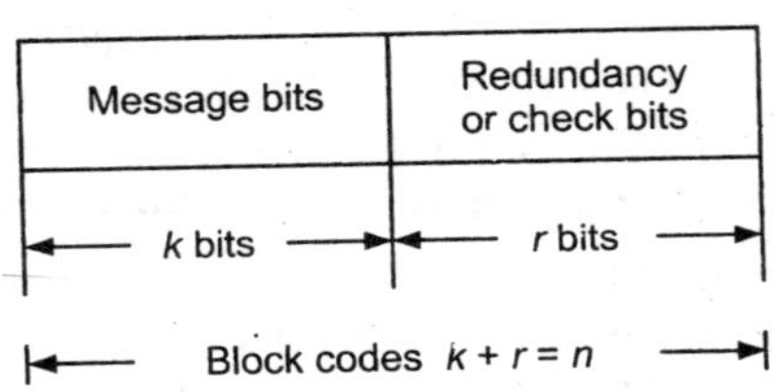

Fig. 5.53

5.9.1. Binary Code Space

In this section we will study some basic concept that is used in block codes.

Codeword	Weight
10010101	4
01110101	5
110000	2
00000	0

Weight: Weight of binary code means number of '1' in that Binary code.

Hamming Distance: It is distance between two codewords in which they differ.

Let $A = 1010100$, $B = 1000000$, $C = 0100000$

Then the distance D between A and B is

$$D(A, B) = 2$$

$$D(B, C) = 1$$

$$D(A, C) = 4$$

Mathematically it can be expressed as

$$D(A, B) = \sum_{k=1}^{n} (a_k \oplus b_k)$$

where $A = a_1, a_2, a_3 \ldots\ldots a_n$

$B = b_1\ b_2\ b_3 \ldots\ldots b_n$

$\oplus$ is the sign of $X \cdot OR$ (Exclusive OR)

$$D(A, B) = ((1 \oplus 1) + (0 \oplus 0) + (1 \oplus 0) + (0 \oplus 0) + (1 \oplus 0) + (0 \oplus 0) + (0 \oplus 0))$$

$$= (0 + 0 + 1 + 0 + 1 + 0 + 0) = 2$$

$$D(A, B) = ((1 \oplus 0) + (0 \oplus 1) + 0 \ldots.$$

$$= 1 + 1 = 2$$

$$D(A, C) = 1 \oplus 0 + 0 \oplus 1 + 1 \oplus 0 + 0 + 1 \oplus 0 + 0 \ldots.$$

$$= 1 + 1 + 1 + 1 = 4$$

The minimum distance of a block codes is defined as the smallest distance between any pair of codewords.

This distance between codes is responsible to correct the error and detect the error. For example if a block code with minimum distance 2 with two digits. Then possible codes are 00, 01, 10, 11. But let in code book of receiver if digits are 10 and 11. Now due to single error 10 may be received as 00 or 01 but it is not in code book so that error is detected but it is difficult to make decision that 10 or 11 are transmitted. Because both are at equal distance from 00 or 01. So that error cannot be detected. To extend this concept follow table may be useful.

Table 5.6.

Minimum distance	Description of coding
1	Error cannot be detected
2	Single error detection
3	Single error correction
4	Single error correction plus double error detection
5	Double error correction
6	Double error correction with three error detection

In general, if n is the minimum distance of block code, then

1. $\frac{n-1}{2}$ errors can be corrected if n is odd.

2. $\frac{n-2}{2}$ errors can be corrected and $\frac{n}{2}$ errors can be detected if n is even.

5.10. CONVOLUTIONAL CODES

We have already discussed block codes for error correction and other one widely used is convolutional codes.

Convolutional code is more efficient than block codes particularly when continuous error correction required or large number of bits get corrupted. A convolutional code is defined by three parameters m, n, p. An (m, n, p) code processes input data n bits at a time and produces an output of m bits for each incoming n bits. So far this is the same as the block codes. In case of a convolutional code m, n are generally quite small numbers.

The main difference is that convolutional codes have memory which is characterized by the constant factor P.

Convolutional codes can be better understood by following example. There are two figures to explain convolutional codes. Figure 5.54 (*a*) explains implementation of codes and Fig. 5.54 (*b*) explains decoding process.

For a specific code (m, n, p), shift register contains the most recent $P \times n$ input bits. The register is initialized to all zeros.

The encoder produces m output bits after which the oldest n bits from the register are discarded and n new bits are shifted in.

Although the output of m bits depend on $P \times n$ bits, the rate of encoding is m output bits per n input. The most commonly used binary encoders have $n = 1$ and hence a shift register length of P.

Our example is a (2, 1, 3) code. In this example encoder converts an input bit U_m into two output V_{m_1}, V_{m_2} using the 3 most recent bits.

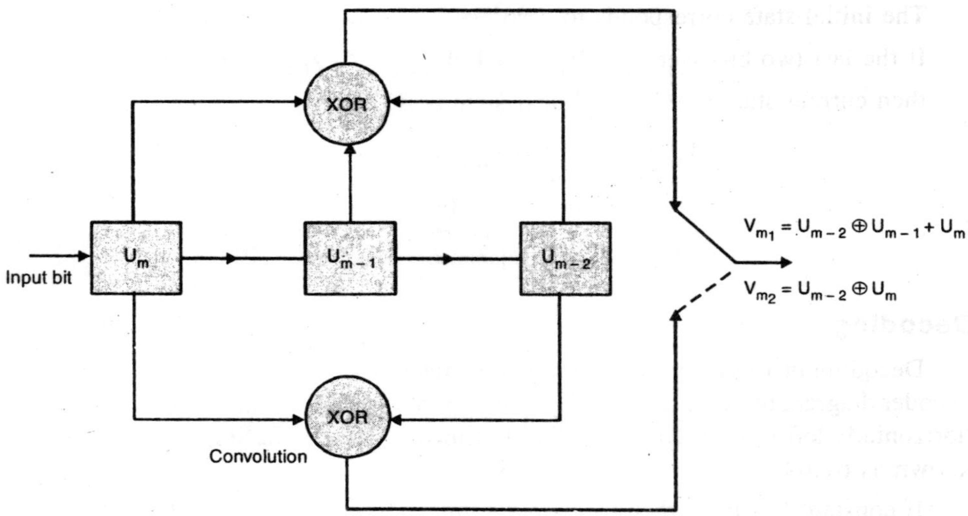

Fig. 5.54(*a*). Encoder [convolution code generator].

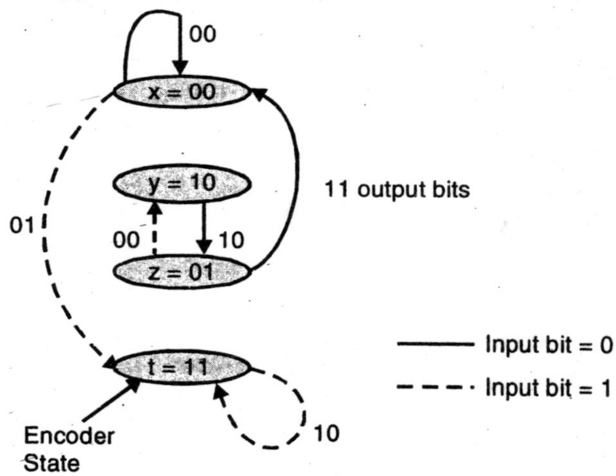

Fig. 5.54(*b*). Encoding of bit.

First output bit produced is from the upper logic circuit

$$V_{m_1} = V_m \oplus V_{m-1} \oplus V_{m-2}$$

And the second output bit from the lower logic circuit

$$V_{m_2} = U_m \oplus U_{m-2}$$

For any given input of *n* bits there are $2^{n(P-1)}$ different functions that map the n input bits into m output bits. Which function is used depends on the history of the last $(P-1)$ input block of n bits each. We can therefore represent a convolutional code using a finite state machine.

The machine has $2^{n(P-1)}$ states and the transition from one state to another is determined by the most recent n input and m outputs.

The initial state corresponds to zero.

If the last two bits were 10 ($V_{m-1} = 1$, $V_{m-2} = 0$) and the next bit is 1.

then current state is $y(10)$ and next state is $t(11)$.

$$V_{m_1} = U_{m-2} \oplus U_{m-1} \oplus U_m$$

$$= 0 \oplus 1 \oplus 1 = 0$$

$$V_{m_2} = 0 \oplus 1 = 1$$

Decoding

Decoding process can be understood by state diagram in Fig. 5.55. Note that if encoder diagram represents vertically as in Fig. 5.55(*b*) then state diagram represents horizontally left to right corresponding to time, or data input. State diagram is also known as trellis.

If constant *k* is large then trellis diagram becomes unwidely. In this case 2^{P-2} simplified trellis parts can be used to depict the transitions. Figure 5.56 shows the transitions for a code (2, 1, 7). Any valid output is defined by a path through the trellis. In our example the $x - y - z - y - z - y - x - x$.

Produces the output 11 10 00 01 01 11 00 and

was generated by the input 1011000

If an invalid path occurs such as $x - y$ then the decoder attempts error correction. In essence, the decoder must determine what data input was most likely to have produced the invalid output.

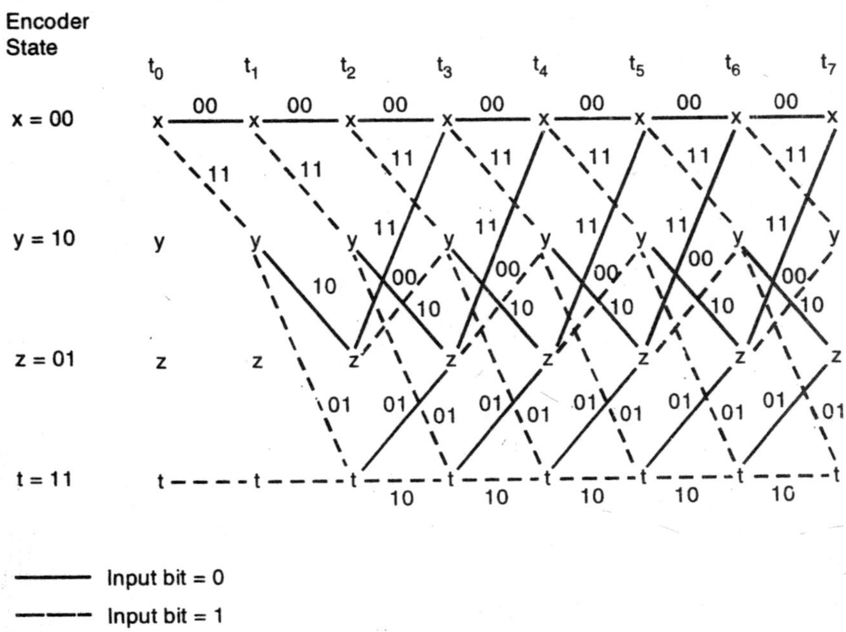

Fig. 5.55. State diagram for Fig. 5.54(*b*).

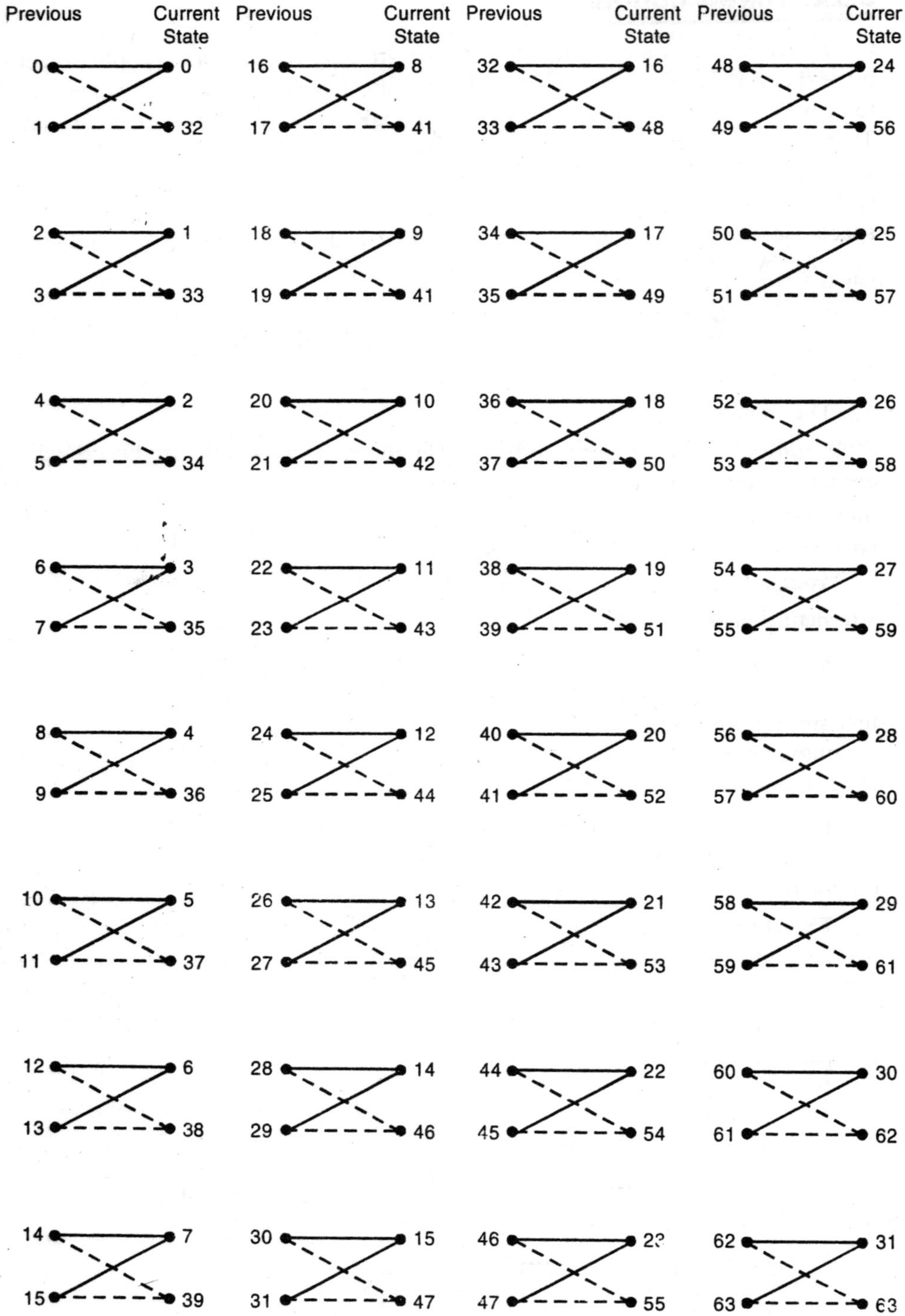

Fig. 5.56. State (Trellis diagram for encoder with mnp = (2, 1, 7).

5.11. TURBO CODING

As we know, wireless communication is growing very fast and people expecting more speed and more services. Then at same time it becomes challenge for engineers to provide error free communication in high speed data transmission environment. Turbo coding becomes very popular coding technique in third generation (3G) wireless system. It provides very good performance in terms of bit error probability and it is very close to Shannon limit and can be efficiently implemented for high speed use. There are so many types of turbo encoding and decoding technique and most of them depend upon convolutional encoding.

Figure 5.61(*a*) shows turbo encoder technique. In this technique one copy of the encoder receives a stream of input bits and produces a single output check bit C_1 for each input bit.

The input to the other encoder is an interleaved version of the input bit stream, producing a sequence of C_2 check bits. The initial input bit plus the two check bits are then multiplexed to produce the sequence $I_1 C_{11} C_{21} I_2 C_{12} C_{22}, \ldots$ that is, the first input bit followed by the first bit from encoder one, followed by the first bit from encoder 2, and so on. The resulting code rate will be 1/3.

A code rate of 1/2 can be achieved by taking only half of the check bits, alternating between outputs from two encoders.

Rates of 1/3 and 1/2 are both found in 3G system. Note that each encoder only produces a single check bit for each bit and that the input bit is preserved. While bits are not preserved in convolutional encoders and there are multiple output bits (*m* output bits for *n* inputs). For turbo coding, a variation of the convolutional code, known as a recursive systematic convolutional code (RSC) is used.

In a (2, 1, *P*) RSC encoder, which is typical, one of the two check bit calculations is fed back to the shift register and the other check bit calculation procedure an output bit.

The output of the encoder consists of alternating input and check bits. The RSC encoder has the same trellis structure as a conventional encoder and similar statistical properties.

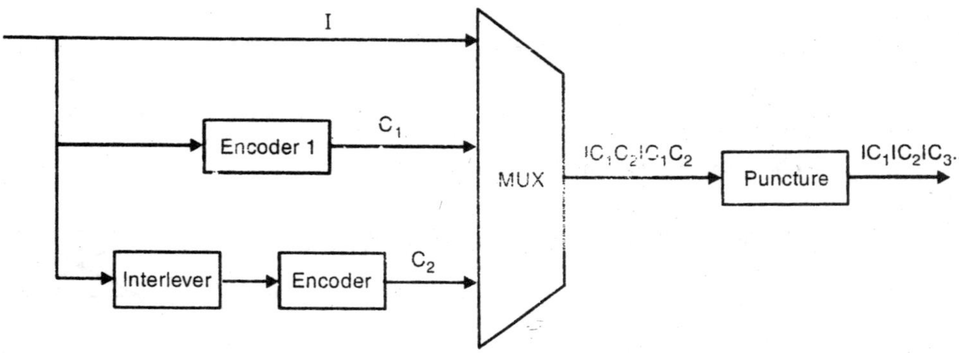

Fig. 5.57. (*a*) Turbo encoder.

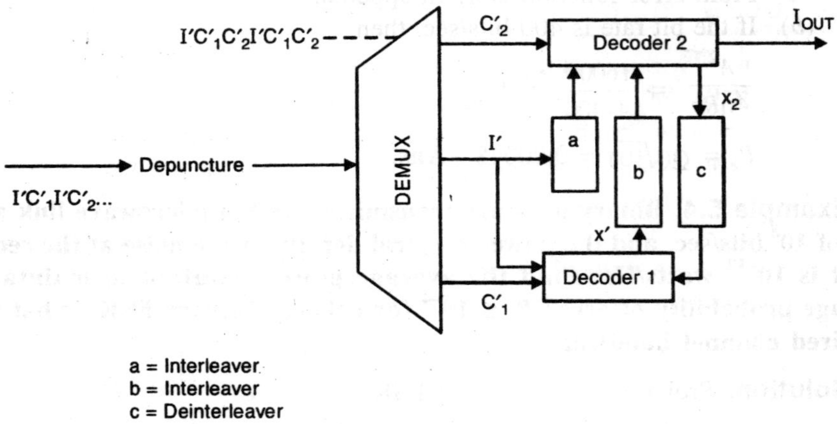

Fig. 5.57(*b*). Turbo decoder.

Example 5.3. Binary data has to be transmitted over a communication link that has a bandwidth of 4 kHz and a maximum achievable signal to noise ratio of 10 dB at its *O/P*.

Determine

(*a*) The maximum signalling rate and P_e if a coherent ASK scheme is used for transmitting binary data through this channel.

(*b*) If the data rate is maintained at 400 bits/sec. Calculate the error probability.

Solution.

(*a*) Average signal power = $\frac{A^2}{4}$

Noise power = $2 \cdot \frac{\eta}{2} \cdot 4000$

S/N ratio = 10 dB ⇒ 1

$$S/N = \frac{\text{Average signal power}}{\text{Noise power}} = 1$$

$$\Rightarrow \frac{A^2}{4 \times 4000\eta} = 1 \text{ or } \frac{A^2}{\eta} = 16000$$

$$\text{Bit rate} = \frac{4 \times 10^3}{3} \Rightarrow 1.33 \times 10^3$$

or 1.33 kbps

(*Note:* Here Bandwidth of ASK is taken 3 R_b.)

Therefore, signalling rate is 1.33 kbps

and probability of error $P_e = Q\left(\sqrt{\frac{A^2}{4\eta R_b}}\right)$

$$\Rightarrow \frac{A^2}{4\eta R_b} \Rightarrow \frac{16000}{4 \times 1.33 \times 10^3} \Rightarrow 3$$

$$P_e = Q(\sqrt{3}) = 0.0401$$

From error function table in appendix.

(*b*) If the bit rate is 400 bits/sec, then

$$\frac{A^2}{4\eta Rb} \Rightarrow \frac{16000}{4 \cdot 400} \Rightarrow 10$$

$$P_e = Q(\sqrt{10}) = 0.00048 \quad \textbf{Ans.}$$

Example 5.4. Binary data are transmitted over a microwave link at the rate of 10^8 bits/sec. and the power spectral density of the noise at the receiver input is 10^{-12} watts/Hz. Find the average power required to maintain an average probability of error $P_e \leq 10^{-4}$ for coherent binary FSK. What is the required channel bandwidth?

Solution. Probability of error for FSK.

$$P_e = \frac{1}{2} \text{ erfc } \left(\sqrt{\frac{E_b}{2N_o}} \right)$$

here

$$P_e = 10^{-4} \Rightarrow \text{erfc } \left(\sqrt{\frac{E_b}{2N_o}} \right) = 2 \times 10^{-4}$$

$$\text{erfc } (V) = \frac{\exp(-V^2)}{\sqrt{\pi} V}$$

$$\Rightarrow \exp(-E_b/2N_o) = 2 \times 10^{-4} \times \sqrt{\pi} \times \left(\sqrt{\frac{E_b}{2N_o}} \right)^{\frac{1}{2}}$$

$$\Rightarrow \frac{E_b}{2N_o} = 2.3$$

Given

$$\frac{N_o}{2} = 10^{-12} \Rightarrow N_o = 2 \times 10^{-12}$$

$$E_b = 2.3 \times 2N_o$$

$$= 2.3 \times 2 \times 2 \times 10^{-12}$$

$$= 9.2 \times 10^{-12} \text{ Joules.}$$

Data rate = 10^8 bits/sec.

$$\frac{1}{T_b} = 10^8 \Rightarrow T_b = 10^{-8} \text{ sec.}$$

Average power or carrier power will be,

$$p_{av} = \frac{E_b}{T_b} \cong \frac{9.2 \times 10^{-12}}{10^{-8}} \Rightarrow 9.2 \times 10^{-4} \text{ W}$$

Hence for $P_e \geq 10^{-4}$, we should have,

$$P_{av} = 9.2 \times 10^{-4} \text{ or } 0.92 \ m \text{ Watt.}$$

$$\text{Bandwidth} = \frac{1}{T_b} \Rightarrow 108 \Rightarrow 10 \text{ MHz.}$$

Example 5.5. An FSK system transmits binary data at the rate of 2.5 × 10^6 bits/sec. During the transmission, white Gaussian noise of zero mean and power spectral density 10^{-20} watts/Hz is added to the signal. In the absence of

noise, the amplitude of the received sinusoidal wave for digit 1 or 0 is 1 μ volt. Determine the av. probability of symbol/error, assuming coherent detection.

Solution: $T_b = \dfrac{1}{2.5 \times 10^6} \Rightarrow 0.4 \times 10^{-6}$ sec.

$$\text{Signal Energy per bit } (E_b) = \frac{1}{2} A_c^2 \cdot T_b$$

$$= \frac{1}{2} \times \left(10^{-6}\right)^2 \times 0.4 \times 10^{-6}$$

$$= 2 \times 10^{-19} \text{ Joules}$$

Average Probability of error for coherent FSK is,

$$P_e = \frac{1}{2} \operatorname{erfc}\left(\sqrt{\frac{E_b}{2N_o}}\right)$$

$$= \frac{1}{2} \operatorname{erfc}\left(\sqrt{\frac{2 \times 10^{-19}}{2 \times 2 \times 10^{-20}}}\right)$$

$$= \frac{1}{2} \operatorname{erfc}\left(\sqrt{5}\right)$$

By using approximation,

$$\operatorname{erfc}(V) = \frac{\operatorname{erfc}(-V)^2}{\sqrt{\pi V}}$$

$$\text{We get } P_e = \frac{\exp(-5)}{2\sqrt{5\pi}} \Rightarrow \frac{6.73 \times 10^{-3}}{14}$$

$$= 0.84 \times 10^{-3} \quad \textbf{Ans.}$$

Example 5.6. Assume odd parity, find the parity bit for each of the following data units.

(a) 1001011 (b) 11010110 (c) 1101011 (d) 101101101

Solution. Odd Parity number of '1' must be odd.

(a) 1001011<u>1</u> (b) 11010110<u>0</u> (c) 1101011<u>0</u> (d) 101101101<u>0</u>

Example 5.7. Using Vertical Redundancy patterns, check that received bit has error or not of 1010001.

Solution. **(a) If even parity is used** then error is there because number of '1' is odd.

(b) If odd parity is used then, there is no error because number of '1' is odd.

Example 5.8. Using Horizontal Redundancy check pattern find the LRC following block of data.

10011001 01101111. Use even parity concept.

Solution. 10011001 01101111 can be placed as

1 0 0 1 1 0 0 1	even 1s hence
0 1 1 0 1 1 1 1	parity is zero.
1 1 1 1 0 1 1 0	parity

Data is to be transmitted.

10011001 01101111 11110110

Example 5.9. Given a 10 bit sequence 1001001010 and divisor of 1011. Find the CRC. And also check your answer.

Solution: To find CRC divide the data bit by divisor.

```
             101011000
           ____________
      1011 ) 1001001010000   --> Appended zeros
             1011
             ----
              0100
              0000
              ----
               1000
               1011
               ----
                0111
                0000
                ----
                 1110
                 1011
                 ----
                  1011
                  1011
                  ----
                   0000
                   0000
                   ----
                    0000
                    0000
                    ----
                     0000
                     0000
                     ----
                      0000
                      0000
                      ----
                       000
```

CRC will be 000

Data is to be transmitted 1001001010000.

To check the result, again divide the result by divisor, if result comes zero, it means CRC is correct.

```
             101011
           ______________
      1011 ) 1001001010000 0
             1011
             ----
              0100
              0000
              ----
               1000
               1011
               ----
                0111
                0000
                ----
                 1110
                 1011
                 ----
                  1011
                  1011
                  ----
                   0000
                   0000
                   ----
                      0
```

Remainder is zero, it means received data is correct.

Example 5.10. **Given a remainder of 111, a data unit of 10110011, and a divisor of 1001, is there an error in the data unit.**

Solution. Data receiver is 10110011111

```
            10100111
      ____________
1001 ) 10110011111
       1001
       ----
        0100
        0000
        ----
         1000
         1001
         ----
          0011
          0000
          ----
           0011
           0000
           ----
            1111
            1001
            ----
             1101
             1001
             ----
              1001
              1001
              ----
               000
```

Remainder is zero, it means receiver data is correct.

Example 5.11, **Find the binary equivalent of polynomial $x^8 + x^4 + x^3 + x + 1$.**

Solution. Binary equivalent of polynomial is

100011011

Example 5.12. **Find the polynomial equivalent of 10001110011.**

Solution. The highest power will be 10, and polynomial is

$$x^{10} + x^6 + x^5 + x^4 + x + 1$$

Example 5.13. **Find the check sum for the following bit sequence. Assume a 16 bit segment size.**

1001001110010011

1001100001001101

Solution.

```
      1 0 0 1 0 0 1 1 1 0 0 1 0 0 1 1
      1 0 0 1 1 0 0 0 0 1 0 0 1 1 0 1
      -------------------------------
Sum   0 0 1 0 1 0 1 1 1 1 1 0 0 0 0 0
                                    1
      -------------------------------
      0 0 1 0 1 0 1 1 1 1 1 0 0 0 0 1
```

Take 1^s complement, check sum = [1 1 0 1 0 1 0 0 0 0 0 1 1 1 1 0]

Example 5.14. **Find the minimum number of redundancy bit needed to correct one single bit error for following data size.**

***(a)* 12** ***(b)* 16**

Solution. To find the redundancy bits, we use following condition.

(a) $2^r \geq m + r + 1$

Let $r = 5$

$2^5 \geq 12 + 5 + 1 = 32 \geq 18$ Satisfied, hence 5 bits required.

(b) $2^5 \geq 16 + 5 + 1$ = Satisfied, hence $r = 5$

Example 5.15. Construct the Hamming code for the bit sequence 10011001.

Solution. First find the redundancy bit, required no. of data bit is 8.

$$2^4 \geq 8 + 4 + 1$$

So $r = 4$

and the position of these 4 bits will be

$2^0, 2^1, 2^2, 2^3$ so that 1, 2, 4, 8 position.

12	11	10	9	8	7	6	5	4	3	2	1
1	0	0	1	r_8	1	0	0	r_4	1	r_2	r_1

	b_0	b_1	b_2	b_3
0	0	0	0	0
1	0	0	0	1
2	0	0	1	0
3	0	0	1	1
4	0	1	0	0
5	0	1	0	1
6	0	1	1	0
7	0	1	1	1
8	1	0	0	0
9	1	0	0	1
10	1	0	1	0
11	1	0	1	1
12	1	1	0	0

r_1 will take care of those bits in which from right position at first position first bit is '1'.

r_1 will take care of 1, 3, 5, 7, 9, 11

r_2 will take care of 2, 3, 6, 7, 10, 11

r_4 will take care of 4, 5, 6, 7, 12

r_8 will take care of 8, 9, 10, 11, 12

Now find the parity bit of redundancy. Here 1 and considering even parity.

$r_1 = 1$

$r_2 = 0$

$r_4 = 0$

$r_8 = 0$

Redundancy bit $r_8\, r_4\, r_2\, r_1$ is 0001

And Hamming code is 1 0 0 1 0 1 0 0 0 1 0 1

Example 5.16. A sender sends 0 1 1 1 0 0 0 1, the receiver receives 0 1 0 0 0. If only VRC is used, how can receiver detect the error.

Solution. No, because VRC checks only parity, parity is same in transmitted and receiver data.

Example 5.17. The following block uses even parity LRC, which bits are in error.

10010101 01001111 11010000 11011011

Solution. LRC is 11011011 (Transmitted with data) and data bit is

10010101 01001111 11010000

```
  10010101
  01001111
  11010000
  --------
  00001010
```

It can be written as LRC is not same, it means data is corrupted.

By LRC, we can detect only error but at which position, it cannot be found.

Example 5.18. If a divisor is 101101, how many bits long is the CRC ?

Solution. CRC is always 1 bit less than divisor.

Example 5.19. A receiver receives the code 11001100111 when it uses Hamming encoding algorithm, the result is 0101, which bit is computed and what is correct code.

Solution. Received bit is 11001100111

Data bit is of 11 bit, it means 4 redundancy bits at position $r_1\, r_2\, r_4\, r_8$.

11	10	9	8	7	6	5	4	3	2	1
1	1	0	0	1	1	0	0	1	1	1
			r_8				r_4		r_2	r_1

The redundancy bits given in question are 0101

$r_8 \quad r_4 \quad r_2 \quad r_1$

$(0 \quad 1 \quad 0 \quad 1)_2 = (5)_{10}$

It means 5th bit is corrupted and correct data is

[1 1 0 0 1 1 1 0 1 1 1]

Example 5.20. Apply Shannon-Fanno coding for the following message probability. And find Efficiency. Take $m = 2$.

Message	**A**	**B**	**C**	**D**	**E**	**F**
	0.25	**0.25**	**0.125**	**0.125**	**0.125**	**0.0625**

Solution. Sum all the probability

$$0.25 + 0.25 + 0.125 + 0.125 + 0.125 + 0.0625 = 0.9375 \neq 1$$

Hence there will be one more message code. G. The probability of G = 1 – 0.9375 = 0.625.

Now arrange in descending order

			Codes	n
A	0.25		11	2
		2		
B	0.25		10	2
		1		
C	0.125		011	3
		4		
D	0.125		010	3
		3		
E	0.125		001	3
		5		
F	0.0.625		0001	4
		6		
G	0.0625		0000	4

$$\text{Efficiency } (\eta) = \frac{n(x)}{\bar{L}\log_2 m}$$

$$\Rightarrow \qquad n(x) = -3.32[0.25 \log 0.25 + 0.25 \log 0.125 + 0.125 \log 0.125 + 0.125 \log 0.125 + 0.0625 \log 0.0625 + 0.0625 \log 0.0625]$$

$$= +3.32[0.3 + 0.33 + 0.15] = 2.58 \text{ bits/message.}$$

$$\bar{L} = 0.25 \times 2 + 0.025 \times 2 + 0.125 \times 3 + 0.125 \times 3 + 0.125 + 0.125 \times 3 + 0.625 \times 4 + 0.0625 \times 4$$

$$= 1 + 1.25 + 0.5 = 2.625$$

$$\eta = \frac{2.58}{2.68} \times 100 = 98.2\%$$

Example 5.21 Find the Huffman code for example 5.20 and find the efficiency.

Solution. Arrange in descending order.

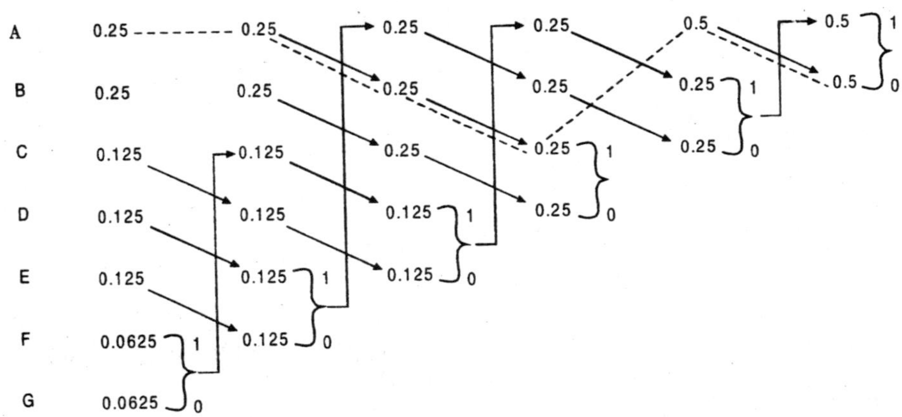

Codeword for A, See the dashed line

0.25	0.25	0.25	0.25	0.5	0.5	1
x	x	x	1	x	0	1

Code is 101

After reversing again 101

Code for B	0.25	0.25	0.25	0.25	0.5	0.5	1
	x	x	x	0	x	0	1

Code is 100

Code for C	0.125	0.125	0.125	0.25	0.25	0.5	1
			0	x	1	1	1

$\Rightarrow$ 1110

Code for D	0.125	0.125	0.25	0.25	0.25	0.5	1
	x	1	x	x	0	1	1

$\Rightarrow$ 1101

Code for E	0.125	0.125	0.25	0.25	0.25	0.5	1
	x	0	x	x	0	1	1

Code for E is 1100

Code for F	0.0625	0.125	0.125	0.25	0.25	0.5	1
	0	x	1	x	1	1	1

Code for F 11111

Code for G	0.0625	0.125	0.125	0.25	0.25	0.5	1
	0	x	1	x	1	1	1

$\Rightarrow$ 11110

Message	Probability	Code	Length
A	0.25	101	3
B	0.25	100	3
C	0.125	1110	4
D	0.125	1101	4
E	0.125	1100	4
F	0.0625	11111	5
G	0.0625	11110	5

Efficiency $(\eta) = \dfrac{H(x)}{\bar{L}\log_2^m}$

$$H(x) = \sum_{i=1}^{n}(x_i)\log_2 P(x_i)\ ,\ \bar{L} = \Sigma P_i n_i$$

$$H(x) = -\ [0.25 \log 0.25 + 0.25 \log 0.25 + 0.125 \log 0.125 + 0.125 \log 0.125 + + 0.125 \log 0.125 + 0.125 \log 0.125 + 0.0625 \log 0.0625 + 0.0625 \log 0.0625] \quad ...(1)$$

Note: Base of log is 2.

Therefore, by applying logarithmic identity.

$$\log_n^m = \frac{\log_{10}^m}{\log_{10}^2}$$

$$\Rightarrow \qquad \log_2^m = \frac{\log_{10}^m}{\log_{10}^2} = \frac{1}{0.3010}\log_{10}^m = 3.32 \log_{10}^m$$

REVIEW QUESTIONS

1. Sketch an ASK waveform for the bit 10101101.
2. Sketch the ASK, PSK, FSK waveform for the following bits :
   (*a*) 10110101 (*b*) 1101101
3. Sketch the QPSK waveform for the bit 11010101.
4. Sketch the QPSK, OQPSK for the bit 11010101.
5. Sketch the MSK waveform for 110101010.
6. Sketch the DPSK waveform for 11010101 with reference bit 1 and 0.
7. Explain the PSK and DPSK and compare them.
8. Explain the MSK and QPSK and compare them.
9. Explain QAM with block diagram.
10. Describe the term bandwidth efficiency.
11. Explain the probability bit error and find the Pb for polar, bipolar signaling.
12. Find the probability of bit error for ASK, FSK, PSK.
13. Explain block coding.
14. Explain turbo coding.
15. What do you mean by OFDM.
16. What are different digital modulation techniques compare QPSK, and BPSK in terms of bandwidth and data rate.
17. Write short notes on OFDM and Turbo coding.
18. What is difference between convolution code and linear block codes ? Make block diagram of a convolution encoder.
19. What is Hamming distance and hamming weight ? Explain from first principles how a (15, 11) hamming code can be used to detect and correct single random error.
20. Write a short note on convolutional codes and explain how these codes differ from block codes.
21. Write a short note on concatenated Turbo codes.
22. Explain how orthogonality and autocorrelation properties are helpful in designing the spreading sequences without interference n CDMA systems.
23. Explain the linearity property of block codes.

❑❑❑

6

CHAPTER

Multiple Access Techniques

6.0. INTRODUCTION

There are various multiple access techniques that need to use as per application like TDMA, FDMA, CDMA. TDMA and FDMA are widely used multiple technique in GSM technology. while CDMA is another technique used in wireless technology. There are various advantages of TDMA and FDMA as per application. We will discuss in detail in this chapter.

INSIDE THIS CHAPTER

6.0. Introduction
6.1. Multiple Access Techniques
6.2. Multiple Access Techniques For Wireless Communication
6.3. Capacity of a Cellular System
6.4. Comparison of Spectrum Efficiency for FDMA and TDMA Systems
6.5. Summary of FDMA, TDMA and CDMA
6.6. Wireless TDMA Design Consideration for Mobile Phone
6.7. Space Division Multiple Access (SDMA)
6.8. Code Division Multiple Access (CDMA)
6.9. Bluetooth
6.10. Comparison Between Wireless Communication System

6.1. MULTIPLE ACCESS TECHNIQUES

In wireless communication system, it is always desirable to allow the user to send the simultaneous information to the base station while receiving the information from the base station.

For example talking on cell phone, at same time user transmits information and receives information, this concept is known as Duplexing. Generally it is classified in two part frequency division duplexing and time division duplexing.

Analog cellular systems are the first generation of cellular systems. Several incompatible analog standards have been developed and implemented world wide. These systems use Analog Frequency Modulation (FM) and have a Frequency Division Multiple Access (FDMA) based Media Access Control (MAC) architecture. The available radio spectrum is shared by a relatively large number of FM modulated signal.

2G cellular systems primarily use digital modulation and processing techniques and have several incompatible air interface standards. The first goal, to use digital techniques, has been achieved, whereas the second goal, to have a single digital cellular wireless global standard, is unlikely to be achieved. Numerous incompatible digital mobile wireless and cellular standards have been developed. Some of these techniques use FDMA, TDMA etc.

Third generation systems such as several spread spectrum (SS), Collision Sense Multiple Access (CSMA), Direct Sequence (DS), Frequency Hoped (FH) spread spectrum have also been standardized.

6.1.1. Frequency Division Duplexing (FDD)

Frequency division duplexing provides two distinct bands of frequencies for every user. The forward band provides traffic from the mobile to BTS and the reverse band provides traffic from the BTS to the mobile.

In FDD, any duplex channel actually consists of two simplex channels (a forward and reverse), and a device called a duplexer is used inside each subscriber unit and base station to allow simultaneous bidirectional radio transmission and reception for both the subscriber unit and the base station on the duplex channel pair.

Their must be a frequency separation between both channels and it must be constant throughout the process.

In India all GSM and CDMA operators are using Frequency Division Duplexing (FDD) technique.

In GSM 900, uplink frequency from mobile to BTS is 890-915 MHz while in Downlink from BTS to mobile we use different frequency (935 MHz–960 MHz) rather than using same frequency in uplink and downlink both. And Duplex distance between uplink and downlink is (935 MHz – 890 MHz = 45 MHz). It means if a signal is transmitted in uplink at 890.2 MHz then corresponding downlink frequency is 935.2 MHz at a separation of 45 MHz.

While in TDD, we use same frequency in uplink and downlink at time sharing basis.

Therefore, GSM is a combination of TDMA + FDMA + FDD.

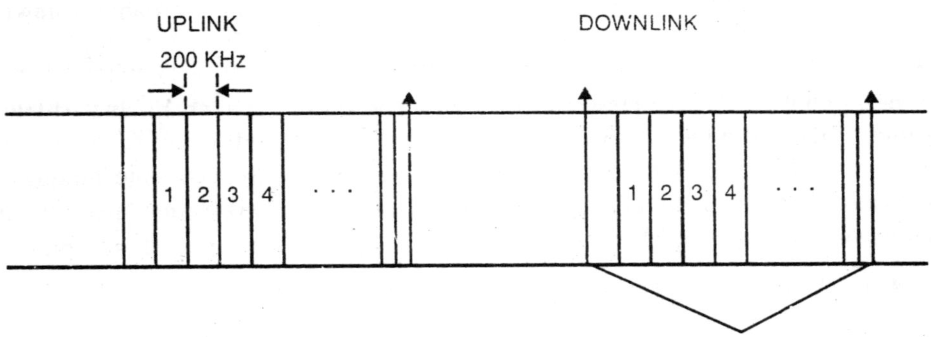

Guard band

GSM 900 : 890 MHz-915 MHz — 935 MHz–960 MHz
Channel Bandwidth = 25 MHz
Duplex Bandwidth = 45 MHz
Carrier Bandwidth = 200 KHz
Total number of channels = 124

GSM 1800 : 1710 MHz-1785 MHz — 1805 MHz–1880 MHz
Channel Bandwidth = 75 MHz
Duplex Bandwidth = 95 MHz
Carrier Bandwidth = 200 KHz
Total number of channels = 374

Fig. 6.1. Frequency division duplexing.

6.1.2. Time Division Duplexing (TDD)

It uses time instead of frequency to provide both forward and reverse link.

In TDD, multiple users share a single radio channel by taking turns in the time domain. All users are allowed to access the channel in pre-specified time and each duplex channel has both a forward time slot and a reverse time slot.

If the time separation is small then both way communication occurs simultaneously but it is not too small so that signal become interfering to each other. TDD requires a single communication channel which make equipment simpler while FDD required two dedicated communication channels [Two simplex].

TDD enables each transceiver to operate as either a transmitter or receiver, due to that a time delay is created between individual users but it is very less, cannot be observed, but as compared to FDD, where no time delay but two separate channels are required which are working at two different frequencies. So the equipment becomes costly than TDD enabled equipment.

TDD generally is limited to cordless phone or short range portable access.

It is more effective for fixed wireless access when all users are stationary so that propagation delays do not vary in time among the users.

6.2. MULTIPLE ACCESS TECHNIQUES FOR WIRELESS COMMUNICATION

- Multiple access schemes are used to allow many mobile users to share simultaneously a finite amount of radio spectrum.
- For high quality communications, this must be done without severe degradation in the performance of the system.

There are three major access techniques used to share the available bandwidth in a wireless communication system. They are Frequency Division Multiple Access (FDMA), Time Division Multiple Access (TDMA), and Code Division Multiple Access (CDMA). These techniques can be grouped as narrowband and wideband systems, depending upon how the available bandwidth is allocated to the users. In addition to FDMA, TDMA, and CDMA, two other multiple access techniques are in use in wireless communications. These are packet radio (PR) and space division multiple access (SDMA).

6.2.1. Principles of Operation of an FDMA

In FDMA each user is allocated a unique frequency band or channel (Fig. 6.2). These channels are assigned on demand to users who request service. In Frequency division duplexing (FDD) systems, the channel has two frequencies—forward channel and reverse channel. During the period of the call, no other user can share the same frequency band.

If the FDMA channel is not in use, then it will be idle and cannot be used by other users to increase or share capacity. This is a wasted resource. Some of the features of FDMA are as follows :

1. The bandwidth of FDMA channels is narrow (30 KHz) since it supports only one call/carrier.

2. Inter Symbol Interference (ISI) is low since the symbol time is large compared to average delay spread. No equalization is required.
3. FDMA systems are simple than TDMA systems, but modern DSP is changing this factor.
4. FDMA systems have higher cell site system costs as compared to TDMA systems, because of the single carrier per channel design, and the need to use due costly bandpass filters to eliminate spurious radiation at the base station.
5. The FDMA mobile unit uses Duplexers since both the transmitter and the receiver operate at the same time. This results in an increase in cost of FDMA subscriber units and base stations.
6. FDMA requires tight RF filtering to minimize ACI. Some of the advantages and disadvantages of FDMA are:

Advantages :

- Very Simple to design
- Narrowband (no Inter Symbol Interference)
- Synchronization is easy
- No interference among users in a cell

Disadvantages :

- Narrowband interference
- Static spectrum allocation
- Frequency reuse is a problem
- High Q analog filters or
- Large guard band required

6.2.1.1. Number of Channels Supported by FDMA System

The first Advanced Mobile Phone System (AMPS) was based on FDMA/FDD. A single user occupies a single channel while the call is in progress, and the single channel is actually two complex channels which are frequency duplexed with a 45 MHz split. When a call is complete, or when a handoff occurs, the channel is vacated so that another mobile subscriber may use it. Multiple or simultaneous users are accommodated in AMPS by giving each user a unique channel. Voice signals are sent on the forward channel from the mobile to base unit, and on the reverse channel from the base unit to the mobile unit. In AMPS, analog narrowband frequency modulation (NBFM) is used to modulate the carrier. The number of channels, N, that can be simultaneously supported in a FDMA system, is given by

$$N = \frac{(B_t - 2B_g)}{B_c} \quad \text{...(6.1)}$$

where B_t is the total spectrum allocation, B_g is the guard band allocated at the edge of the allocated spectrum band and B_c is the channel bandwidth.

Example 6.1. AMPS System as example of FDMA

If an AMPS cellular operator is allocated 12.5 MHz for each simplex band, and if B_t is 12.5 MHz, B_{guard} is 10 kHz, and B_c is 30 kHz, find the number of channels available in an FDMA system.

Solution. The number of channels available in the FDMA system is given as

$$N = \frac{12.5 \times 10^6 - 2(10 \times 10^3)}{30 \times 10^3} = 416$$

In the AMPS, each cellular carrier is allocated 416 channels,

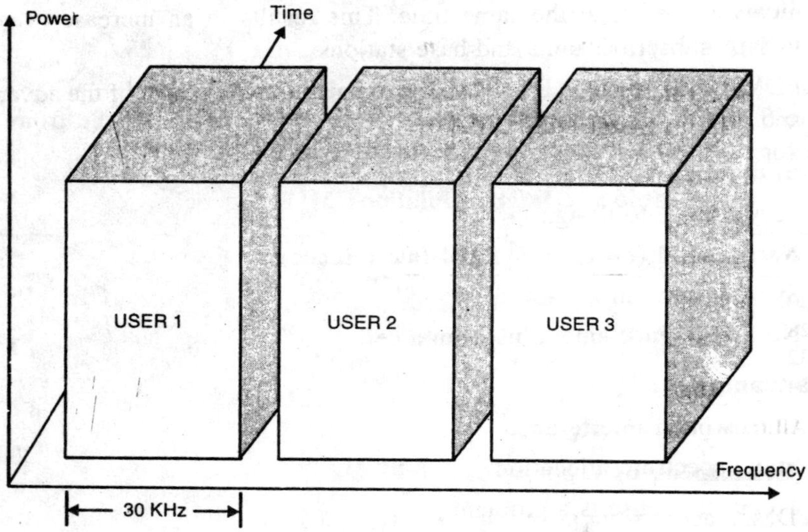

Fig. 6.2. FDMA schemes (AMPS).

6.2.1.2. Non-linear Effects in FDMA

In FDMA system, many channels share the same antenna at the base station. The power amplifiers, when operated at or near saturation for maximum power efficiency, non-linear response, and this non-linearity cause signal spreading in the frequency domain and generate inter-modulation (IM) frequencies. These IM frequencies can interfere with other channels in the FDMA systems.

Spreading of the spectrum results in adjacent channel interference.

Intermodulation is the generation of undesirable harmonics. These harmonics generated outside the mobile radio band cause interference to adjacent services, while those present inside the band cause interference to other users in the wireless system.

The first generation cellular phone (AMPS) was based on FDMA/FDD.

In AMPS, analog narrowband frequency modulation is used to modulate the carrier. The number of channels that can be simultaneously supported in FDMA system is given by

$$N = \frac{B_t - 2B_g}{B_c} \quad ...(6.2)$$

B_t → Total spectrum (Bandwidth allocation)

B_g → Guard band allocated at the edge of the allocated spectrum band,

B_c → Channel bandwidth.

B_t and B_c may be specified in terms of simplex bandwidth.

Intermodulation distortion products occur at frequencies $(mf_1 + nf_2)$ where m and n are integers.

$$m < \infty, \quad n < \infty$$

and f_1 and f_2 are base station carrier frequencies.

Some of the possible intermodulation frequencies produced by a non-linear device are as follows :

$$(2n + 1)f_1 - 2nf_2,\ (2n + 2)f_1 - (2n + 1)f_2,\ (2n + 1)f_2 - 2nf_1,$$
$$(2n + 2)f_2 - (2n + 1)f_1, \text{ etc. for } n = 0, 1, 2, 3, \ldots$$

Table 6.1 shows some of the intermodulation frequencies derived from above formula for base station carrier frequencies 1930 MHz, 1932 MHz.

Table 6.1. Intermodulation (IM) Frequencies.

$n = 0$	$n = 1$	$n = 2$
1930	1926	1922
1928	1924	1920
1932	1936	1940
1934	1938	1942* → Outside the band

Note : All frequencies are in MHz.

6.2.2. Principles of Operation of TDMA

In TDMA, access time is divided into frames are again divided into time slots. A basic channel is formed by a particular time slot inside every frame. Usually same frame structure is used. In order to avoid simultaneous transmission and reception of a user, the corresponding time slots for the forwards and return links are separated in time.

The TDMA scheme is illustrated in Figures 6.3of 6.4.

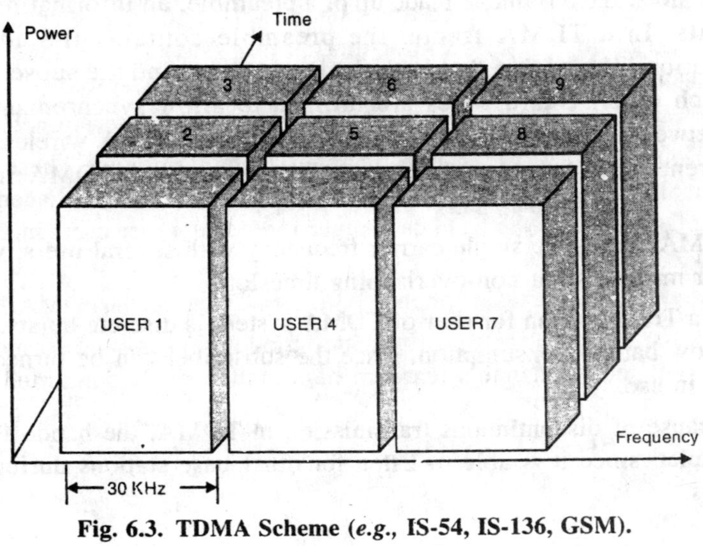

Fig. 6.3. TDMA Scheme (*e.g.*, IS-54, IS-136, GSM).

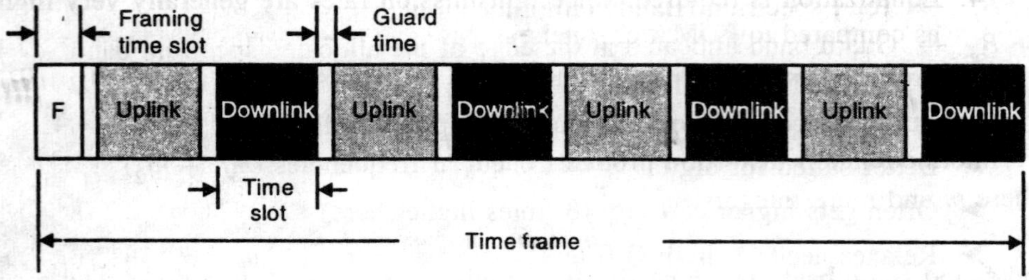

Fig. 6.4.

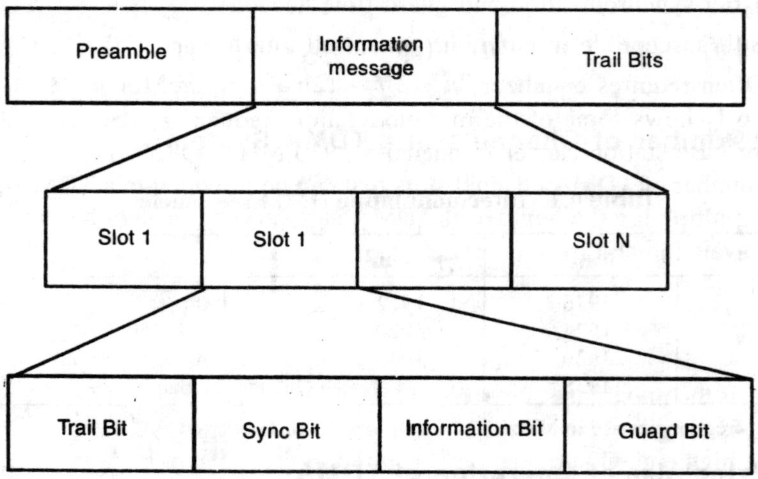

Fig. 6.5. TDMA frame structure.

TDMA systems transmit data in a buffer and burst method, thus the transmission for any user is noncontinuous. This implies that unlike in FDMA systems which accommodate analog FM, digital data and digital modulation must be used with TDMA. The transmission from various users is interlaced into a repeating frame structure as shown in Fig. 6.6. It can be seen from Fig. 6.6 that a frame consists of a number of slots. Each frame is made up of a preamble, an information message, and tail bits. In a TDMA frame, the preamble contains the address and synchronization information that both the base station and the subscriber use to identify each other. Guard times are utilized to allow synchronization of the receivers between different slots and frames. Different TDMA wireless standards have different TDMA frame structures. The features of TDMA include the following :

1. TDMA shares the single carrier frequency with several users, where each user makes use of non-overlapping timeslots.
2. Data Transmission for user of TDMA system is discrete bursts. The result is low battery consumption, since the subscriber can be turned off when not in use.
3. Because of discontinuous transmission in TDMA, the handoff process is simpler, since it is able to listen for other base stations during idle time slots.

4. Equalization is required, since transmission rates are generally very high as compared to FDMA channels.

Some of the advantages and disadvantages of TDMA are :

Advantages :

- Better suited for digital
- Often gets higher capacity (3 times higher here)
- Relaxes need for high Q filters

Disadvantages :

- Strict synchronization and guard time needed
- Still susceptible to jamming, other cell interference
- Often requires equalizer

6.2.2.1. Number of Channels in a TDMA System

The number of TDMA channel slots that can be provided in a TDMA system is found by multiplying the number of TDMA slots(m) per channel by the number of channels available and is given by :

$$N = \frac{m(B_{tot} - 2B_{guard})}{B_c} \quad ...(6.3)$$

where B_{tot} is the maximum number of TDMA users supported on each radio channel and B_{guard} is the guard band, one at the low end of the allocated frequency band and one at the high end. Guard band are required to ensure that users at the edge of the band do not "bleed over" into an adjacent radio service where B_c is channel bandwidth.

6.2.2.2. Efficiency of TDMA

The efficiency of a TDMA system is a measure of the percentage of transmitted data that contains information as opposed to providing overhead for the access scheme. The frame efficiency, η_f, is the percentage of bits per frame which contains transmitted data. Note that the transmitted data may include source and channel coding bits, so the raw end-user efficiency of a system is generally less than η_f. The frame efficiency can be found as follows :

The number of overhead bits per frame is

$$b_{OH} = N_r b_r + N_t b_p + N_t b_g + N_r b_g \quad ...(6.4)$$

where
N_r = Number of reference bits per frame
N_t = Number of traffic bits per frame
b_r = Number of overhead bits per reference burst
b_p = Number of overhead bits per preamble in each slot
b_g = Number of equivalent bits in each guard time interval

The total number of bits per frame, B_T is

$$B_T = T_f R \quad ...(6.5)$$

where T_f is the frame duration, and R is the channel bit rate.

The frame efficiency η_f is given as

$$\eta_f = \frac{\text{Number of bits / frame containing transmitted data}}{\text{Total number of bits / frame}}$$

$$\eta_f = \left(1 - \frac{b_{OH}}{b_T}\right) \times 100\% \qquad \text{...(6.6)}$$

Example 6.2. Consider Global System for Mobile, which is a TDMA/FDD system that uses 25 MHz for the forward link, which is broken into radio channels of 200 kHz. If 8 speech channels are supported on a single radio channel, and if no guard band is assumed, find the number of simultaneous users that can be accommodated in GSM.

Solution. The number of simultaneous users that can be accommodated in GSM is given as 25 MHz

$$N = \frac{25 \text{ MHz}}{\dfrac{(200 \text{ kHz})}{8}} = 1000$$

Thus, GSM can accommodate 1000 simultaneous users.

Example 6.3. GSM System uses a TDMA/FDD system.

The GSM System uses a frame structure where each frame consists of 8 time slots, and each time slot contains 156.25 bits, and data is transmitted at 270.833 kbps in the channel. Find :

1. **Time duration of a bit**
2. **Time duration of a slot**
3. **Time duration of a frame and**
4. **How long must a user occupying a single slot wait between two simultaneous transmissions ?**

Solution. The time duration of bit,

$$T_b = \frac{1}{\text{BitRate}} = \frac{1}{270.833 \times 10^3} = 3.692 \ \mu\text{s}$$

The time duration of a slot T_{slot} = 156.25 × T_b = 0.577 ms

The time duration of frame, T_f = 8 × T_{slot} = 4.615 ms

A user has to wait 4.615 ms, the arrival time of a new frame, for its next transmission.

Example 6.4. If a normal GSM timeslot consists of 6 trailing bits, 8.25 guard bits, 26 training bits, and 2 traffic bursts of 58 bits of data, find the frame efficiency.

Solution. A time slot has 6 + 8.25 + 26 + 2/58 = 156.25 bits.

A frame has 8 * 156.25 = 1250 bits/frame

The number of overhead bits per frame is given by

$$b_{OH} = 8(6) + 8(8.25) + 8(26) = 322 \text{ bits}$$

$$\text{Frame efficiency, } \eta_f = \frac{(1250 - 322)}{1250} = 74.24\%$$

6.3. CAPACITY OF A CELLULAR SYSTEM

Channel capacity of a radio system is defined as the maximum number of channels or users that can be provided in a fixed frequency band. For a cellular system Radio Capacity is given by

$$m' = \frac{B_t}{(B_c)(N)} \quad \text{...(6.7)}$$

where B_t = Total allocated spectrum for the system

B_c = Channel bandwidth

N = Number of cells in frequency reuse pattern

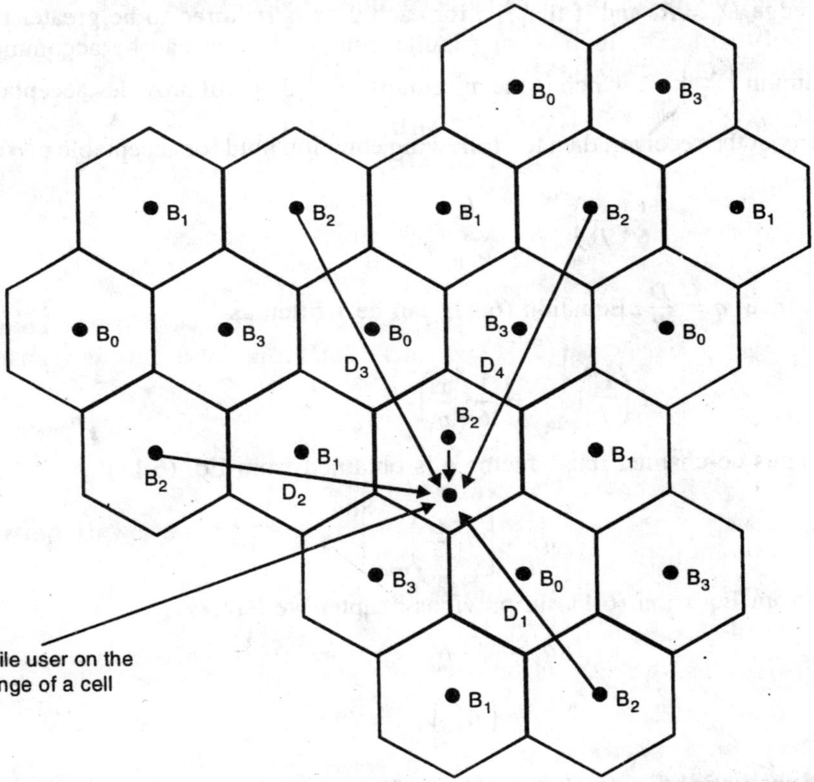

Fig. 6.7. Illustration of forward channel interference for a cluster size of $N = 4$. Shown here are four co-channel base stations which interfere with the sewage base station. The distance from the serving base station to the user is D_0, and interferes are a distance D_k from the user.

Parameter, m' is determined by the required carrier-to-interference (C/I) ratio and the channel bandwidth B_c. As shown in Fig. 6.67, the M closest co-channel cells may be considered as first order interference, in which case (C/I) is given by (See Eq. 1.10)

$$\frac{C}{I} = \frac{D_0^{\alpha_0}}{\sum_{i=1}^{M} D_i^{-\alpha_i}} \quad \text{...(6.8)}$$

where α_0 is the path exponent in the desired cell, D_0 is the distance from the desired base station to the mobile, D_i is the distance of the ith cell from the mobile, and α_i is the path loss exponent to the ith interfering base station. If only the sixth closest interfering cells are considered, and all are approximately at the same distance D and have similar path loss exponents equal to that in the desired cell, then $\frac{C}{I}$ is given by

$$\frac{C}{I} = \frac{D_0^{-\alpha}}{6\,D^{-\alpha}} \quad \text{...(6.9)}$$

Now if it is assumed that maximum interference occurs when mobile is at the cell edge $D_0 = R$, and if the $\frac{C}{I}$, for each user is required to be greater than some minimum $\left(\frac{C}{I}\right)_{\min}$, which is the minimum $\left(\frac{C}{I}\right)$ that still provides acceptable signal quality at the receiver, then the following equation hold for acceptable performance :

$$\frac{1}{6}\left(\frac{R}{D}\right)^{-\alpha} \geq \left(\frac{C}{I}\right)_{\min} \quad \text{...(6.10)}$$

Since $q = \frac{D}{R}$. Equation (6.47) can be written as

$$\left(\frac{C}{I}\right)_{\min} = \frac{1}{6}\left(\frac{1}{q}\right)^{-\alpha} \quad \text{...(6.11)}$$

Thus co-channel reuse factor q is obtained from Eq. (6.11).

$$q = \left[6\left(\frac{C}{I}\right)_{\min}\right]^{(1/\alpha)} \quad \text{...(6.12)}$$

From Equation (6.11) in previous chapter we have

$$q = \sqrt{3}\,N$$

$$N = \left(\frac{q^2}{3}\right) \quad \text{...(6.13)}$$

Substituting Eq. (6.13) into Eq. (6.7) we get following expression for cellular system capacity

$$m = \frac{B_t}{B_c\left(\frac{q^2}{3}\right)} \quad \text{...(6.14)}$$

From Eqs. (6.12) and (6.14) we obtain

$$m = \frac{B_t}{B_c\left[\frac{6}{3^{\alpha/2}}\left(\frac{C}{I}\right)_{\min}\right]^{2/\alpha}} \quad \text{...(6.15)}$$

For example when $\alpha = 4$, the cellular capacity is given by

$$m = \frac{B_t}{B_c \sqrt{\frac{2}{3}\left(\frac{C}{I}\right)_{min}}} \text{ Radio channels/cell} \quad ...(6.16)$$

In order to provide the same voice quality, $\left(\frac{C}{I}\right)_{min}$ may be lower in a digital system when compared to an analog system. Typically, the minimum required $\frac{C}{I}$ is about 12 dB for narrowband digital systems and 18 dB for narrowband analog FM systems, although exact values are determined by subjective listening tests in real-world propagation conditions. Each digital wireless standard *e.g.*, FDMA, TDMA, CDMA has a different $\left(\frac{C}{I}\right)_{min}$, and in order to compare different systems, an equivalent $\frac{C}{I}$ must be used. Lower $\left(\frac{C}{I}\right)_{min}$ values imply more capacity. If B_t and m are kept constant in Equation (6.48), then it is clear that B_c and $\left(\frac{C}{I}\right)_{min}$ are related by

$$\left(\frac{C}{I}\right)_{eq} = \left(\frac{C}{I}\right)_{min} \left(\frac{B_c}{B_c'}\right)^2 \quad ...(6.17)$$

where B_c is the bandwidth of a particular system, $\left(\frac{C}{I}\right)_{min}$ is the tolerable value for the same system, B'_c is the channel bandwidth for a different system, and $\left(\frac{C}{I}\right)_{eq}$ is the minimum $\frac{C}{I}$ value for the different system when compared to the $\left(\frac{C}{I}\right)_{min}$ for a particular system. Notice that for a constant number of users per radio channel, the same voice quality will be maintained in a different system, if $\left(\frac{C}{I}\right)_{min}$ increases by a factor of four when the bandwidth is halved. Equation (6.16) indicates that maximum radio capacity occurs when $\left(\frac{C}{I}\right)_{min}$ and B_c are minimized, yet Equation (6.17) shows that $\left(\frac{C}{I}\right)_{min}$ and B_c are inversely related.

In a digital cellular system, $\frac{C}{I}$ can be expressed as

$$\frac{C}{I} = \frac{E_b R_b}{I} = \frac{E_c R_c}{I} \quad ...(6.18)$$

where R_b is channel bit rate, E_b is the energy per bit, R_c is the rate of the channel code, and E_c is the energy per code symbol. From Eqs. (6.17) and (6.18), $\alpha = 4$, the ratio of $\frac{C}{I}$ to $\left(\frac{C}{I}\right)_{eq}$ of a different system is given as

$$\frac{\left(\frac{C}{I}\right)}{\left(\frac{C}{I}\right)_{eq}} = \frac{\frac{E_c R_c}{I}}{\frac{E_c' R_c'}{I'}} \quad ...(6.19)$$

The relationship between R_c and B_c is always linear, and if the interference level I is the same in mobile environment for two differential digital systems, then Equation (6.19) can be rewritten as

$$\frac{E_c}{E_c'} = \left(\frac{B_c'}{B_c}\right)^3 \quad ...(6.20)$$

Equation (6.20) shows that if B_c is reduced by half, then the energy code symbol increase eights times. This gives the relationship $\frac{E_b}{N_0}$ and B_c in a digital cellular system.

6.4. COMPARISON OF SPECTRUM EFFICIENCY FOR FDMA AND TDMA SYSTEMS

6.4.1. FDMA

In FDMA, the total bandwidth B_t is divided into M channels each with bandwidth B_c. Therefore, the cellular capacity for FDMA is given by (Eq. 6.16 above)

$$m' = \frac{M}{\sqrt{\left(\frac{2}{3}\right)\left(\frac{C}{I}\right)}} \quad ...(6.21)$$

Carrier and Interference terms can be written as

$$C = E_b R_b \quad ...(6.22)$$

$$I = I_0 B_c \quad ...(6.23)$$

where E_b is the energy per bit, R_b is the transmission rate of FDMA system and I_0 represents the interference power per Hz.

6.4.2. TDMA

Assume that multichannel FDMA system occupies the same spectrum as a single channel TDMA system with multiple time slots. For this TDMA case,

$$C' = E_b R_b' \qquad ...(6.24)$$

$$I' = I_0 B_c' \qquad ...(6.25)$$

where R_b' is transmission rate of TDMA system.

Example 6.5. Consider a FDMA system with 3 channels, each having a bandwidth of 10 kHz and transmission rate of 10 kbps. A TDMA system has 3 time slots, channel bandwidth of 30 kHz, and a transmission rate of 30 kbps.

Solution. For the TDMA multiple access, the received $\left(\frac{C}{I}\right)$ for a single user is measured for $\left(\frac{1}{3}\right)$ of the time the channel is in use. We will compare the cellular capacity of the FDMA and TDMA schemes.

For FDMA system from Eqs. (6.17) and (6.18), we get

$$C = E_b R_b = (E_b)(10^4)$$

$$I = I_0 B_c = (I_0)(10 \text{ kHz}) = (I_0)(10^4)$$

So
$$\frac{C}{I} = \left(\frac{E_b}{I_0}\right) \times \frac{10^4}{10^4} = \left(\frac{E_b}{I_0}\right) \qquad ...(6.26)$$

For TDMA system from Eqs. (6.24) and (6.25)

$$C' = E_b R_b' = \frac{E_b \times 10^4}{0.333}$$

C', I' can be measured in 333.3 ms in 1 second.

$$I' = I_0 B_c' = (I_0)(30)(10^3)$$

So
$$\frac{C'}{I'} = \left(\frac{E_b}{I_0}\right)\left(\frac{10^4}{.333} \times \frac{1}{30 \times 10^3}\right)$$

$$= \left(\frac{E_b}{I_0}\right) \qquad ...(6.27)$$

Equations (6.26) and (6.27) show that the received $\left(\frac{C'}{I'}\right)$, for a user in TDMA system is the same as $\left(\frac{C}{I}\right)$ for a user in the FDMA scheme. Therefore, for this example, FDMA and TDMA have the same radio capacity and consequently the same spectrum efficiency. However, the required peak power for TDMA is 10 log (K) higher than FDMA, where K is the number of time slots in a TDMA system of equal bandwidth.

Table 6.2. Multiple Access Technique used in Different Wireless Communication Systems.

Cellular System	Multiple Access Technique
Advanced Mobile Phone System (AMPS)	FDMA/FDD
Global System for Mobile (GSM)	TDMA/FDD
US Digital Cellular (USDC)	TDMA/FDD
Pacific Digital Cellular (PDC)	TDMA/FDD
CT2 (Cordless Telephone)	FDMA/TDD
Digital European Cordless Telephone (DECT)	FDMA/TDD
US Narrowband Spread Spectrum (IS-95)	CDMA/FDD
W-CDMA (3GPP)	CDMA/FDD CDMA/TDD
CDMA 2000 (3GPP2)	CDMA/FDD CDMA/TDD

Table 6.3. Comparison of AMPS with Digital Cellular (TDMA).

Parameter	AMPS	GSM
Bandwidth (MHz)	25	25
Voice Channel	833	1000
Frequency Re-use (cluster size)	7	4 or 3
Channels/Site	119	250 or 333
Traffic (Erlangs/sq. km)	11.9	27.7 or 40
Capacity Gain	1.0	2.3 or 3.4

6.4.3. Capacity of Digital Cellular CDMA

For evaluating the capacity of a CDMA system, first consider a single cell CDMA system. The cellular network consists of a large number of mobile users communicating with a base station. The cell site transmitter consists of linear combiner which adds spread signal of individual users for the forward channel. For a single cell system under consideration, these weighting factors can be assumed to be equal. A pilot signal is also included in the cell-site transmitter and is used by each mobile to set its own power control for the reverse link. For a single cell system with power control, all the signals on the reverse channel are received at the same power level at the base station.

Consider a single cell CDMA system. Let the number of users be N and the signal power from each of N users be S. Thus, the Signal-to-Noise (SNR) is,

$$SNR = \frac{S}{(N-1)S} = \frac{1}{(N-1)} \qquad \text{...(6.28)}$$

In addition to SNR, the bit energy to noise ratio is $\frac{E_b}{N_0}$ an important factor in communication systems as is given as

$$\frac{E_b}{N_0} = \frac{S/R}{(N-1)S/W} = \frac{W/R}{N-1} \quad ...(6.29)$$

where R is the baseband information bit rate, and W is the total RF bandwidth. The term W/R is the processing gain of the CDMA system. If background thermal noise n, is also considered the above equation becomes,

$$\frac{E_b}{N_0} = \frac{W/R}{(N-1)+(n/S)} \quad ...(6.30)$$

Number of users that can access the system is thus given as

$$N = 1 + \frac{W/R}{E_b/N_0} - \frac{n}{S} \quad ...(6.31)$$

Here, $\frac{E_b}{N_0}$ is the value required for adequate performance of the demodulator/decoder and for digital voice transmission, this implies a BER of 10^{-3} or better.

6.4.3.1. An Example on Comparison of CDMA, FDMA, TDMA Systems

At this stage, using the above equation, we can do a simple comparison of the CDMA system with the other multiple-access systems. Consider a bandwidth (W) of 1.25 MHz and a bit rate (R) of 8 kpbs using voice coders. Let's assume that a minimum $\frac{E_b}{N_0}$ of 5 (7 dB) is required to achieve adequate performance (BER of 10^{-3}). Ignoring the effect of the spurious interference and thermal noise, the number of users in the CDMA system (in 1.25 MHz bandwidth) works out to be

$$N = 1 + \frac{1.25\text{ MHz}/8\text{ kHz}}{5} = 32\text{ users} \quad ...(6.32)$$

On the other hand for a (single-celled) AMPS system (FDMA) operating over the same bandwidth, the number of users is given by :

$$\frac{125\text{ MHz}}{30\text{ kHz}} = 42\text{ users.}$$

For a D AMPS based 3-slot digital TDMA system.

The number of user will be 126 users.

Till now, the CDMA capacity is much less than that of other conventional systems (since the number of users is much less than the processing gain (W/R) of the system). However, it is important to consider the fact that we still haven't taken attributes like Voice Activity Detector (VAD), Sectoring, Frequency Reuse, etc. into account yet (which, as shown later, will increase the capacity by orders of magnitude). Note that, in a multi-cell AMPS system (with a frequency reuse factor of 7), the number of users per cell reduces from 42 to 6 (and a reduction from 126 to 18 in 3-slot TDMA) and thus the CDMA will show a capacity increase when compared to these systems.

One way of improving the CDMA capacity is the use of complicated modulation and channel coding schemes that reduce the E_b/N_0 requirement and increase capacity as shown by the Eqn. (6.5). The other way is to reduce the interference, which translates to an increase in the capacity according to Eqns. (6.3) and (6.4). We will now discuss the VAD (Voice Activity Detector) and antenna sectoring in reducing the effect of mutual interference in a CDMA system.

Antenna Sectorization : Any spatial isolation of users in a CDMA system translates directly into an increase in the system capacity. Consider an example where three directional antennas having 120° effective beamwidths are employed. Now, the interference sources seen by any of these antennas are approximately one-third of those seen by the omni-directional antenna. This reduces the interference term in the denominator of Eq. (6.4) by a factor of 3 and the number of users (N) is approximately increased by the same factor. Consider N_s to be the number of users per sector and thus the interference received by the antenna in that particular sector is proportional to N_s. The number of users per cell is approximately given by $N = 3N_s$.

Monitoring or Voice Activity Detection : The human voice activity cycle is 35 percent. When users assigned to a cell are not talking, VAD will allow all other users to benefit due to reduced mutual interference. Thus interference is reduced by a factor of 65 percent. CDMA is the only technology that takes advantage of this phenomenon. Voice Activity monitoring is a feature present in most digital vocoders where the transmission is suppressed for that user when no voice is present. Consider the term, voice activity factor (β), to be 3/8 (corresponding to the human voice activity cycle of 35-40 percent). The interference term in the denominator of Eq. (6.4) is thus reduced from (N – 1) to (N_s – 1) (In reality, the net improvement in the capacity will be reduced from 8/3 to 2 due to the fact that with a limited number of calls per sector, there is a non-negligible probability that an above average number of users are talking at once).

Thus, with VAD and Sectorization, the new average value of (E_b/N_0) within a sector is given as,

$$\frac{E_b}{N_0} = \frac{\frac{W}{R}}{(N_s - 1)\alpha + \left(\frac{\eta}{S}\right)} \quad \text{...(6.33)}$$

N_s is given by

$$\left[1 + \frac{1}{\alpha}\left\{\frac{\frac{W}{R}}{\frac{E_b}{N_0}} - \frac{q}{S}\right\}\right] \quad \text{...(6.34)}$$

The number of users is now :

$$N = 3N_s = 3\left[1 + \frac{1}{\alpha}\left\{\frac{\frac{W}{R}}{\frac{E_b}{N_0}} - \frac{n}{S}\right\}\right] \quad \text{...(6.35)}$$

For the same conditions and assumption discussed previously, the capacity of the CDMA system is now,

$$N = 3\times\left[1+\frac{8}{3}\left\{\frac{\frac{1.25\text{ MHz}}{8\text{ kbps}}}{5}\right\}\right] = 253\text{ users} \qquad \ldots(6.36)$$

That's works out to be a 8-fold capacity increase when compared to the previous case (Eq. 6.7) (without VAD and sectoring). In reality, due to the variability of E_b/N_0, the capacity increase has to be backed off to 5 or 6 times. Even this capacity increase is enough to bring the number of users much closer to the processing gain (W/R) of the system. This makes the CDMA capacity comparable to the TDMA and FDMA capacity. Again, it's important to note that these calculations are for a single-celled system, where frequency reuse considerations are not taken into account at all. The biggest advantage of CDMA comes from the fact that it can reuse the same frequencies in all the cells (unlike TDMA and FDMA). To take this into account, the CDMA capacity (for both forward and reverse links) has to be calculated for the multicell case, where additional interference is caused by the users in the adjacent cells.

Example 6.6. If $W = 1.25$ MHz, $R = 9600$ bps, and a minimum acceptable E_b/N_0 is 10 dB, determine the maximum number of users that can be supported in a single cell CDMA system using

(*a*) omni-directional base station antennas and no voice activity detection

(*b*) 3 sectors at base station and $P = 3/8$. Assume the system is interference limited.

Solution. (*a*) Using Eq. (6.5)

$$N = 1+\frac{\frac{(1.25\times10^6)}{9600}}{10} = 1+13.02 = 14$$

(*b*) Using Eq. (6.8) (Note $\eta = 0$)

$$N_s = 1+\frac{1}{\beta}\left\{\frac{\frac{W}{R}}{\frac{E_b}{N_0}}\right\} = 1+\frac{1}{\frac{3}{8}}\left\{\frac{\frac{1.25\times10^6}{9600}}{10}\right\} = 35.7$$

Total number of users is given by $N = 3\ Ns$ (since 3 sectors exist within a cell),

Therefore, $N = 3 \times 35.7 = 107$ users/cell

6.5. SUMMARY OF FDMA, TDMA AND CDMA

A comparison of FDMA, TDMA and CDMA multiple techniques is discussed below :

1. FDMA, TDMA and CDMA yield conceptually the same capacity
2. However, in wireless communications CDMA has improved capacity due to

(*a*) statistical multiplexing

(*b*) graceful degradation

3. Performance can still be improved by adaptive antennas, multiuser detection, Forward Error Correction (FEC), and multi-rate encoding.
4. TDMA allocates a time instant for a user

   FDMA allocates a frequency band for a user

   CDMA allocates a code for user
5. CDMA-system can be *synchronous* or *asynchronous* :
   - Synchronous CDMA cannot be used in multipath channels that destroy code orthogonality.
   - Therefore, in wireless CDMA-systems as in IS-95, cdma2000, WCDMA and IEEE 802.11 user are asynchronous.
6. Code classification :
   - Orthogonal, as Walsh-codes for orthogonal or near-orthogonal systems
   - Near-orthogonal and non-orthogonal codes :
     - Gold-codes, for asynchronous systems
     - Maximal length codes for asynchronous systems
7.

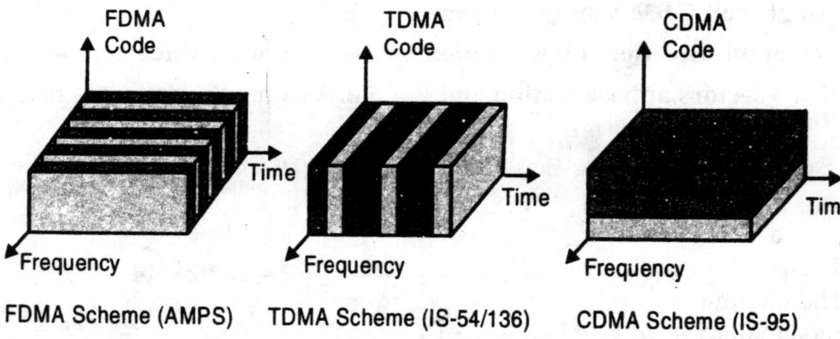

Fig. 6.6.

6.6. WIRELESS TDMA DESIGN CONSIDERATION FOR MOBILE PHONE

Before designing a TDMA structure, let us consider following set of requirements.

1. **Number of logical channels [number of time slot in TDMA frame]** 8, this appears to be the minimum to justify the additional costs of multiplexing.
2. **Maximum cell Radius (R)** 35 km, to give a sufficiently high traffic level in rural areas.
3. **Frequency** (900 MHz), this is commonly allocated to mobile radio applications.
4. **Maximum vehicle speed (V_m)** 250 km/hr, or 69.4 m/s, to accommodate mobile units on high-speed trains.
5. **Maximum coding delay** approximately 20 ms, to avoid adding unduly to delays within the fixed network, which may involve satellite links.

6. **Maximum delay spread (Δ_m)** 10 s (in mountainous regions); this is the difference in propagation delay among different multipath signals arriving at the same antenna.
7. **Bandwidth.** Not to exceed 200 kHz, corresponding to 25 kHz per channel (the current spacing for analog FM cellular systems in Europe).

Figure 6.9 suggests the step to be considered in designing the TDMA time slot. We use this as a guide in the following discussion.

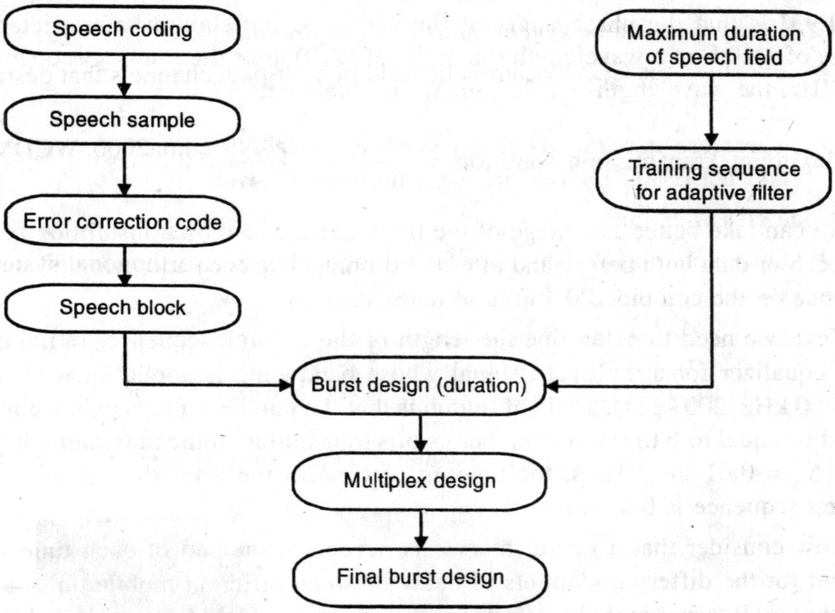

Fig. 6.9. Steps in design of TDMA time slot.

The speech coder must provide satisfactory speech quality at minimum data rate. The traditional form of speech coding to produce a digital bit stream is pulse code modulation (PCM), results in a data rate of 64 kbps. This rate is undesirably high for use in cellular radio. With current technology, a data rate of 12 kbps is reasonable for producing good-quality speech reproduction.

If we restrict the coding delay to 20 ms, then it would be acceptable to form the encoded speech into blocks of 20 ms duration, or speech samples of 240 bits.

Data at 12 kbps could also be blocked in 240-bits units. Error corrections can then be applied to the 240-bit blocks. For 2G systems, convolutional error connecting codes are commonly used with a code rate of 1/2.

This overhead raises the number of bits in a block to 480. In addition, there is constraint factor of *S*, meaning that 4-bits must be added to the data block to account for the length of the shift register.

This brings the speech block length to 488-bits. With the parameter choosen so for, The minimum bit rate for an 8 channel system is

$$= \frac{8 \text{ channels} \times 488 \text{ bits / channel}}{20 \times 10^{-3} \text{ s}} = 195.2 \text{ kbps.} \quad ...(6.37)$$

In fact, the bit rate will be somewhat higher to take care of other design consideration, discussed subsequently. This means that a data rate of greater than 200 kbps will need to be carried in the available bandwidth of 200 kHz. In practice, such data rates cannot be achieved without the use of adaptive equalization. As we know, in a mobile environment, adaptive equalization will require the inclusion of a new training sequence each time the mobile unit moves a distance sufficient to potentially cause changes in transmission path characteristics. Let us assume that a training sequence is included in each time slot. A rough criterion suggested in [JUNE93] is that the phase angle of the carrier signal should be restricted to a change of 1/20th of a wavelength (an angle of $\pi/10$) after the training sequence. At 900 MHz, the wavelength is 0.333 m. We can calculate

$$\text{Maximum Transmission Duration} = \frac{\lambda/20}{V_m} = \frac{0.333/20}{69.4} = 0.24 \text{ ms.} \quad \text{...(6.38)}$$

We can take better advantage of the training sequence by transmitting 0.24 ms of speech or data both before and after the training sequence and using the training sequence on the combined 0.48 ms of data.

Next, we need to determine the length of the training sequence. In the design of an equalizer for a multipath signal whose bandwidth is about equal to the bit rate (200 kHz, 200 kpbs), a rule of thumb is that the number of taps on the equalizer should be equal to 6 times the number of bits transmitted in the maximum dispersal time (Δ_m = 0.01 ms). Thus, the amount of time in the time slot devoted to the training sequence is 0.06 ms.

Now consider that a guard interval is needed at the end of each time slot to account for the differing amounts of delay between different mobile units and the base station. Because eight mobile units share the same TDMA frame, it is necessary to adjust the timing of the transmission of the mobile units so that the transmission from one unit does not interfere with adjacent time slots. It is the responsibility of the base station to provide synchronization information to each mobile unit to adjust relative delays to enforce the time slot structure of the TDMA frame. However, the mobile units may be moving relative to the base station and relative to each other, so a guard time is inserted in each time slot to account for these discrepancies. When a mobile first makes a connection through the base station, the base station can provide the mobile unit with the timing information based on the current propagation delay between the mobile unit and the base station. We would also like to add a guard time sufficient to avoid the need to frequently update this synchronization information. We can calculate the guard time as follows. The average telephone call is about 130 seconds, so the radial distance toward or away from the base station that a mobile could cover is (130s) × (69.4 m/s) = 9022 m. The change in propagation delay caused by a movement of this distance is 9022/ (3×10^8 m/s) = 0.03 ms.

Figure 6.10(*a*) shows the tentative time slot design. The next step is to fit a coded data block into a convenient number of time slots, together with the training sequence and guard bits. We have a maximum duration of a time slot of approximately 0.57 ms. With 8 time slots per frame, that gives a frame time of about 4.6 ms. We said that we wanted to send data with a coding delay of 20 ms, so if we round the frame time down to 4 ms (time slot = 0.5 ms), then we could conveniently send a block of speech in five successive slots on the same channel. A speech block consists of 488 bits, so each time slot would need to hold 488/5 or about 98 data bits. This yields a bit rate of 98/0.4 = 245 kbps. At this data rate, the minimum number of training bits required is (0.06 ms) × (245 kbps) = 14.7, which on rounding becomes 15 bits. Similarly, the minimum number of guard bits is (0.03 ms) × (245 kbps) = 7.35, which on rounding becomes 8 bits.

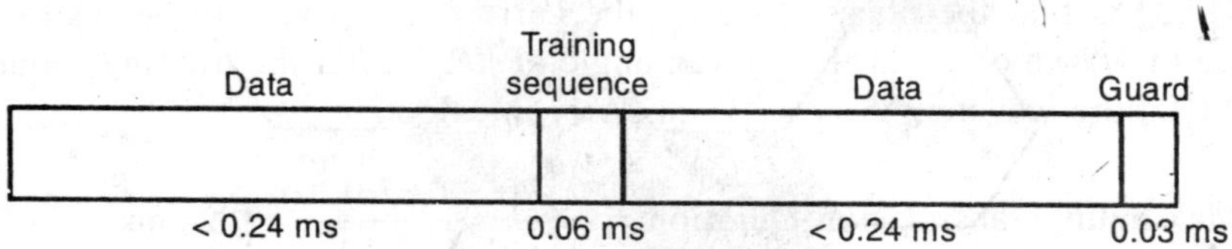

(*a*) Approximate field duration.

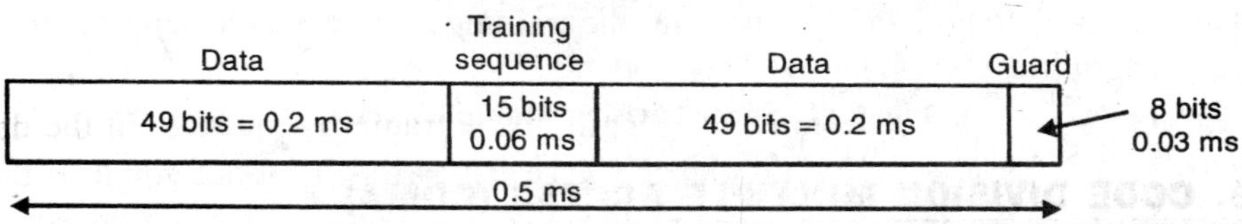

(*b*) Approximate field sizes.

Fig. 6.10.

The resulting frame structure is shown in Fig. 6.10(*b*), we have 121-bits transmitted in 0.5 ms for a channel bit rate of 242 kbps.

6.7. SPACE DIVISION MULTIPLE ACCESS (SDMA)

In 1946, Bell Telephone System planners started submitting proposals for a large-scale system that would satisfy the growing customer demand for more wireless access. The idea behind the proposals was to break a huge geographical region into smaller areas called "cells." Each cell would use a frequency different than those of its nearest neighbours to prevent any interference.

That is the idea behind the Space Division Multiple Access (SDMA), the same frequency can be used multiple times in the same geographical region.

The advantage to this technology is increased network capacity. The easiest way for FDMA broadcasters to increase their coverage area is to increase their transmitting power. However, increased power causes interference problems and increases the distance before a frequency can be reused. SDMA can increase coverage by adding more cells. Modern cellular uses higher frequencies and lower power. This causes less interference and reduces the frequency reuse distance. This technology emerged with the offering of Advanced Mobile Phone System (AMPS) in the early 1980s.

Although this was a big capacity improvement, it soon ran into its limits. The network planners made a few modifications to this design to increase capacity. One solution was to reduce the cell size even further and to add more cells to fill in the newly created uncovered areas. A second alternative was to add another frequency to the existing cell, so that two calls could be placed from the same cell. Both of these solutions, however, did not overcome the basic limit of one call per frequency.

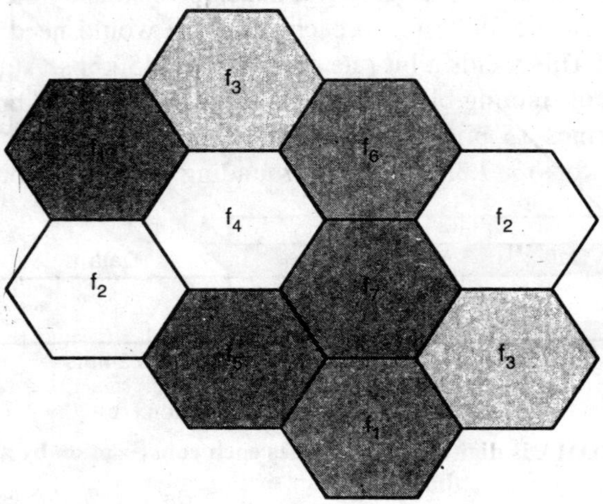

Fig. 6.11. Space Division Multiple Access.

6.8. CODE DIVISION MULTIPLE ACCESS (CDMA)

Code Division Multiple Access (CDMA) also uses digital format. In CDMA systems, several transmissions via the radio interface take place simultaneously on the same frequency bandwidth. The user data is combined at the transmitter's side with a code, then transmitted. On air, all transmission get mixed. At the receiver's side, the same code is used as in the transmitter's side. The code helps the receiver to filter the user information of the transmitter from the incoming mixture of all transmissions on the same frequency band and same time. This is often represented by layers, as can be seen in the figure below.

In contrast to classical FDMA and TDMA systems, the same carrier frequency band can be used in neighbouring cells. Frequency reuse factor in CDMA is one.

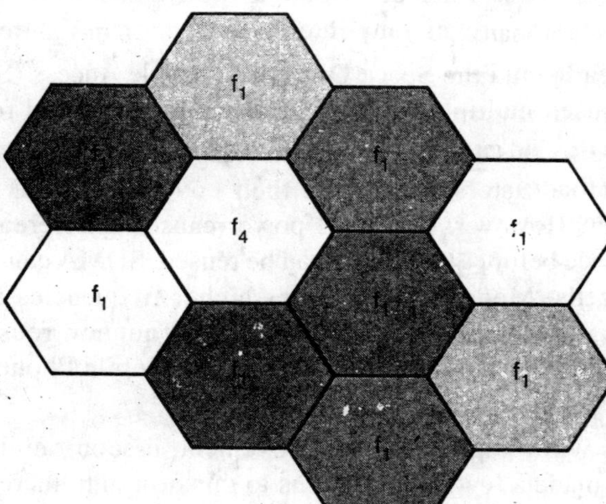

Fig. 6.12. Code Division Multiple Access.

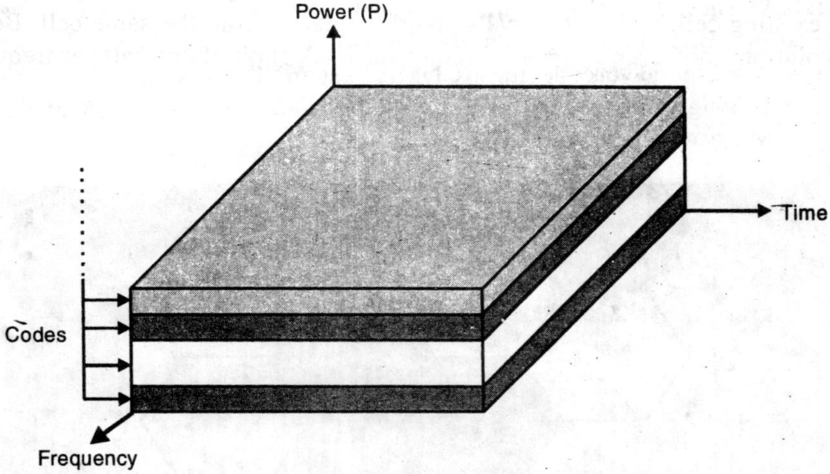

Fig. 6.13. CDMA is digital and identifies each conversation by a code rather than frequency or time slice.

6.8.1. CDMA Background

Code Division Multiple Access (CDMA) is a type of spread-spectrum; a family of digital communication methods that the military has used for some time dating back to World War II. It is particularly useful to the military for two reasons :

- It provides protection from enemy jamming, because the spread signal is difficult to interfere with.
- It can conceal that any communication is taking place.

Even though CDMA was hypothetically possible in the late 1940s, it was not available to the civilian market for another four decades. A primary reason for this was that low cost, high-density digital integrated circuits had to be developed to keep the cost and the weight of the units down.

6.8.2. Principles of CDMA

CDMA can be compared to a room with people speaking different languages.

Let us imagine that a corporate CEO is hosting a large multinational gathering. Our host, having mastered many languages, is primarily the one making the conversation. Our host demands that his guests speak in their native tongues. Our host, a true mediator, is able to interpret the conversations between guests if they wish to talk with each other; he can fluently follow several conversations at the same time. He can understand different speakers, all talking at the same time, because they speak in different languages.

He occasionally has to tell some guests, who tend to get carried away, to speak a little softer; and he has to ask the soft speakers to talk more loudly so that he can hear them better.

The party starts to mature and many more guests arrive. The overall volume begins to rise, because there are more people speaking at the same time. The host asks the guests nearest to him to speak more softly, while he asks the ones further away to please speak up.

CDMA functions are much like our party. The CEO hosting the party is our Base Station (BS) and the guests are the Mobile Stations (MS). The different languages correspond to codes in a CDMA system. The BS can tell the mobiles apart, even though they are transmitting at the same time, by the codes that they use. Each MS uses a separate code.

Fig. 6.14. The CDMA multinational gathering.

6.8.3. CDMA Information, Theory and Codes

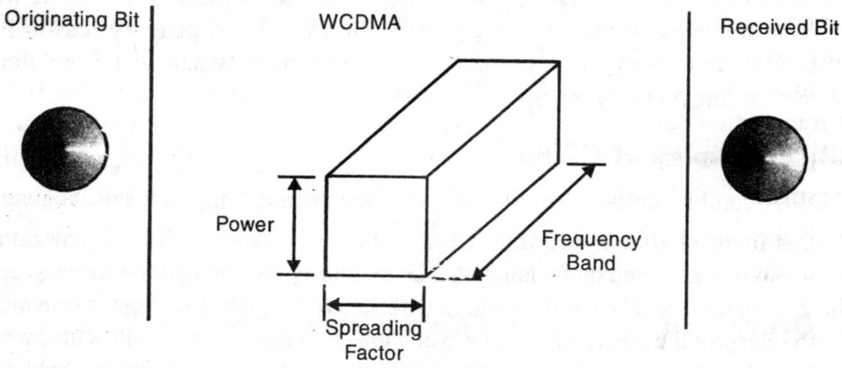

Fig. 6.15. General transmission principle.

In direct sequence CDMA, the transmission takes place continuously. If one user data bit has to be transmitted from the transmitter (*e.g.*, the mobile phone) to the receiver (*e.g.*, the base station), a certain amount of energy is required. The amount of energy depends on the distance of the transmitter from the receiver, the obstacles in the transmission path, etc. The energy can be represented like a "box" having specific *volume*. The energy/volume is constant—but the dimensions of the box can be change. As can be seen above, the box's volume is made of the frequency band, transmitter power, and time for the transmission. In UMTS, the frequency band is constant. The other two dimensions, power and duration for the transmission, are subject to change. A high data rates means many bits in one second, so the duration for each information bit is short. Consequently, the output power for each bit must be high to keep the box's volume at a specific, constant level. If the data rate goes down, less information bits are transmitted in one second, and therefore, the duration of one information bit is longer. If the energy for the transmission of the information bit has not changed, the volume of the box is the same. Consequently, less output power is required at the transmitter's side.

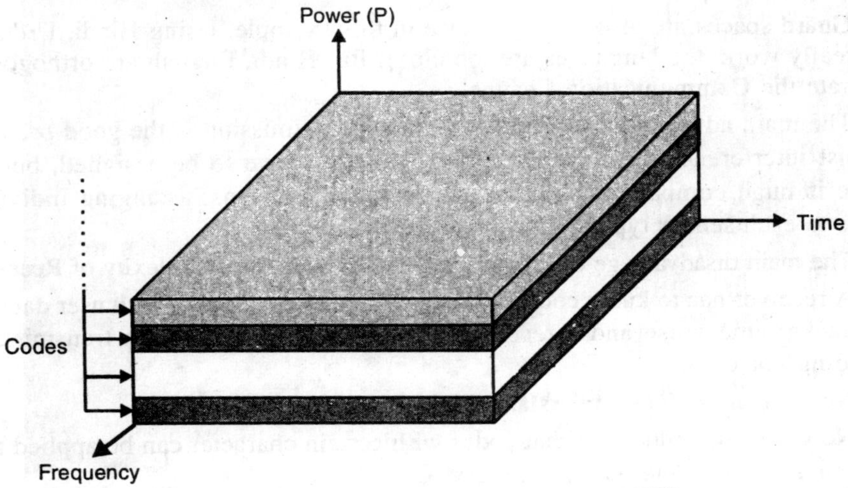

Fig. 6.16. CDMA (Code Division Multiple Access).

In direct sequence CDMA technology based systems (like WCDMA), every user is assigned a code/codes varying per transaction, *i.e.*, the different users use separate codes. These codes are called **spreading codes**. It should be noted that one user may also use several spreading codes in certain situations. The user data is directly multiplied with his code. This process is called spreading. Then the user data is transmitted via the common frequency band.

- If the originating bit rate is low, the power required for transmission is small. This kind of case can be seen as a narrow layer in Fig. 6.16.
- If the originating bit rate is high, is higher. This kind of case can be seen as a thick layer in Fig. 6.16.

6.8.4. Generation of Code for CDMA : Code Division Multiple Access

For better understanding of CDMA, there is need to understand CDM :

Code Division Multiplexing : Code division multiplexing (CDM) is a relatively new scheme in commercial communication systems. First used in military application due to its inherent security features. It is now also introduced in many civilian applications.

The typical everyday example of CDM is a function with many participants came from different countries around the world.

Many participants establish communication channels, they talk to each other, using the same frequency range (approx 300 to 6000 Hz) depending upon a Person's voice, at same time. If everybody speaks the same language, SDM (Space Division

Multiplexing) is needed to be able to communicate (standing in groups) talking with limited transmitting power.

But as soon as another code, another language is used one can tune into this language and clearly separate communication in this language from all other languages (the other languages appear as background noise). This example explains why CDM has built in security. If the language is unknown, the signals can still be received, but they are useless, by using a secret code (or language), a secure channel can be established in a 'hostile' environment. (At function this may cause some confusion).

Guard spaces are also of importance in this example. Using Hindi, Urdu does not really work, the languages are too close. But Hindi, English are orthogonal to separate the Communication Channel.

The main advantage of CDM for Wireless transmission is the good protection against interference and tapping. Different codes have to be assigned, but code space is huge compared to the frequency space and thus, assinging individuals codes to each sender typically does not cause problem.

The main disadvantage of CDM is that it is having too complexity of Receiver.

A receiver has to know code and must separate the channel with user data from the background noise and a receiver must be synchronised with transmitter for decoding correctly.

Now we are able to understand CDMA.

Now we can understand that codes with certain character can be applied to the transmission to enable the use of CDMA.

CDMA systems use exactly these codes to separate different users in code space and to enable access to a shared medium without interference.

The main problem is how to find 'good' codes and how to separate the signal from noise generated by other signals and the environment.

To better understanding of CDMA, let us elaborate an example :

- Two senders X and Y want to send data, CDMA assigns the following key sequences.

  Key $X_K = 01001$

  $Y_K = 11010$

  Sender X wants to send the bit $X_d = 1$, sender Y sends $Y_d = 0$.

  Let us assume that we code a binary 0 as –1 and 1 as +1.

- Now both senders spread their signals using their key as sequence. After coding

  Sender X sends $X_s = X_d * X_K$

  $= +1 * (-1, +1, -1, -1, +1)$

  $= (-1, +1, -1, -1, +1)$

  Sender Y sends $Y_s = Y_d * Y_K$

  $= -1 * (+1, +1, -1, +1, -1)$

  $= (-1, -1, +1, -1, +1)$

Both signals are then transmitted at the same time using the same frequency, thus the signals superimpose in space.

Neglecting interference, noise from other sources for simplicity of example and assuming signal received at receiver of same strength,

So at receiver $R = X_s + Y_s = (-2, 0, 0, -2, +2)$

The receiver now wants to receive data from sender X and therefore, tunes into the code of X, so applies X's code for despreading

$$R * X_K = (-2, 0, 0, -2, +2) * (-1, +1, -1, -1, +1)$$
$$= 2 + 0 + 0 + 2 + 2 = 6$$

It is much larger than zero so the receiver detects a binary 1.

For detecting Y

$$\Rightarrow \quad R * Y_K = (-2, 0, 0, -2, +2) * (+1, +1, -1, +1, -1)$$
$$= (-2, -2, -2) = -6$$

It is very much less than zero so, on receive, it will detect to be zero.

6.9. BLUETOOTH

Bluetooth is a type of Local Area Network with a very limited coverage area. Generally this type of network is used to connect small electronic devices without wire. Many of today's devices already offer an Infrared Data Association (IrDA) interface with transmission rates of 115 Kbit/sec.

Problems of IrDA involve its very limited range (typically upto 5 meter like T.V. remote) and line of sight communication between the interfaces and another limitation at a time only two devices can connect to each other.

Mean IrDA Communication support only point to point communication. And also IrDA has no ability to work with internet, or to access the media devices.

One big advantage is also there, the low cost of IrDA mechanism, so it is found in almost any mobile devices (Cellular Phones, PDAs, Laptops).

After doing some R&D and taken initiative by Ericsson to form a group of Five companies (Ericsson, Intel, IBM, Nokia, Toshiba) to do work in the direction of achieving multi-communication link technologies, simply a technology that can remove the limitation of IrDA Communication.

Later this group named SIG (Special Interest Group) and founded the Bluetooth consortium. Bluetooth represents a single chip, low cost, radio based wireless network technology. Many other companies and research institutions joined the special interest group around Bluetooth (1998), whose goal is the development of mobile phones, laptops, Notebooks, Headsets, Palmtops etc., including Bluetooth technology by the end of 1999. Main aim of SIG to create a technology which is standarised globally, so we can say Bluetooth is one of that technology which established and standardised by industry and promoted by the Bluetooth consortium.

Today User Scenarios: Nowadays, most devices are connected to a Desktop computer via wires, likely keyboard, mouse, speaker earphones etc.

This type of connection involves various disadvantages. Each device has its own type of cable, different plugs are needed and somewhere space is also a problem. But if you are using a wireless technologies then there is no need of wire, plugs and upto an extent space problem can also be solves. Just try to imagine in a seminar or conference. Several people coming together, discussing issues, exchanging data.

The persons who are giving seminar using a laptop (Bluetooth enabled) and if the audians having an electronic Bluetooth device they can exchange their data same time and simultaneously.

Using wireless Piconets, a mobile phone can be connected to a PDA or laptop in a simple way. Mobile phones will not have full WLAN adapters built in, but could have a bluetooth chip.

Then the mobile phone can access a bridge between the local piconet and the GSM network (see Fig. 6.17). For instance, on arrival at an airport, a Person's mobile phone could already receive e-mail via GSM and forward it to the laptop which is still in the suitcase or via a Piconet, a file server could update local information stored on a laptop or PDA while the person is walking into office. When comparing Bluetooth with other WLAN technology we have to keep in mind that one of the goals of Bluetooth was to provide local wireless access at very low costs.

From a technical point of view, WLAN technologies like those presented above could also be used, however, WLAN for IEEE 802.11, have been designed for higher bandwidth and larger range and are thus more expensive.

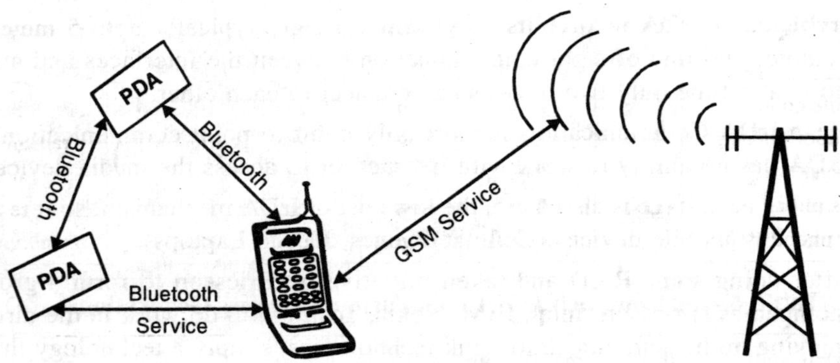

Fig. 6.17. PDA (Personal Digital Assistants).

6.10. COMPARISON BETWEEN WIRELESS COMMUNICATION SYSTEM

Table 6.4(*a*) and 6.4(*b*) illustrate the type of services, if we analyse critically, we see that all five mobile systems use a fixed base system.

For example, cordless telephones use fixed base stations so that they may be plugged into the telephone line supplied by the phone company.

The radio link between the cordless phone base-station and the Portable handset is designed to behave identically to the coiled cord, connecting a traditional wired telephone handset to the telephone carriage. But note that each wireless system has its own application and every wireless system has its infrastructure design. For the case of low power like hand-held cellular phones, a large number of base stations are required to ensure that every phone must be in close range to a base station within area.

If base stations were not within close range, a great deal of transmitter power would be required of the phone, which limited the battery life and made the service useless for hand-held users.

Table 6.4(*a*). Comparison of Mobile Communication Systems for Mobile Station.

Service	Coverage range	Required infra-structure	Complexity	Hardware cost	Carrier frequency	Functionality
TV remote control	Low	Low	Low	Low	Infrared	Transmitter
Garage door opener	Low	Low	Low	Low	<100 MHz	Transmitter
Paging system	High	High	Low	Low	<1 GHz	Receiver
Cordless phone	Low	Low	Moderate	Low	1-3 GHz	Transceiver
Cellular phone	High	High	High	Moderate	<2 GHz	Transceiver

Table 6.4(*b*). Comparison of Mobile Communication Systems for Base Station.

Service	Coverage range	Required infra-structure	Complexity	Hardware cost	Carrier frequency	Functionality
TV remote control	Low	Low	Low	Low	Infrared	Receiver
Garage door opener	Low	Low	Low	Low	<100 MHz	Transmitter
Paging system	High	High	High	High	<1 GHz	Transmitter
Cordless phone	Low	Low	Low	Moderate	1-3 GHz	Transceiver
Cellular phone	High	High	High	High	<2 GHz	Transceiver

Table 6.4(*c*). Comparison between Various Wireless Generations.

Generation	Standard Name	Bandwidth	Speed	Access Technology
2G	GSM	25 MHz	≃ 9.6 Kbps	TDMA
2.5G	GPRS	25 MHz	≃ 53 Kbps	TDMA
2.75G	EDGE	25 MHz	≃ 473 Kbps	TDMA
3G	WCDMA	60 MHz	≃ 2 Mbps	CDMA

GSM (Global System for Mobile Communication)

GPRS (General Packet Radio Service)

EDGE (Enhanced Data for Global Evolution)

WCDMA (Wideband Code Division Multiple Access)

As we know from previous discussion 1G was an analog system but when demand increase and technology evolved then second generation came into picture which was based on TDMA and FDMA and it was a complete digital technology. 2G was very much compatible with voice transmission but not with data services like, Fax, Picture, Video or Internet browsing. Then 2.5G came into picture to resolve this limitation but when Internet became very much popular then people started to demand more speed to access internet or phone and to see live program on mobile phone. Again cellular system went for major changes and new technology has been implemented over existing architecture of GSM. This is called 2.75G generation and named EDGE. But again speeds of accessing data was not much satisfactory particularly for internet browsing, then authorities, who are involved with development of wireless technology, decided to implement new technologies over GSM architecture and called 3G, which is named WCDMA.

Here it is important to note that CDMA-2000 also comes in 3G but network operator of CDMA is not using all features of 3G and secondary bandwidth of CDMA-2000 is only 1.25 MHz/channel.

But WCDM is a 3rd Generation technology which is based on CDMA but implemented over GSM architecture not by replacing existing GSM architecture and having bandwidth of 5 MHz/channel.

Here we can note that various generations evolved only to increase data speed otherwise speed (9.6 Kbps) for voice is very much sufficient. Speed for voice transmission and for data depends upon mobile handset and network parameter.

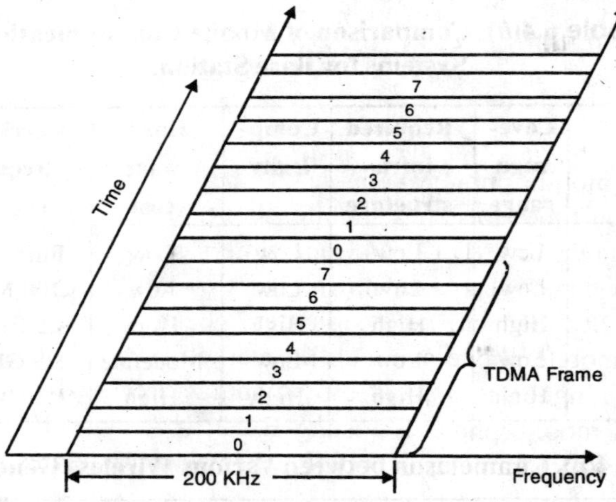

Fig. 6.18. TDMA channel.

As we know GSM is based on TDMA and FDMA, firstly whole bandwidth of GSM is divided into channels and then channels are divided into Time slots. How many subscribers can talk simultaneously in a network it depend upon availability of channel and time slot.

Time slot is parameter of mobile handset and coding scheme is parameter of network operator which decide speed of transmission and reception for voice and data that you can get.

Let us take GSM 900 which is having bandwidth of 25 MHz.

Uplink frequency 890–915 MHz

Downlink frequency 935–960 MHz

Both uplink and downlink frequency can divide in bandwidth of 200 KHz.

$$\text{Total number of channels} = \frac{25 \text{ MHz}}{200 \text{ KHz}} = 125 \text{ channels}$$

Each channel is divided into 8 time slots.

$$\text{TDMA frame} = 8 \text{ time slots}$$

One time slot duration = 0.577 ms

Total duration of TDMA frame= 8 × 0.577 ms = 4.615 m second.

Time slot is a type of physical channel on which signal is transmitted and received. If a person gets second time slot by network then it will be continued until and unless conversation ends. Then this will be released and allocated to any other user.

If all 125 channels and all time slots are allocated to a single network operator and operator is not using concept of frequency re-use then in a cell

Maximum number of subscribers that can talk simultaneously

= 125 × 8 = 1000 subscribers

But practically only 7 time slots allocated for traffic (voice, picture etc.). One time slot is reserved for signalling purposes.

As we know speed to access data or voice depends upon time slot and coding scheme and it is classified as follows :

Mobile class	Supported time slot in (D.L. + U.L.)
Class 1	2 + 1
Class 2	2 + 2
Class 3	3 + 1
Class 4	3 + 2
Class 5	4 + 1
Class 6	4 + 2

Technically mobile phone available in market is classified by mobile class which is based on time slots. If a mobile phone is of class 1 then it can support 2 time slots in down link (D.L.) and one time slot in uplink (U.L.). It means if your mobile phone is of class 1 then if network operator allocate to you more than 2 time slots in D.L. that will be meaningless because your mobile phone cannot support. Latest mobile phone available in market can support (4 + 1) time slots. But class may be higher, it is decided by manufacturer.

Well known mobile phone Samsung E250 is of class 10 but it can support (3 + 1) only. Second parameter is coding scheme which relates to speed directly.

Coding scheme	Speed
CS-1	9.6 kbps
CS-2	13.4 kbps
CS-3	15.6 kbps
CS-4	21.4 kbps

If network operator implemented CS-2 and your mobile phone supports (4 + 2) then

Speed in downlink is = 13.4 × 4 = 53.6 kbps

And in uplink = 13.4 × 2 = 26.8 kbps

But if network operator implements CS-4 and your mobile phones of class (8 + 8) then

Maximum speed that you can get

8 × 21.4 = 171.2 kbps

In India Network operators are using CS-1 and CS-2 only.

If Network operator is providing EDGE services then some another coding scheme is required called MCS1–MCS9. Where MCS stands for modified coding scheme. CS1–CS4 can be implemented for upto GPRS services only. In MCS9, we can get 59.2 kbps per time slot.

So maximum speed that you can get

$$59.2 \times 8 = 473 \text{ kbps}$$

Note that for voice conversion. We require CS-1 and one time slot only. So for voice conversation speed is equal to = 9.6 kbps × 1 = 9.6 kbps

Table 6.5. Multiple Access Technique for Different Traffic.

Type of Traffic	Multiple Access Technique
Bursty, short messages	Contention protocols
Bursty, long message,	Reservation protocols with fixed
Small number of users	TDMA reservation channel
Stream or Deterministic (Voice)	FDMA, TDMA, CDMA

Table 6.6. Comparison of SDMA, TDMA, FDMA and CDMA Mechanisms.

S.No.	Approach	SDMA	TDMA	FDMA	CDMA
1.	**Idea**	Segment space into cells/sectors	Segment sending time into disjoint time-slots, demand driven or fixed patterns	Segment the frequency band into disjoint sub-bands	Spread the spectrum using orthogonal codes
2.	**Terminals**	Only one terminal can be active in one cell/one sector	All terminals are active for short periods of time on the same frequency	Every terminal has its own frequency uninterrupted	All terminals can be active at the same place at the same moment, uninterrupted
3.	**Signal separation**	Cell structure/ directed antennas	Synchronization in the time domain	Filtering in the frequency domain	Code plus special receivers
4.	**Advantages**	Very simple, increases capacity per km^2	Established, fully digital, very flexible	Simple established, robust	Flexible, less planning needed soft handover
5.	**Disadvantages**	Inflexible, antennas typically fixed	Guard space needed (multipath propagation), synchronization difficult	Inflexible, frequencies are a scarce resource	Complex receivers, needs more complicated power control for senders
6.	**Comment**	Only in combination with TDMA, FDMA or CDMA useful	Standard in fixed networks, together with FDMA/SDMA used in many mobile networks	Typically combined with TDMA (frequency hopping patterns) and SDMA (frequency reuse)	Still faces some problems, higher complexity, lowered expectations; will be integrated with TDMA/FDMA.

If your mobile phone is GPRS enabled and network operator is also providing this GPRS service then you can send receive MMS (Multimedia Messaging Service), and you can browse Internet but with maximum speed of 53 kbps but if EDGE enabled phone and operator providing this service then these services can be accessed with 473 kbps.

It is important to note in uplink (mobile phone to network), we require less time slots because we just send request to network to access a particular service while in downlink (network to mobile phone), we require more time slots because we load these MMS services from network.

Example 6.7. In a GSM system which uses a TDMA/FDD system that uses 30 MHz for the forward link, and is broken into radio channels of 150 kHz. If 7 speech channels are supported on a single radio channel, and if no guard band is assumed, find the number of simultaneous users that can be accommodated in GSM cellular system.

Solution. Given data,

Forward link frequency = 30 MHz (f_l)

Channel bandwidth = 150 kHz (f_c)

No. of speech channels = 7 (n)

Number of simultaneous users

$$= \frac{nf_l}{f_c} \Rightarrow \frac{7.30 \times 10^3 \times 10^3}{150 \times 10^3}$$

= 1400 users. **Ans.**

Example 6.8. If a GSM cellular operator is allocated 25 MHz for each simplex band and if B_{guard} = 20 kHz, B_t is 25 MHz, B_C is 50 kHz, and 1000 users supported on each radio channel. Find number of TDMA channel.

Solution. Given : $B_g = 20$ kHz, $B_t = 25$ MHz, $B_C = 50$ kHz, $m = 1000$

$$N = \frac{m(B_t - 2B_g)}{B_C}$$

$$= \frac{1000(25 \times 10^6 - 2 \times 20 \times 10^3)}{50 \times 10^3}$$

$= 499 \times 10^3$ channels. **Ans.**

SUMMARY

1. In this chapter, we became familier with different access techniques used for Mobile Communication like TDMA, FDMA, CDMA.
2. Frequency Division Duplexing provides two different frequencies of two different users. As number of users increases, number of channels increases.
3. Time Division Duplexing provides user to access the system in specified time slots. So in TDD, multiple users share a single radio channel by using different time slots in same channel.

4. FDD provides two simplex channels at same time while TDD provides two simplex time slots on the same frequency.
5. FDMA, TDMA, CDMA are the three major access techniques used to share the available bandwidth in a wireless communication system.
6. CDMA allows all of the transmitters to access the channel at the same time.
7. Instead of TDMA, FDMA, CDMA, two other multiple access technique packet radio and space division multiple access (SDMA) also used for accessing a cellular system.
8. Spread spectrum multiple access (SSMA) uses signals which have a transmission bandwidth that is several orders of magnitude greater than the minimum required RF bandwidth.
9. SSMA is classified in two parts, frequency hopped multiple access (FH) and direct sequence multiple access. DS is also called CDMA.
10. Space division multiple access works on the principle to control the energy radiated by antenna in different directions.

REVIEW QUESTIONS

1. What geometric shape is used in cellular system design and why ?
2. What is the principle of frequency re-use in the context of a cellular network ?
3. Explain the paging function of a cellular system.
4. List five ways of increasing the capacity of a cellular system.
5. As a mobile unit in communication with base station moves, what factors determine the need for power, Control the amount of power adjustment ?
6. What are the advantages and disadvantages of CDMA cellular systems ?
7. What are some key characteristics that distinguish third generation cellular system from second generation cellular systems ?
8. What are the different techniques used to access the cellular system described in brief ?
9. What do you mean by term spread spectrum. What are their features ?
10. Explain the spread spectrum, with its application working etc.
11. Make a list comparing the different access technique—TDMA, CDMA, FDMA.
12. Why we prefer TDMA in digital cellular system ?
13. What do you mean by space division multiple access, how is it different with respect to other access system ? Give its one application.

NUMERICAL BASED PROBLEMS

1. For a cellular system, FDMA spectral efficiency is defined as $\eta_a = \dfrac{B_c N_t}{B_w}$,

   where $B_c \Rightarrow$ Channel bandwidth

   $B_w \Rightarrow$ Total bandwidth in one direction.

   $N_t \Rightarrow$ Total number of voice channel in the covered area.

   (*a*) What is an upper bound of η_a ?
   (*b*) Determine η_a for the system.
2. Consider four different cellular systems that share the following characteristics. The frequency bands are 825 to 845 MHz for mobile unit transmission and 870 to 890 MHz for base station transmission.

A duplex circuit consists of one 30 kHz channel in each direction. The systems are distinguished by the re-use factor, which is 4, 7, 12 and 19, respectively.

(*a*) Suppose that in each of the systems, the cluster of cells (4, 7, 12, 19) is duplicated 16 times. Find the number of simultaneous communications that can be supported by each system.

(*b*) Find the number of simultaneous communications that can be supported by a single cell in each system.

(*c*) What is the area covered, in cells, by each system ?

(*d*) Suppose the cell size is the small in all 4 systems and a fixed area of 100 cells is covered by each system. Find the number of simultaneous communications that can be supported by each system.

3. Consider a 7-cell system covering an area of 3100 km^2. The traffic in the seven cell is as follows :

Cell number	1	2	3	4	5	6	7
Traffic (Erlangs)	30.8	66.7	48.6	33.2	38.2	37.8	32.6

Each user generates an average of 0.03 Erlangs of traffic per-hour, with a mean holding time of 120 sec. The system consists of a total of 395 channels and is designed for a grade of service of 0.02.

(*a*) Determine the number of subscribers in each cell.

(*b*) Determine the number of cells per hour per subscriber.

(*c*) Call per-hour in each cell.

(*d*) Determine the total number of subscribers.

(*e*) Determine the average number of subscribers per channel.

(*f*) Determine the subscriber density per km^2.

(*g*) Determine the total traffic (total Erlangs)

(*h*) Determine the Erlangs per km^2.

(*i*) What is radius of each cell ?

4. The GSM TDMA system uses a 270. 833 kbps data rate to support eight users per frame.

(*a*) What is the raw data rate provided for each user ?

(*b*) If guard time, ramp up time, and synchronization bits occupy 10.1 kbps, determine the traffic efficiency for each user.

5. The US Digital cellular TDMA system uses a 48.6 kbps data rate to support three users per frame. Each user occupies two of the six time slots per frame. What is the raw data rate provided for each user ?

❑❑❑

7

CHAPTER

Second Generation, Digital and Wireless Systems

7.0. INTRODUCTION

In previous chapters, we have discussed many types of wireless techniques and standards. In this chapter, we will discuss some more wireless devices, which belong to 1 G and 2 G and 3 G systems. As we know AMPS belongs to 1 G technologies. Here we will discuss more parameter about AMPS.

Wireless system and standards are related to our life very much because if we decide to purchase a wireless device then it must be standardized world-wide, so that we can use everywhere. In this chapter we are going to discuss AMPS, GSM, CDMA, IS-95 and various International standards related to these devices.

INSIDE THIS CHAPTER

7.0. Introduction
7.1. AMPS and ETACS
7.2. United States Digital Cellular (IS-54 and IS-136)
7.3. Introduction to GSM, the Global System for Mobile Communication
7.4. Setup Channels
7.5. Frame Structure of GSM
7.6. Signal Processing in GSM
7.7. Interim Standard 95 (IS-95)
7.8. Difference Between GSM and CDMA
7.9. Personal Communication Access System (PACS)
7.10. Personal Digital Cellular/Pacific Digital Cellular (PDC)
7.11. Personal Handy-Phone System
7.12. The Industrial, Scientific and Medical (ISM Band)
7.13. Wireless Personal Area Network (WPAN)
7.14. Network Database
7.15. Spectrum Allocation
7.16. Frequency Hopping
7.17. Walsh Code
7.18. CDMA (Code Division Multiple Access)
7.19. Design Concept of CDMA
7.20. Direct Sequence CDMA (DS-CDMA)
7.21. 3G (WCDMA) Architecture
7.22. Overview of Handover in Mobile Networks
7.23. Vocoders
7.24. Dropped Call Rate

7.1. AMPS AND ETACS

AMPS is refered as Advance Mobile Phone Service or Analog Mobile Phone Service. It is first mobile cellular service belonging to first generation technology. In the late 1970, AT&T Bell Laboratories developed a cellular system called AMPS. AMPS was first implemented in the chicago, U.S.A. Initially FCC (Federal Communication Commission) allocated 40 MHz of Bandwidth in the 800 MHz band but when number of subscribers increased then in 1989 FCC allocated an additional 10 MHz for AMPS communications.

The first AMPS cellular system used large cells and ominidirectional antenna at the base stations to minimize the initial equipment needs.

AMPS system was based on $N = 7$ cell reuse pattern for cell splitting and sectoring purposes to increase the cellular capacity. It was found that the AMPS 30 kHz channel required a signal to interference ratio (SIR) of 18 dB for satisfactory system performance. $N = 7$ was the smallest reuse factor which satisfied the requirement of 18 dB (SIR) by using 120 degree directional antennas.

ETACS (European Total Access Communication System) was developed in the mid-1980 and technology of ETACS was identical to the AMPS. The difference is that, it is implemented in Europe with 25 kHz channel oppose to 30 kHz channel of some another parameter of AMPS and ETACS as shown in Table 7.1.

7.1.1. Frequency Bands

AMPS cellular service operates in the 800 MHz "Cellular" FM band. For each market area, the Federal Communications Commission (FCC) allowed two licensee (networks) known as "A" and "B" carriers. Each carrier within a market uses a specified "block" of frequencies consisting of 21 control channels and 395 voice channels. Originally, the B (wireline) side license was usually owned by the local phone company such as a "Baby Bell" (Ameritech), and the A (non-wireline) license was made available to private companies such as Cellular One. At the inception of cellular in 1983, the FCC had granted each carrier within a market 333 channels (666 channels total). By the late 1980's, the cellular industry's subscriber base had grown into the millions across America and it became necessary to add channels for additional capacity. In 1989, the FCC granted carriers an expansion from the current 666 channels to the now 832 (416 per carrier). The additional frequency was available in the upper 800 MHz band which also was home to UHF channels 70-83. This meant that these UHF channels could no longer be used for UHF TV transmission as these frequencies were to be used for AMPS transmission.

The anatomy of each channel is composed of 2 frequencies. 416 of these are in the 824-849 MHz range for transmissions from mobile stations to the base stations, paired with 416 frequencies in the 869-894 MHz range for transmissions from base stations to the mobile stations. Each cell site will use a subset of these channels, and must use a different set than neighbouring cells to avoid interference. This significantly reduces the number of channels available at each site in real-world systems. Each AMPS frequency is 30 kHz wide.

7.1.2. Introduction of Digital TDMA

Later, many AMPS networks were partially converted to D-AMPS, often referred to as TDMA (though **TDMA** is a generic term that applies to many cellular systems). D-AMPS is a digital, 2G standard used mainly by Cingular Wireless and U.S. Cellular.

7.1.3. Introduction of GSM and CDMA

AMPS and D-AMPS ("TDMA") are now being phased out in favour of either CDMA2000 or GSM which allow for higher capacity data transfers for services such as WAP, Multimedia Messaging System (MMS), and wireless Internet access. There are some phones capable of supporting AMPS, D-AMPS and GSM all in one phone (using the GAIT standard).

7.1.4. The Future of AMPS

In 2002, the FCC decided to no longer require A and B carriers to support AMPS service. Since the AMPS standard is analog technology, it suffers from an inherently inefficient use of the frequency spectrum. All AMPS carriers have converted most of their consumer base to a digital standard such as CDMA2000 or GSM and continue to do so at a rapid pace. Digital technologies such as GSM and CDMA2000 support multiple voice calls on the same channel and offer enhanced features such as two-way text messaging and data services.

7.1.5. Analog System in Europe

Total Access Communication System or TACS is the European version of AMPS. ETACS was an extended version of TAGS with more channels. TAGS and ETACS are now obsolete in Europe, having been replaced by GSM system. In the United Kingdom, the last ETACS service operated by Vodafone was discontinued on 31 May 2001, after sixteen years of service. ETACS is, however, still in use in a handful of countries elsewhere in the world. NMT is another analog cellular standard that was widely used in Europe, mainly in the Nordic countries, which has now been fully replaced by GSM except for limited use in rural areas due to its superior range.

7.1.6. AMPS and ETACS Specifications

There are various parameters of AMPS and ETACS, which are different to each other. AMPS and ETACS are same technology but their implementation in two different areas is different.

Table 7.1. AMPS and ETACS Specifications.

Parameter	ETACS	AMPS
1. Channel bandwidth	25 kHz	30 kHz
2. Forward channel frequency	935 - 960 kHz	869 - 894 kHz
3. Reverse channel frequency	890-915 kHz	824-849 kHz
4. Number of channel	1000	832
5. Voice modulation	FM	FM (Frequency Modulation)
6. Access technique	FDMA	FDMA
7. Spectral efficiency	0.33 bps/Hz	0.33 bps/Hz
8. Duplexing technique	FDD	FDD
9. Data rate on control channel	8 Kbps	10 Kbps
10. Traffic channel per RF channel	1	1

7.2. UNITED STATES DIGITAL CELLULAR (IS-54 AND IS-136)

As we know from previous discussion, AMPS was a 1 G system based on Analog technology. AMPS was not designed to support the increasing subscribers day by day. Therefore, there is need for improvement in existing technology.

After extensive research and comparison of various parameters in cellular system and keeping view of industry requirements, in late 1980 s, the United States Digital Cellular System (USDC) was developed to support more users in a fixed spectrum location.

USDC accessing technique was based on TDMA that support three full rate users or six half rate users on each AMPS channel. In such a way USDC supports six time more capacity than AMPS system. The dual mode USDC/AMPS system was standardized as Interim Standard 54 (IS-54) in 1990 by Electronic Industry association and (EIA) and by Telecommunication Industry association (TIA).

In coming year, with some improvement it is known as IS-136. USDS is installed in Canada and it is also known as NADC. (North American Digital Cellular).

USDC system was supposed to use same standard of AMPS like Frequency reuse plan and base stations standard. It became beneficial for subscribers also because base station and subscriber unit were equipped with AMPS and USDC both.

7.2.1. IS-54 and IS-136 (D-AMPS)

IS-54 and **IS-136** are second-generation (2G) mobile phone systems, known as **Digital AMPS** (D-AMPS). It is used throughout America, particularly in the United States and Canada. D-AMPS is considered end-of-life, and existing networks are in the process of being replaced by GSM/GPRS and CDMA2000 technologies.

Although this system is most often referred to as **TDMA,** this name is based on the acronym for time division multiple access, a common multiple access technique which is used by multiple protocols, including GSM, as well as in IS-54 and IS-136. The two different uses of this term can be confusing. TDMA (the technique) is also used in the GSM standard. However, TDMA (the standard, *i.e.,* IS-136) has been competing against GSM and systems based on Code division multiple access (CDMA) for adoption by the network carriers, although it is now being phased out in favour of GSM technology.

D-AMPS uses existing AMPS channels and allows for smooth transition between digital and analog systems in the same area. Capacity was increased over the preceding analog design by dividing each 30 kHz channel pair into three time slots (hence *time division)* and digitally compressing the voice data, yielding three times the call capacity in a single cell. A digital system also made calls more secure because analog scanners could not access digital signals. Calls were encrypted, although the algorithm used (CMEA) was later found to be weak.

IS-136 added a number of features to the original IS-54 specification, including text messaging, circuit switched data (CSD), and an improved compression protocol. SMS and CSD were both available as part of the GSM protocol, and IS-136 implemented them in a nearly identical fashion.

Large IS-136 networks include Cingular Wireless and U.S. Cellular in the United States, and Rogers Wireless in Canada. Cingular (who purchased AT&T

Wireless in October 2004) and Rogers Wireless have upgraded their existing IS-136 networks to GSM/GPRS, while US Cellular is migrating most of their network to CDMA2000. Rogers Wireless removed all 1900 MHz IS-136 in 2003, and is slowly doing the same with their 800 MHz spectrum as the equipment fails. It is not unusual for a Rogers D-AMPS customer to have coverage in a specific location one day, and then none in that same area the next.

IS-54 is the first mobile communication system which had provision for security, and the first to employ TDMA technology.

The technical specifications can be summarized as below:

Table 7.2. IS Specifications.

Mobile Frequency Range	Rx: 869-894 kHz; Tx: 824-849 kHz
Multiple Access Method	TDMA/FDM
Duplex Method	FDD
Number of Channels	832 (3 users per channel)
Channel Spacing	30 kHz
Modulation	DQPSK
Channel Bit Rate	48.6 Kb

7.2.2. IS-136

A pragmatic effort was launched to improve IS-54 that eventually added an extra channel to the IS-54 hybrid design. Unlike IS-54, IS-136 utilizes time division multiplexing for both voice and control channel transmissions. Digital control channel allows residential and in-building coverage, dramatically increased battery standby time, several messaging applications, over the air activation and expanded data applications. IS-136 systems needed to support millions of AMPS phones, most of which were designed and manufactured before IS-54 and IS-136 were considered. IS-136 added a number of features to the original IS-54 specification, including text messaging, circuit switched data (CSD), and an improved compression protocol. IS-136 TDMA traffic channels use 7t/4-DQPSK modulation at a 24.3-kband channel rate and gives an effective 48.6 kbit/s data rate across the six time slots comprising one frame in the 30 kHz channel.

7.2.3. IS-95

As the demand for mobile communication went up still, the need for a more sophisticated system arose. In 1994 Qualcomm, Inc. proposed a cellular system and standard based on spread spectrum technology to increase capacity. It was and still is called IS-95. It uses the AMPS protocol as a default, but in normal operation operates quite differently than analog cellular or the more advanced IS-54. IS-95 is the first Code Division Multiple Access (CDMA)-based digital cellular standard. The brand name for IS-95 is **CDMAONE.** IS-95 is also known as TIA-EIA-95.

7.2.4. IS-2000

IS-95 is now being supplanted by IS-2000 (CDMA2000), a later CDMA-based standard. It is used in the USA, South Korea, Canada, Mexico, Israel, Australia,

Venezuela and China. CDMA2000 is a family of 3G mobile telecommunications standards and is an incompatible competitor of the other major 3G standard W-CDMA (UMTS). CDMA2000 is one of the approved radio interfaces for the ITU's IMT-2000 standard and a successor to 2G CDMA (IS-95, branded cdmaOne). It is standardized by 3GPP2. CDMA2000 is a registered trademark of the Telecommunications Industry Association (TIA-USA) in the United States, not a generic term like CDMA. There are 4 major versions of CDMA2000, namely:

- CDMA2000 1x
- CDMA2000 1xEV-DO (1x Evolution-Data Optimized)
- CDMA2000 1xEV-DV (1x Evolution-Data and Voice)
- CDMA2000 3x

7.3. INTRODUCTION TO GSM, THE GLOBAL SYSTEM FOR MOBILE COMMUNICATION

7.3.1. Introduction: The Evolution of Mobile Telephone Systems

Cellular is one of the fastest growing and most demanding telecommunications applications. Today, it represents a continuously increasing percentage of all new telephone subscriptions around the world. Currently there are more than 45 million cellular subscribers worldwide, and nearly 50 percent of those subscribers are located in the United States. It is forecasted that cellular systems using a digital technology will become the universal method of telecommunications. By the year 2005, forecasters predict that there will be more than 100 million cellular subscribers worldwide. It has even been estimated that some countries may have more mobile phones than fixed phones by the year 2000.

The concept of cellular service is the use of low-power transmitters where frequencies can be reused within a geographic area. The idea of cell-based mobile radio service was formulated in the United States at Bell Labs in the early 1970s. However, the Nordic countries were the first to introduce cellular services for commercial use with the introduction of the Nordic Mobile Telephone (NMT) in 1981.

Cellular systems began in the United States with the release of the advanced mobile phone service (AMPS) system in 1983. The AMPS standard was adopted by Asia, Latin America, and Oceanic countries, creating the largest potential market in the world for cellular.

In the early 1980s, most mobile telephone systems were analog rather than digital, like today's newer systems. One challenge facing analog systems was the inability to handle the growing capacity needs in a cost-efficient manner. As a result, digital technology was welcomed. The advantages of digital systems over analog systems include ease of signalling, lower levels of interference, integration of transmission and switching, and increased ability to meet capacity demands. Table 7.3 charts the worldwide development of mobile telephone systems. Brief history of GSM is shown in Table 7.3.

Table 7.3.

S. No.	Date	Activity
1.	1982-1985	Conference European Post Telecommunications (CEPT) begin specifying a European digital telecommunications standard in the 900 MHz frequency band. This standard later became Global system for mobile communication (GSM).
2.	1986	Field tests were held in Paris to select which digital transmission technology to use. The choice was Time Division Multiple Access (TDMA) or Frequency Division Multiple Access (FDMA).
3.	1987	A combination of TDMA and FDMA was selected as the transmission technology of GSM. Operators from 12 countries signed a memorandum of understanding introducing GSM by 1991.
4.	1988	CEPT began producing GSM specifications for a phased implementation. Another five countries signed Memorandum of Understanding (MoU).
5.	1989	European Telecommunication Standards Institute (ETSI) took over responsibilities for GSM specification.
6.	1990	Phase one specifications were frozen and allowed manufacturers to develop network equipment.
7.	1991	➤ The GSM 1800 standard released. ➤ An addendum was added to the MoU allowing countries outside CEPT to sign.
8.	1992	➤ Phase 1 specifications were completed. First commercial phase one GSM networks were launched. ➤ The first international roaming Agreement was established between Telecom Finland and Vodafone in UK.
9.	1993	➤ Australia became the first non European country to sign the MoU. ➤ The MoU now had a total of 70 signatories. ➤ GSM networks were launched in Norway, Austria, Ireland, Hong Kong, Australia. ➤ The number of GSM subscribers reached one million.
10.	1994	➤ The MoU now had over 100 signatories covering 60 countries. ➤ More GSM networks were launched. ➤ Total number of GSM subscribers exceeded 3 million.
11.	1995	➤ Another type of GSM, Personal Communication Services (PCS) was developed in USA and operates at 1900 MHz.
12.	1998	At the beginning of 1998 the MoU has a total of 253 members in over 100 countries and there are over 70 million GSM subscribers worldwide.

7.3.2. GSM (Global System For Mobiles Communication)

GSM stands for Global System for Mobiles Communication. This is a world-wide standard for digital cellular telephony, or as most people know them Digital Mobile Telephones. GSM was created by the Europeans, and originally meant "Groupe Special Mobile", but this didn't translate well, so the now more globally appealing name was adopted. GSM is a published standard by ETSI, and now enjoys widespread implementation in Europe, Asia, and increasingly in America.

Throughout the evolution of cellular telecommunications, various systems have been developed without the benefit of standardized specifications. This presented many problems directly related to compatibility, especially with the development of digital radio technology. The GSM standard is intended to address these problems.

From 1982 to 1985 discussions were held to decide between building an analog or digital system. After multiple field tests, a digital system was adopted for GSM. The next task was to decide between a narrow or broadband solution. In May 1987, the narrowband time division multiple access (TDMA) solution was chosen.

7.3.3. GSM Specifications

GSM specifications do not specify the actual hardware requirements, in specifying the network functions and interfaces in detail.

This allows hardware designers to be creative that, how they provide the actual functionality, but at the same time makes it possible for operators to buy equipment from different suppliers.

The GSM recommendations consist of 12 series which are listed in the table below. These series were written by different working parties and a number of expert groups. A Permanent nucleus was established in order to co-ordinate the working parties and to manage the editing of the recommendations.

Series	Content
1.	General
2.	Service aspects
3.	Network aspects
4.	MS-BSS interface and protocol
5.	Physical layer on the radio path
6.	Speech coding specification
7.	Terminal Adaptor for MS (Mobile Station)
8.	BSS (Base Service Station)—MSC (Mobile Service Switching Centre) interface
9.	Network interworking
10.	Service interworking
11.	Equipment and type Approval specifications
12.	Operation and maintenance.

7.3.4. GSM Phases

In the late 1980, the groups involved in developing the GSM standard realized, that within the given time frame they could not complete the specifications for the entire range of GSM Services and features as originally planned. So it is decided that GSM will be released in different phases.

Phase one contains following services:

- Voice Telephony
- International Roaming
- Basic fax/Data Services (upto 9.6 Kbits/sec)
- Call forwarding

- Call barring
- Short Message Services (SMS)

Phase Two (Additional Features):

- Advice of Charges (Parameters on which charging will depend)
- Calling line identification [CLIP]
- Call Waiting
- Call hold
- Conference calling
- Closed user groups
- Additional data communication capabilities.

Phase 2^+:

- Multiple Service Profiles
- Private Numbering Plans
- Access to Centres Services
- Internetworking with Different GSM standards like GSM1800, GSM1900, DECT (Digital Enhanced Cordless Telecommunication Standard).

7.3.5. GSM Geographical Network Structure

Every telephone network needs a specific structure to route incoming calls to the correct exchange and then on to the subscriber. In a mobile network, this structure is very important because the subscribers are mobile. As subscribers move through the network; these structures are used to monitor their location.

A cellular network is the user of multiple low power transmitters.

Generally, these transmitters are of very less power. So, naturally the range of such transmitters will be very small, therefore, to overcome these difficulties whole network area is divided into small areas called Cells.

Each cell has its own transmitter section with Antenna, receiver, control unit named as Base station with specific band of frequency.

Adjacent cells are assigned different frequencies to avoid interference or crosstalk.

However, cell sufficiently at longer distance can use same band of frequencies. Whole network area can be represented a set of square cells.

But it is not ideal. Because, if the width of a cell is r then 4 adjacent cells will be at distant r, while 4 another adjacent cells will be at $\sqrt{2}r$.

As a mobile user within a cell moves towards the cell boundaries, it is better if all the cells are at equidistant. And it also simplified, the task of determining when to switch the user to an adjacent Antenna and which Antenna to choose. So, if we choose cells of Hexagonal shape, this provides equidistant centre of all the cells. For a cell radius R, the distance between the cell centre is $\sqrt{3}R$. As shown in Fig. 7.1(a) and (c). And the area of a hexagonal cell = 2.59 R^2.

7.3.5.1. Cell

A cell (Hexagonal shape) is the basic unit of a cellular system and is defined as the area of radio coverage given by one BS Antenna system.

Each cell is assigned a unique number called Cell Global Identity (CGI). In a complete network covering on entire country, the number of cells can be quite high.

Note. In practice, a precise hexagonal pattern is not used, due to topographical limitations, local signal propagation conditions, and practical limitation on sitting antennas.

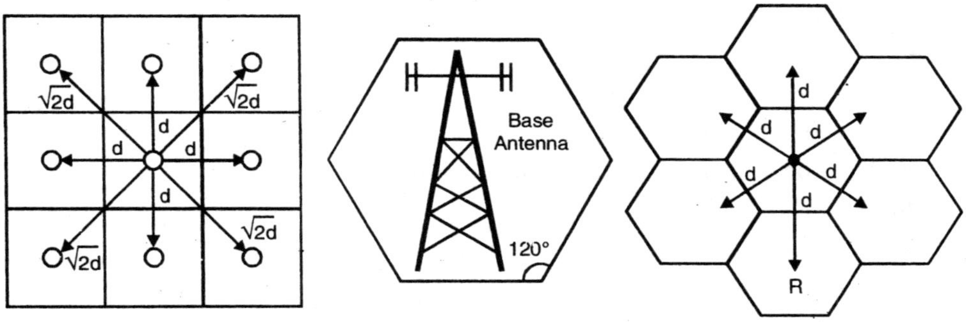

(*a*) Square shape (*b*) (*c*) Hexagonal pattern

Fig. 7.1. Cell.

7.3.6. GSM Frequency Ranges

GSM frequency ranges or frequency bands are the radio spectrum frequencies designated by the ITU for the operation of the GSM system for mobile phones.

GSM Frequency Bands

There are five frequency bands used by GSM mobile phones: GSM-900, GSM-1800, GSM-850, GSM-1900, and GSM-400. GSM Frequency band usage is defined in ETSI 05.05.

System	Band	Uplink	Downlink	Channel Number
GSM 400	450	450.4-457.6	460.4-467.6	259-293
GSM 400	480	478.8-486.0	488.8-496.0	306-340
GSM 850	850	824.0-849.0	869.0-894.0	128-251
GSM 900 (P-GSM)	900	890.0-915.0	935.0-960.0	1-124
GSM 900 (E-GSM)	900	880.0-915.0	925.0-960.0	975-1023, (0, 1-124)
GSM-R (R-GSM)	900	876.0-880.0	921.0-925.0	955-973
GSM 1800	1800	1710.0-1785.0	1805.0-1880.0	512-885
GSM 1900	1900	1850.0-1910.0	1930.0-1990.0	512-810

Note: The table shows the extents of the band and not center frequency.

GSM-900 and GSM-1800

GSM-900 and GSM-1800 are used in most parts of the world.

➤ GSM-900 uses 890-915 MHz to send information from the Mobile Station to the Base Transceiver Station (uplink) and 935-960 MHz for the other direction (downlink), providing 124 RF channels (channel numbers 1 to 124) spaced at 200 kHz. Duplex spacing of 45 MHz is used.

➤ In some countries the GSM-900 band has been extended to cover a larger frequency range. This 'extended GSM', E-GSM, uses frequency range 880-

915 MHz (uplink) and 925-960 MHz (downlink), adding 50 channels (channel numbers 975 to 1023 and 0) to the original GSM-900 band. The GSM specifications also describe 'railways GSM', GSM-R, which uses frequency range 876-915 MHz (uplink) and 921-960 MHz (downlink). Channel numbers 955 to 1023. GSM-R provides additional channels and specialized services for use by railway personnel.

- All these variants are included in the GSM-900 specification.
- GSM-1800 uses 1710-1785 MHz to send information from the Mobile Station to the Base Transceiver Station (uplink) and 1805-1880 MHz for the other direction (downlink), providing 374 channels (channel numbers 512 to 885). Duplex spacing is 95 MHz.

  GSM-1800 is also called PCS in Hong Kong and the United Kingdom. Most of the GSM operators in India use the 900 MHz band. Operators like Hutch, Airtel, Idea etc. use 900 MHz in most areas and also use 1500 MHz in some areas.

GSM-850 and GSM-1900

GSM-850 and GSM-1900 are used in the United States. Canada, India and many other countries in the Americas. GSM-850 is also sometimes erroneously called GSM-800.

- GSM-850 uses 824-849 MHz to send information from the Mobile Station to the Base Transceiver. Station (uplink) and 869-894 MHz for the other direction (downlink). Channel numbers 128 to 251.
- GSM-1900 uses 1850-1910 MHz to send information from the Mobile Station to the Base Transceiver Station (uplink) and 1930-1990 MHz for the other direction (downlink). Channel numbers 512 to 810.
- PCS is an initialism for Personal Communications Service and merely represents the original name in North America for the 1900 MHz band. Cellular is the term used to describe the 850 MHz band, as the original analog cellular mobile communication system was allocated in this spectrum. Providers commonly operate in one or both frequency ranges.

GSM-400: Another less common GSM version is GSM-400. It uses the same frequency as and can co-exist with old analog NMT systems. NMT is a first generation (1G) mobile phone system which was primarily used in Nordic countries, Eastern Europe and Russia prior to the introduction of GSM. It operates in either 450.4-457.6 MHz paired with 460.4-467.6 MHz (channel numbers 259 to 293), or 478.8 - 486 MHz paired with 488.8-496 MHz (channel numbers 306 to 340).

Multi-band and multi-mode phones: Nowadays most phones support multiple frequencies used in different countries. These are typically referred to as multi-band phones. Dual-band phones can cover GSM networks in pairs such as 900 and 1800 MHz frequencies (Europe, Asia, Australia and Brazil) or 850 and 1900 (North America). European tri-band phones typically cover the 900, 1800 and 1900 bands giving good coverage in Europe and allowing limited use in North America. There are also multi-mode phones which can operate on GSM systems as well as on mobile-phone systems using other technical standards. Often these phones use multiple frequency bands as well. For example, one version of the Nokia 6340i GAIT phone sold in North America can operate on GSM-1900, GSM-850 and legacy TDMA-1900. TDMA-800, and AMPS-800, making it both multi-mode and multi-band.

7.3.6.1. Key Terms of GSM

During the development of mobile systems, many terms arose which are used to describe the call cases and situations involving MS. The primary terms used are described below. Figure 7.2 shows a overview of GSM architecture which describes various terms used in GSM and their relations.

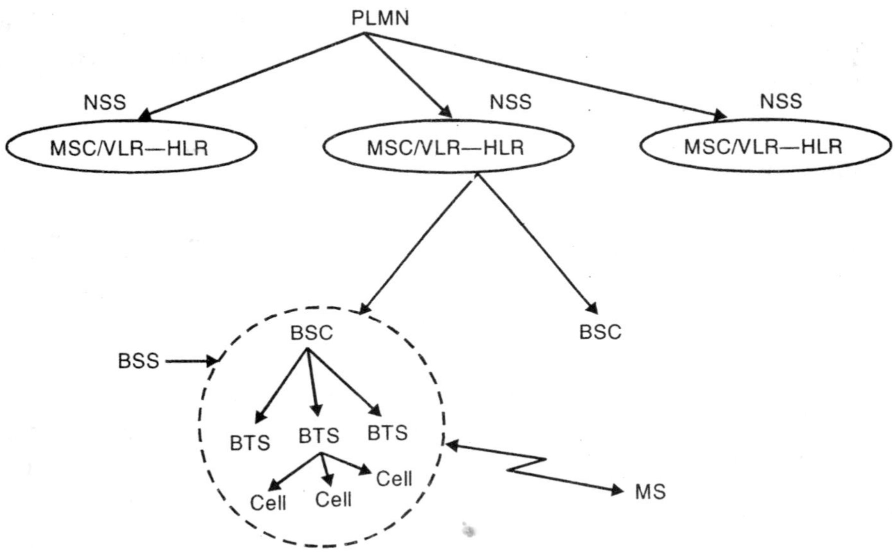

Fig. 7.2. Overview of GSM architecture.

PLMN (Public Land Mobile Network): PLMN is a geographical area which is allocated to a network operator. For example, Delhi is geographical area which is allocated to 4 operators AirTel, Idea, Vodafone, MTNL. But keep in mind whole area is allocated to every operator not shared. Hence in Delhi there are 4 PLMN. PLMN is a operator specific word not area specific. Airtel is a complete one PLMN and to provide service in whole Delhi. And it is distributed in Location Areas (LA) and then it is divided in small area called cell.

NSS (Network Switching Subsystem): NSS is responsible for switching and it consists Home Location Register (HLR), Equipment Identity Register (EIR), Authentication Centre Unit (AUC) and Visitor Location Register (VLR) and all these elements are controlled by MSC (Mobile Switching Center). In a PLMN there may be one MSC or more than one, it depends upon Area and number of subscribers.

BSS (Base Station Subsystem): BSS is responsible for signalling towards Mobile Station (MS) and towards NSS. BSS consists of BSC (Base Station Controller) and BTS (Base Transceiver Station). In a network, more than one BSC is connected with one MSC but some time only one BSC is connected with one MSC. It depends upon Area and Subscriber. In terms of area BSC covers one Location Area and Parameter of

Location area defined in MSC. Generally one BSC area is equal to one Location area. But it is not necessary.

BTS (Base Transceiver Station): BTS is a cabinet which consists of Directional Coupler, Combiner, TRx (Transmit-Receive) equipment and it is controlled by BSC and provides service to BSC and to Mobile Station.

If a BTS consists of 3 TRx, it means BTS has 3 carrier frequencies. One BTS may have 1 to 12 TRx depend upon type of BTS. For example, it is Insight, Metrosite, Ultrasite. To provide service to mobile station BTS is connected to Antenna.

LA (Location Area): Location Area is BSC area but it is defined in MSC.

Cell: Cell is smallest area which is covered by one Antenna.

MS: Mobile Station is mobile phone which is required for Voice/Data Conversation with network. It consists of Mobile Equipment (ME) and SIM (Subscriber Identity Module).

A mobile station can be put in following states.

Idle: MS is on but call is not in progress.

Active: MS is on and voice/data call is in progress.

Detached: MS is OFF.

Table 7.4. Mobile Phone Status.

Mode	Term used in Idle State	Description
Idle	Registration	This is the process in which an MS informs a network that it is attached.
	Roaming	When an MS moves around a network in idle mode, it is referred to as roaming.
	International Roaming	When an MS moves into a network which is not its home network.
	Location Updating	An MS roaming around the network must inform the network when it enters a new LA. This is called location updating.
	Paging	This is a process whereby a network attempts to contact a particular MS. This is achieved by broadcasting a paging message containing the identity of that MS.
Active	Handover	This is the process in which control of a call is passed from one cell to another while the MS moves between cells.
Detach	1. Normal Detach 2. Implicit Detach 3. MS Purging	Manual switch off. But VLR keeps present data (cell identity, location area services etc.) If signal strength is very low or battery strength is very low and periodic registration is not possible with network then MSC/VLR forcefully detach from network to MS. If your mobile phone is in OFF mode for more than 4 hours or out of coverage area then VLR will delete all data stored in normal detach. So whenever you will turn on your mobile, network treats you new subscriber to attach with network.

Note. In idle mode, it is the MS which decides which cell to move into, in active mode, the network makes this decision.

7.3.6.2. MS Registration and Roaming

When an MS is powered off, it is detached from the network. When the subscriber switches power on, the MS scans the GSM frequencies for special channel called Control Channels, when it finds a Control Channel, the MS measures the signal strength, it receives on that channel and records it. When all Control Channels have been measured, the MS tunes to the strongest one. If the location of the MS is noticed to be different from the currently stored location then a location update will also take place. As the MS moves through the network, it continues to scan the control channel to ensure that it is tuned to the strongest possible channel. If the MS finds one which is stronger, then the MS returns to this new control channel. If the new control channel belongs to a new LA, the MS will also inform the network of its new location.

7.3.7. Various Types of Number for Network Identity

Network uses various types of number for establishing releasing, supervision of a call and to keep record of mobile subscriber.

MSISDN (Mobile Station ISDN) Number: It is International mobile subscriber number which is normally called mobile number. It is unique world wide and it consists of

MSISDN = Country Code (CC) + Network Destination Code (NDC) + Subscriber Number (SN)

For example if your Vodafone MSISDN number is 919899638102.

Then CC is 91, NDC is 9899, SN is 638102.

In India MSISDN is a 12 digit format but it may be varied as per country. It is defined in HLR.

MSRN (Mobile Subscriber Routing Number): MSRN is of same format as MSISDN. Some series of MSISDN are reserved by network operator, it is not allocated to subscriber.

MSRN is used during mobile terminate trunk call (call from PSTN or other MSC) to provide location of mobile subscriber.

Generally 2000 MSRN is more than sufficient for Twenty Lacs subscribers. It is defined in MSC and analysed by central memory of MSC.

MSRN = CC + NDC + SN

HON (Hand Over Number): If you are on call and moving also then it become very much annoyed if your call is dropped. To make your call continuing without drop network uses handover concept. HON used by MSC to provide information required to transfer your call from one BSC to another BSC or to another MSC. HON is also same as MSISDN and some series of number reserved by operator to provide this service and not allocated to subscriber.

HON = CC + NDC + SN

IMSI (International Mobile Subscriber Identity Number): It is very important number and world wide unique. It is stored in HLR and in SIM card. Purpose of IMSI is for Location Update and Authentication.

IMSI = MCC + MNC + MSIN

MCC: Mobile country code and it is of 2 to 3 digits. For India it is 404.

MNC: Mobile Network code of 2 digits. For BSNL it is 51, 52.

MSIN: Mobile Subscriber Identification Number (10 digits).

TMSI (Temporary Mobile Subscriber Identity): TMSI is allocated by VLR corresponding to every IMSI and it is used for same purpose as IMSI but once TMSI is allocated, IMSI will not be used for any purpose untill unless TMSI allocated expired due to mobile phone is OFF mode for more than 4 hours.

TMSI is used instead of IMSI to improve security efficiency of network. Because TMSI is of half of IMSI in length so it takes less time to process. And as soon as you change MSC area it will change. Hence it will be very difficult to crack by any Unauthorized User. It is stored in VLR and in SIM Card.

$$\text{TMSI} = \frac{1}{2}(\text{Length of IMSI})$$

IMEI (International Mobile Equipment Identity). It is unique number world wide of mobile equipment and it is stored in mobile equipment and in EIR (Equipment Identity Register). If your mobile phone has been stolen then it may be used to track your mobile or to block permanently your mobile. But unfortunately this facility is not used by network operators in India.

$$\text{IMEI} = \text{TAC} + \text{FAC} + \text{SNR} + \text{Spare digit}$$

TAC (Type Approval Code). Determined by a Central GSM body and it is of 4 to 6 digits.

FAC (Final Assembly Code). It identifies the manufacturer and it is of 2 digits.

SNR (Serial Number). An identical serial number of six digits uniquely identifies all equipment within each TAC and FAC.

Spare Digit. It is to keep for future use, but when transmitted by MS then it is always zero.

To know IMEI of your mobile you can push * # 06 # and Date of manufacturing can be seen by * # 0000 # or by * # 06 # only.

IMEISV (International Mobile Equipment Identity and Software Version) Number. IMEISV provides a unique identity for every MS and also refers to version of software which is installed in MS. If mobile networks need to know the MS speech coding capabilities when a call is being made (Half Rate/Full Rate channel etc.) then it is used. It consists of

$$\text{IMEISV} = \underset{\text{6 digits}}{\text{TAC}} + \underset{\text{2 digits}}{\text{FAC}} + \underset{\text{6 digits}}{\text{SNR}} + \underset{\text{2 digits}}{\text{SVN}}$$

SVN (Software Version Number). Allow the mobile equipment manufacturer to identify software version of a given type approved mobile.

LAI (Location Area Identity). It is temporary network identity, which is required for routing, paging, location update.

$$\text{LAI} = \text{MCC} + \text{MNC} + \text{LAC}$$

MCC (Mobile Country Code) is of 3 digits.

MNC (Mobile Network Code) is of 2-3 digits.

LAC (Location Area Code) is of 16 digits.

One PLMN can be divided into 65536 location areas. But it is maximum number decided by ITU. Practically this number is very less.

CGI (Cell Global Identity). It is used to identify network cell. Maximum number of cells may be in one location area is 65536, while in one PLMN it may be upto 4.3 billion.

But practically this number is very less and depends upon number of parameter.

$$\text{CGI} = \underset{\text{3 digits}}{\text{MCC}} + \underset{\text{2-3 digits}}{\text{MNC}} + \underset{\text{16 digits}}{\text{LAC}} + \underset{\text{16 digits}}{\text{CI}}$$

CI (Cell Identity) used for location update, handover etc. It is of 16 digit.

7.3.8. GSM System Architecture

GSM system architecture consists of various number of component which is onnected to various elements to provide GSM service. But mainly it is divided in our part.

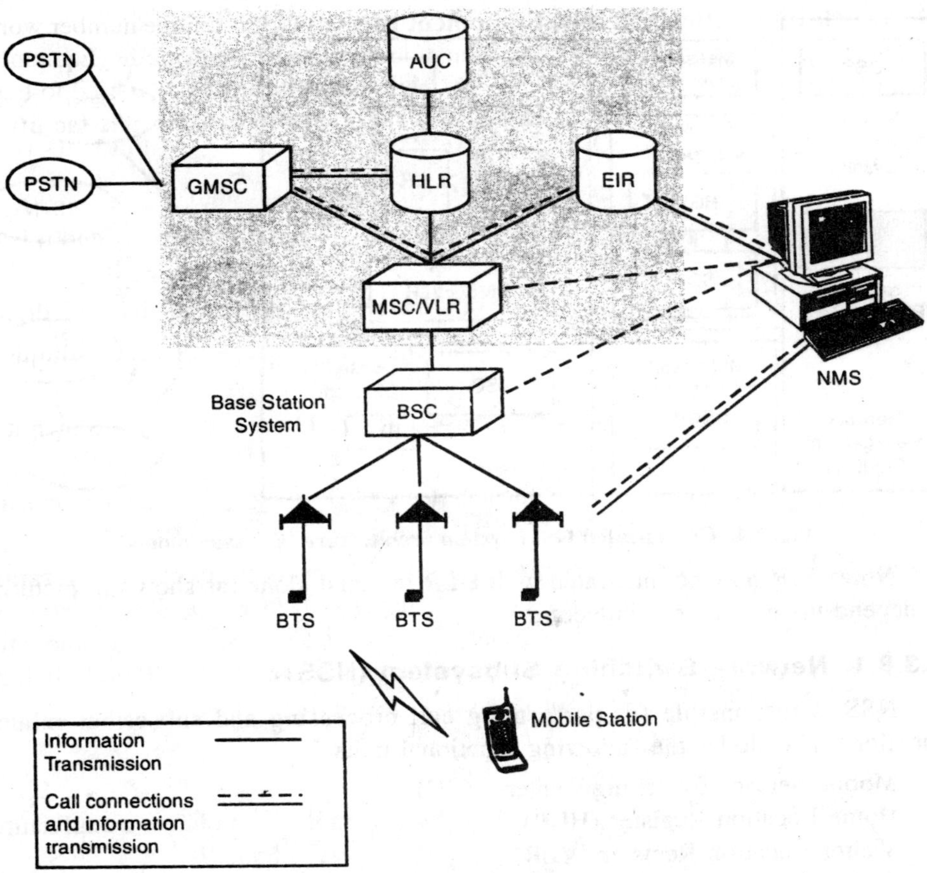

Fig. 7.3. Simple GSM system architecture.

BSS (Base Station Subsystem)

NSS (Network Switching System)

NMS (Network Management System)

MS (Mobile Station)

But note that GSM architecture provides service to MS and it is independent from network operator hence technically it is not part of GSM architecture. But here I am mentioning to just discuss about MS. Simplified structure of GSM architecture is shown in Fig. 7.3. Thick line picture represents to carry traffic and dashed line is to use for signalling only.

Abbreviations:

MI	Mobile Intelligent
MS	Mobile Station
BTS	Base Transceiver Station
BSC	Base Station Controller
MSC	Mobile Service Switching Center
GMSC	Gateway MSC
VLR	Visitor Location Register
HLR	Home Location Register
EIR	Equipment Identity Register

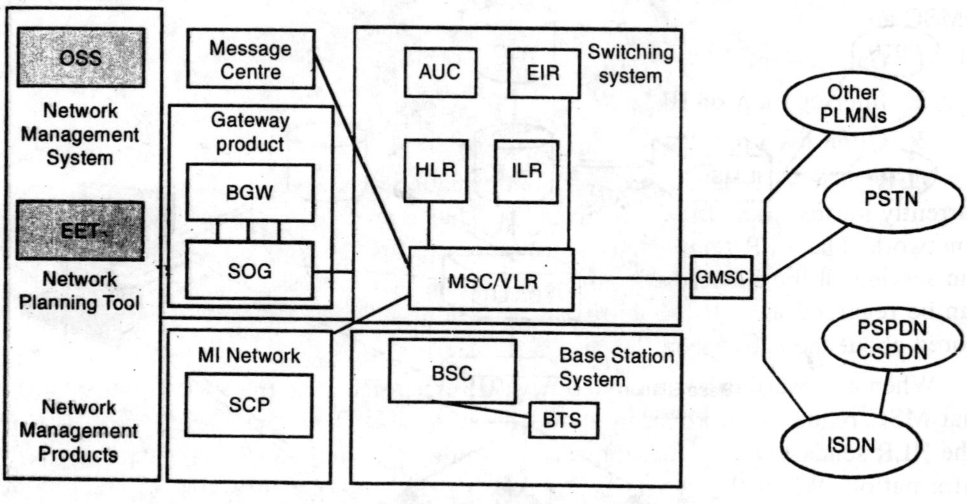

Fig. 7.4. Complicated GSM system architecture (Ericsson model).

Note: EIR may be integrated with HLR or stand alone (as shown in picture). It depend upon vendor architecture.

7.3.8.1. Network Switching Subsystem (NSS)

NSS is responsible for performing call processing and subscriber related functions. It includes the following functional units.

Mobile Service Switching Center (MSC)
Home Location Register (HLR)
Visitor Location Register (VLR)
Authentication Center (AUC)
Equipment Identity Register (EIR)

MSC (Mobile Service Switching Center): MSC is responsible for following :

1. Call establish, supervision, call terminate.
2. Mobility management (to keep record of mobile subscriber).

3. Initiation of Paging (to search a particular subscriber in case of mobile terminated call).
4. Collection of charging information and send to billing gateway for further processing. This charging information is based on various parameters that is called call tarrif.

It is important to note that VLR is a integrated part of MSC, it is not isolated cabinet. VLR is only subscriber data base associated with MSC.

GMSC (Gateway MSC): It is same as MSC but difference is that it is end point of your GSM architecture. If any call is coming from PSTN or from other PLMN then it will reach to your mobile phone and to your MSC through GMSC only.

Hence there is no need of VLR with GMSC. But if MSC is working like GMSC then VLR will be there. Major difference is that GMSC can interrogate HLR directly but MSC cannot. MSC interrogates HLR through VLR only. Other functions of GMSC are

1. Call control
2. Interrogation of HLR
3. Often used for interworking with external networks.

VLR: The VLR database contains information about all the mobile subscribers currently located in an MSC Service Area. Thus there is one VLR for each MSC in a network. The VLR temporarily stores subscription information so that the MSC can service all the subscribers currently visiting that MSC Service Area. The VLR can be regarded as a distributed HLR as it holds a copy of the HLR information stored about the subscriber.

When a subscriber roams into a new MSC Service area, the VLR connected to that MSC requests information about the subscriber from the subscriber's HLR. The HLR sends a copy of the information to the VLR and updates its own location information. When the subscriber makes a call, the VLR will already have the information required for call setup.

MSC/VLR: In GSM networks, the VLR is integrated into the MSC node. This means that signalling between the VLR and the MSC is done internally within the MSC/VLR network node and does not have to be carried over the rest of the network. This has the benefit of reducing the overall signalling load on the network.

VLR stores following Id for various operations in the network.

1. TMSI, IMSI
2. Services for visitor subscriber currently located in MSC.
3. Triplets used for security (Refer to AUC)
4. Location Area Identity Number.

Home Location Register (HLR): The HLR is a centralized network database that stores and manages all mobile subscription information until that subscription is cancelled. The information stored includes:

- Subscriber identity (MSISDN, IMSI)
- Subscriber supplementary services (Call waiting, call forwarding etc.)

- Subscriber location information (Current cell area, location area)
- Subscriber authentication information (Triplets).
- Ki (Authentication key)
- A_3, A_8 Algorithm.

The HLR can be implemented in the same network node as the MSC or as a stand-alone data base. If the capacity of a HLR is exceeded by the number of subscribers, additional HLRs may be added.

AUC: The function of the AUC is to authenticate the subscribers attempting to use a network. In this way, it is used to protect network operations against fraud. The AUC is a data base connected to the HLR which provides it with the authentication parameters and cippering keys used to ensure network security. Authentication is very important step that is required to authorize a user to use a service.

It is set by operator that when authentication is required, for example at every mobile originated call, at every location update etc.

Therefore, whenever call will be originated first it will be authorized (check service of subscriber, identity of user) before connecting call to desired subscriber.

It is important to note that authentication information is generated by authentication centre but stored by HLR and it will be used by VLR.

Let VLR ask to HLR to provide authentication information then HLR instruct to AUC to generate a random number.

Authentic cèntre generates a random number and algorithm, Ki is provided by HLR. Then AUC combine as following:

Random number (RAND) + K_i + A_3 = SRES (Signed Response)

RAND + K_i + A_8 = K_c (Ciphering key used in encryption of data)

Note that K_i, A_3, A_8 is already stored in HLR and Random Number generated by AUC change every time so it is impossible to crack any data from network without completing authentication.

Note: RAND, SRES, K_c Collectively known as Triplets.

EIR: The EIR is a database containing mobile equipment identity information which helps to block calls from stolen, unauthorized, or defective MSS. EIR consists of list of mobile subscribers in three categories.

White list: Those subscriber who authorise to use all services.

Gray list: Subscribers are allowed to use all services but under supervision.

Black List: All services of subscriber have been blocked.

But these services are not used by operators. All subscribers placed in white list. But it is total authority of network operator, network operator can allow to all services of black listed subscriber also.

AUC/EIR: The AUC and EIR are implemented either as stand alone nodes or as combined AUC/EIR node. The UNIX based AUC and the EIR are developed by same Group.

7.3.8.2. Mobile Station (MS)

Mobile station consists two part of as follows.

MS = Mobile Equipment (ME) + SIM (Subscriber Identity Module)

Mobile equipment stores following data:

IMEI (International Mobile Equipment Identity)

A_5 (Algorithm used for coding to speech between mobile and BTS)

SIM stores following data

IMSI/TMSI, A_3, A_8, K_i

Authentication between mobile and network can be explained as follows.

Let mobile originates a call and MSC/VLR require authentication then first MSC/VLR asks to HLR for triplets and HLR asks to AUC to generate random number.

AUC generate authentication parameter and give to HLR and give to VLR as follows.

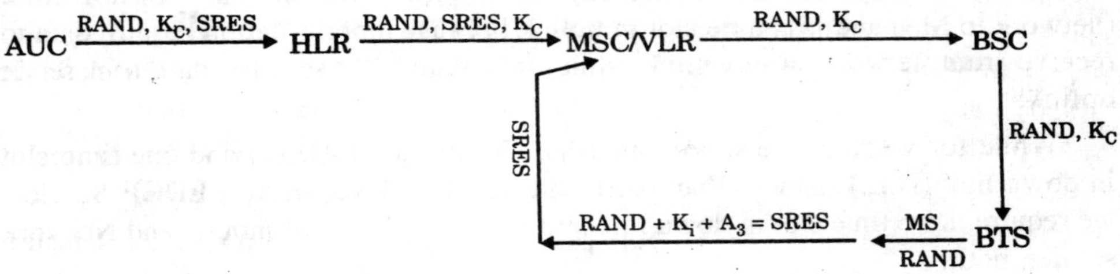

Notes:

(1) It is important to note that (K_C, SRES, RAND). Collectively known as Authentication Triplet.

where, Signed Response (SRES) = $K_i + A_3$ + RAND

Ciphering Key (K_C) = $K_i + A_8$ + RAND

SRES is known as signed response.

RAND is known as random number and it is generated by Authentication Center Unit (AUC) for authentication between mobile subscriber and network.

(2) AUC is integral part of HLR (Home location register unit). AUC also stores K_i and A_3, A_8, IMSI details.

K_i is a unique number for each SIM card.

(3) SIM is known as subscriber identity mobile.

(4) AUC generate RAND number.

(5) AUC generate SRES using RAND and Send to MSC/VLR via HLR alongwith RAND.

(6) MSC/VLR send only RAND to MS and SRES keep with itself.

(7) MS generate SRES using RAND and send to VLR.

(8) MSC/VLR match SRES obtained by HLR and send by MS, if both SRES match together it mean authentication is successful.

(9) MSC/VLR send K_C to BTS and it will remain with BTS for ciphering.

As we know MS has also K_i, A_3, A_8 hence as soon as MS gets RAND number from VLR it generates SRES and send to VLR and VLR match together.

Mobile station available in market of various type but technically it is classified based on time slot. If a mobile phone support more time slot then it can provide more service and provide more speed for voice/data. And time slot is classified in mobile class.

Class	D.L.	+	U.L.
One	2	+	1
Two	2	+	2
Three	3	+	1
Four	3	+	2
Five	4	+	1
Six	4	+	2

Class one mobile is basic mobile which supports 2 time slots in Downlink (Network to MS) and supports one time slot in Uplink (MS to Network).

In Data Service (To access GPRS), we require more time slot in downlink (network to MS) and less time slot in uplink because mobile station require data to receive from network in downlink while only request is sent for data transfer in uplink.

While for voice conversation one time slot in uplink (U.L.) and one time slot in down link (D.L.) is more than sufficient but for GPRS service, EDGE Service, we require more time slot in down link and it increases cost of mobile and Network service both.

7.3.8.3. Base Station System (BSS)

Base station is responsible for all types of signalling and it consists of BSC, BTS and transcoder.

Functions of BSS are as follows:

1. Radio Path Control (Radio channel allocation and quality of radio allocation)
2. BTS and TC (Transcoder control)
3. Connection establishment with MS-NSS.
4. Mobility management, speech transcoding.
5. Collection of statistical data.

7.3.8.3.1. Base Station Controller (BSC)

BSC manages all the radio-related functions of GSM network. It is a high capacity switch that provides functions such as MS handover, radio channel assignment and the collection of cell configuration data. More than one BSC can be controlled by single MSC. Following are the functions of BSC.

1. Connection establishment between the MS and the NSS.
2. Mobility management.
3. Statistical raw data collection.
4. BTS and TC control.
5. Power control of MS, which depends upon measurement report sent by MS to network.

Statistical raw data consists of information from the BTS, TC, and collected in BSC and forwarded to NMS.

7.3.8.3.2. Base Transceiver Station (BTS)

BTS is connected with BSC through TC and it consists of TRx, Directional Coupler, combiner etc. Antenna is directly connected with BTS. BTS also consists of A_5 algorithm. Functions of BTS are as follows.

1. Air interface signalling (towards MS)
2. Ciphering (speech coding)
3. Speech processing (channel coding, interleaving, burst formating)
4. Generation of alarms report.
5. Baseband/radio frequency transformation
6. Modulation of voice/data

Ciphering in BTS. User information and signalling exchanged between BTS and MS is not disclosed to unauthorised individuals hence there is need to code speech.

To cipher a message there is need of TDMA frame number (TFN) and A_5 algorithm.

As we know A_5 is stored in BTS and mobile equipment and TDMA frame number on which this transmission and reception will take place.

At the time of authentication between MS and network, VLR sends K_c and random number where K_c will be kept by BTS and random number forwarded to MS. But note that in Authentication process K_c is not shown but it will be sent by HLR.

And MS already having K_i, A_3, A_8, A_5 to generate K_c and ciphered message. If it is mobile terminated call then ciphering and de-ciphering between BTS and ME (Mobile Equipment) will be as follows.

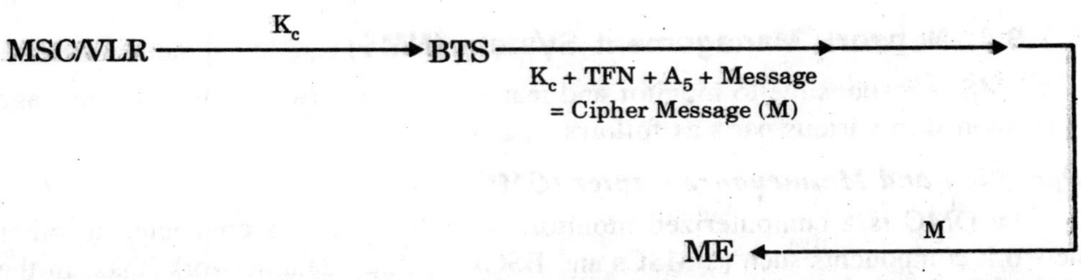

Cipher message (M)-TFN (TDMA Frame Number) – A_5 – K_c = Original Message

(1) MSC/VLR sends K_c to BTS.

(2) BTS is already having A_5 + TFN, hence by adding message and K_c produce cipher message.

(3) ME is already having TFN and A_5 and K_c which will be subtracted from cipher message and obtained original message.

Encrypting of message at BTS is called ciphering while de-crypting message at MS is called De-crypting of message.

But if it is mobile originated call then first mobile generates K_c by using RAND as follows.

$A_8 + K_i + \text{RAND} = K_c$

and then cipher message as follows.

Cipher Message (M)-TFN $- A_5 - K_c =$ Original Message

Note that this ciphering and de-ciphering between MS and BTS only. There is no role of MSC/VLR, they just provide Kc and RAND number.

Note: It is a dummy message (Blank) send by mobile equipment and BTS.

7.3.8.3.3. Trans Coder (TC)

Transcoder is connected at NSS site but it is part of BSS only. Function of converting from PCM information (A/D conversion) to the GSM speech coder information is called Trans Coding. This function is present in both MS and BSS. It also involves the conversion of information arriving from the MSC/VLR at a rate of 64 kbps to rate of 16 kbps. Without this rate of conversion the links to BSC would require 4 times the data rate capabilities which increases cost of network.

By reducing rate to 16 kbps it is possible to use one quarter of the transmission links equipment. It is not shown in figure but it is connected between BSC and MSC as follows.

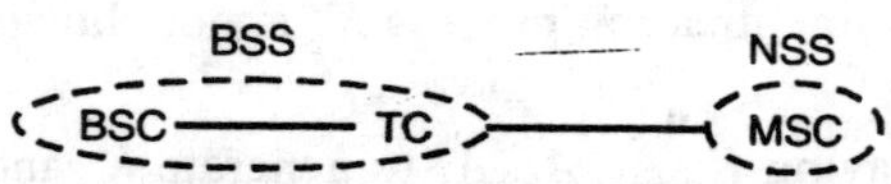

TC is a transcoder. It used for compressing the 4 channel of 16 kbps into 64 kbps of one channel towards MSC and de-compress towards BSC.

7.3.8.4. Network Management System (NMS)

NMS is responsible to monitor and manage whole network. To better manage it is divided in various parts as follows.

Operation and Maintenance Center (OMC)

An OMC is a computerized monitoring center which is connected to other network components such as MSCs and BSCs via X.25 data network links. In the OMC, staff are presented with information about the status of the network and can monitor and control a variety of system parameters. There may be one or several OMCs within a network depending on the network size.

Network Management Center (NMC)

Centralized control of a network is done at a Network Management Center (NMC). Only one NMC is required for a network and this controls the subordinate OMCs. The advantage of this hierarchical approach is that staff at the NMC can concentrate on long term system wide issues, whereas local personnel at each OMC can concentrate on short term, regional issues. OMC and NMC functionality can be combined in the same physical network node or implemented at different locations.

Operation and Support System (OSS)

Operation and Support System (OSS) is to support the activities performed in an OMC and/or NMC. The network operator monitors and controls the network through OSS which offers cost effective support for centralized, regional and total operation and maintenance activities. OSS is based on Telecommunication Management and Operation Support (TMOS) platform.

7.3.8.5. Additional Network Components

A basic GSM network can be enhanced by the addition of some or all of the following functions.

7.3.8.5.1. Additional Nodes

Message Centre (MC): A Message Center (MC) may be added to a GSM network to provide voice/fax mail and handling of the Short Message Service (SMS) and SMS Cell Broadcast (SMSCB) text messages. These services can generate considerable revenue for a network operator, as they are becoming increasingly popular.

Mobile Intelligent Network (MIN) Nodes: Mobile Intelligent Network (MIN) nodes can be added to a basic GSM network to provide value-added services to subscribers.

MIN nodes include:

Service Switching Point (SSP): An SSP acts as an interface between the call control functions of the mobile network and the service control functions of a Service Control Point (SCP).

Service Control Point (SCP): As SCP contains the intelligence of a MIN service or Services. This intelligence is realized of a MIN services. This intelligence is realized in software programs and data.

Service Data Point (SDP): SDP manages the data which is used by a MIN service.

Note: MIN Service includes:

- Free Phone
- Premium rate
- Personal number
- Televolting
- Cellular virtual private network

Post Processing Systems: Post Processing Systems are used by network operators to handle and analyse the large amounts of information which is generated by calls in the network.

Service Order Gateway (SOG): A network operator requires administrative systems to analyze and manage network information such as customer subscriptions, billing information and fraud detection. An operator's administrative systems are normally called Customer Administration Systems (CAS). They are complex systems which are often inflexible and costly to adapt to the specific needs of individual network operators.

The Service Order Gateway (SOG) is a system, which enables CASs to exchange information with network elements which contain service information, such as the HLR.

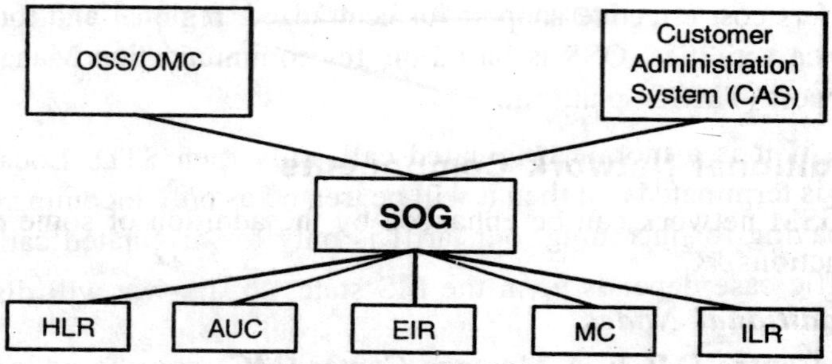

Fig. 7.5. Service order gateway.

Billing Gate Way (BGW): BGW collects billing information or Call Data Record (CDR) files from network elements such as MSCs and forwards them to Post-processing systems that use the files as input. Any internal BGW alarms are forwarded to OSS at an OMC.

A BGW is usually connected to the customer administration and billing systems and is handled by the administrative organization.

The figure below shows some of the possible billing information required when analyzing a specific call.

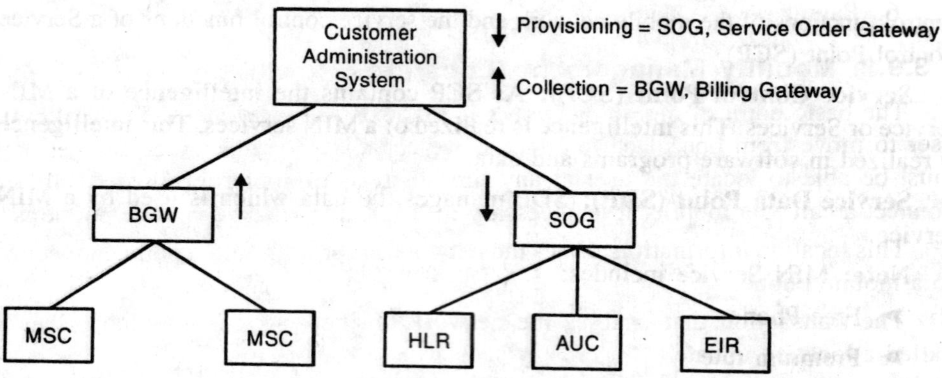

Fig. 7.6. Illustration of billing information.

7.3.9. Traffic Cases

Traffic cases may be of following type.

(*a*) Mobile Originated Call

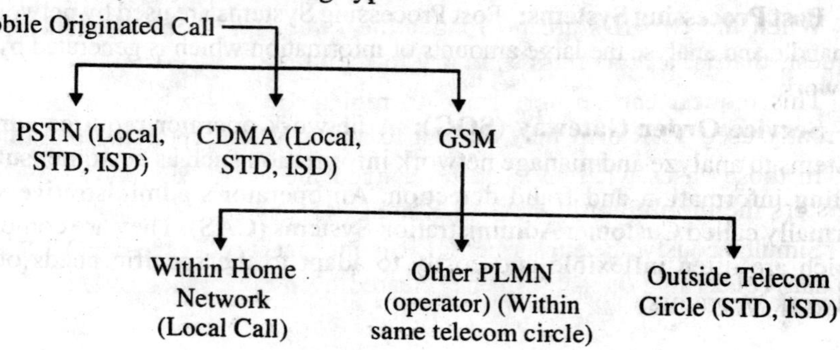

(*b*) Mobile Termined Call

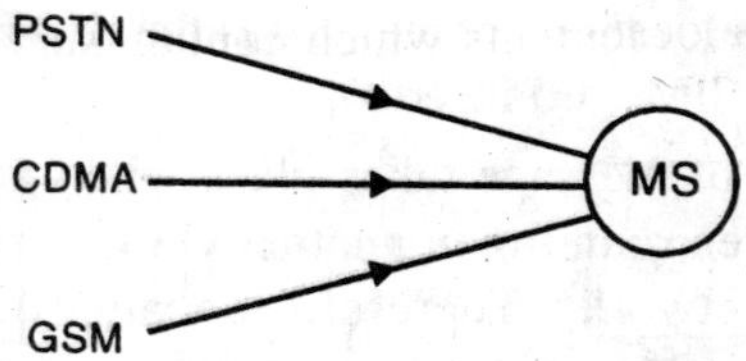

In India, if it is a mobile originated call, only then STD, Local, ISD terms arise but if it is terminated call then it will be treated as only incoming call. Because we are not paying for incoming, our tariff is only for originated call.

This traffic case depends upon the MS state, so first we will discuss various states of MS.

7.3.9.1. MS in Idle State

If MS is in Idle state it means only power is ON, No call/data conversion are taking place. In idle state following terms are defined.

1. Location update
2. Cell-selection
3. Cell re-selection
4. IMSI detach/attach
5. Implicit detach
6. Paging

7.3.9.2. Mobility Management (Location Update)

The basic point of mobile networks is the user ability to move. If we allow a user to move from one place to another, we cause a challenge since the network must be able to locate the user at any time there is a mobile terminated call. To connect a call to a mobile, it is necessary to determine where a mobile phone is.

This location information allows the network to form a point to point connection to a mobile phone.

The transaction that enables the network to keep track of the subscriber is called a location update.

The mobile phone constantly receives information sent by the network. This information includes identification (ID) of the VLR area in which the mobile is currently located. In order to keep track of its location, the mobile stores the ID of the area, the mobile compares this information to the area ID stored in its memory.

When the two IDs are no longer the same, the mobile sends the network a request, that is, a registration inquiry to the area it has just entered.

This request can be performed to register in a new area logically under the currently used VLR or it may be sent to a new VLR depending on situation.

In those cases where VLR are changed the network receives the request and registers the mobile in the new VLR area.

Simultaneously, the subscriber HLR is informed about the new VLR location and the data concerning the subscriber is cleared from the previous VLR.

Figure 7.7 briefly describes the types of messages that are exchanged the very first time a subscriber makes a location update, that is the first time the mobile station (MS) is switched ON with the SIM card inserted.

A similar exchange of message takes place when a subscriber switches on the MS for the first time when roaming in another GSM network then the home PLMN (Public Land Mobile Network). Therefore, Location update can be classified as follows.

1. First Time Location Update (PowerON, IMSI attach): MS will be applicable for this type of location update in two cases.

(*a*) New subscriber, who purchased SIM and inserted into mobile and power turn on.

(*b*) A MS is turning ON power after 4 hours.

2. Periodic Location Update: It is based on a timer set in network, just to check MS is present in network or not.

3. Generic Location Update: Whenever you change MSC/VLR area.

7.3.9.2.1. First Time Location Update (IMSI Attach)

As soon as you turn on Power, first MS will detect own network signal and then read information of network. Because you are new user hence there is need to inform network that I am in this cell area/location area, please update yourself and if any terminated call me then you can page me in this area of network. Hence MS sends a request to network and if authentication is set by network then it also takes place.

Note that if authentication is ON then first authentication will be performed then location area updated.

Moving of MS in the Network

Changing Cells within an LA

MS are constantly moving around in the cellular network. The MS location information stored in the VLR. If an MS changes cells within an LA, the network is not updated. The MS knows that the new cell belongs to the same LA by listening to the control channel frequency in the new cell. This broadcasts the cell's LAI. The MS compares the last LAI received with the new LAI. If they are the same, it means that the MS has not changed LAs and does not need to inform the network.

Location Updating (Same MSC/VLR)

If an MS detects a change in LAI on the control channel, it informs the network. When the MS sends the location updating message, the MSC/VLR determines whether it is an MS which is already registered or an MS visiting from another MSC/VLR. Whole process is shown in Fig. 7.7.

1. Then MS listens to the control channel in the new cell to determine the LAI. The new LAI is compared to the old one, if they differ, a location update is necessary.
2. The MS establishes a connection with the network and authentication of the subscriber and phone is performed.

3. If authentication is successful, the MS sends a location updating request to the system.
4. The system acknowledges location updating.

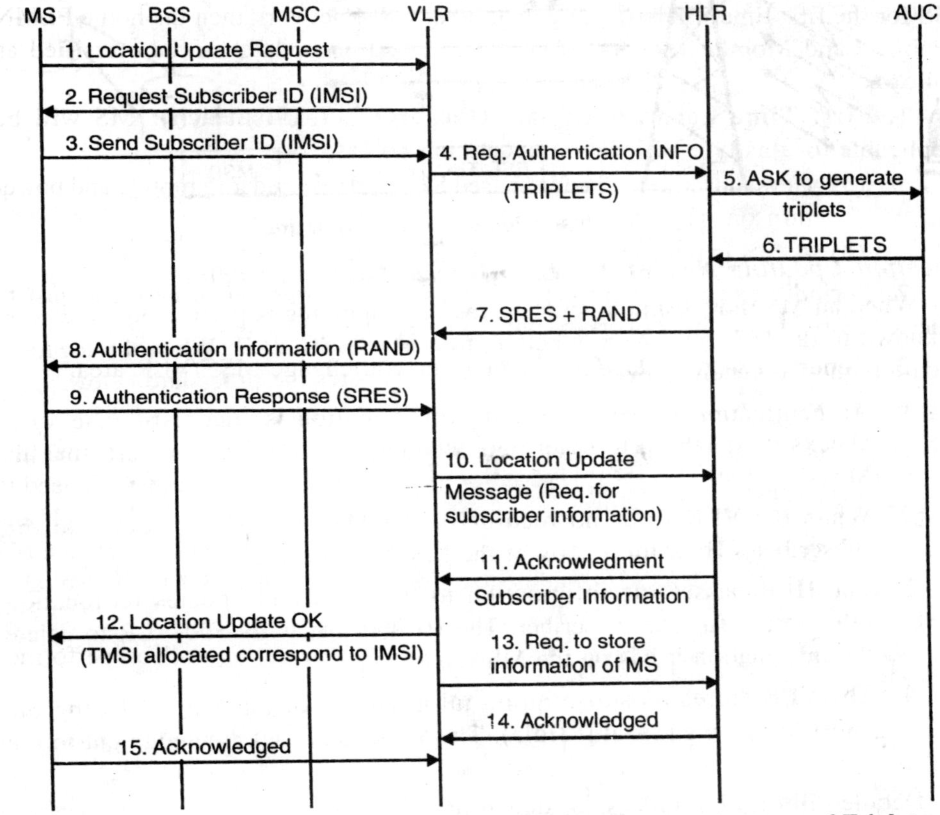

Fig. 7.7. IMSI Attach (K_C is not shown as it is understood that it is a part of Triplets and need to send)

1. MS sends location update request.
2. VLR asks for IMSI.
3. MS sent IMSI.
4. VLR asks for authentication information.
5. HLR asks to AUC to generate triplets.
6. AUC sent triplets to HLR.
7. HLR gives SRES + RAND to VLR.
8. VLR keeps SRES and sends RAND to MS.
9. MS generates SRES and sends to VLR then VLR match together and if matches then authentication successful.
10. VLR asks to HLR to provide subscriber information of MS.
11. HLR provides subscriber information to VLR.
12. VLR allocates TMSI corresponds to IMSI and for further transaction TMSI will be used between MS and MSC/VLR instead of IMSI and stores current location area in data base.
13. VLR also sends a copy of MS location to HLR that will be used in mobile terminated call for MS.
14. HLR acknowledged.
15. MS will acknowledge to VLR.

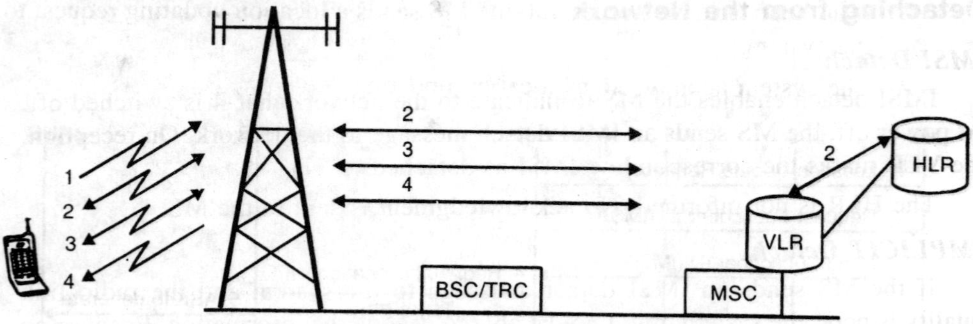

Fig. 7.8. Illustration of location updating.

Location Updating New MSC/VLR (Generic Location Update)

When an MS moves into a new LA, location updating is performed. However, unknown to the MS, the LA may belong to a new MSC/VLR. When the location update request is received by the new VLR, it executes the procedure below.

1. Authentication is performed. If authentication is successful, the VLR checks its database to determine whether or not it has a record for this MS.
2. When the VLR finds no record for the MS, it sends a request to the subscriber's HLR for a copy of the MS's subscription.
3. The HLR passes the information to the VLR and updates its location information for the subscriber. The HLR instructs the old VLR to delete the information it has on the MS.
4. The VLR stores its subscription information for the MS, including the latest location and status (idle). The VLR sends acknowledgment to the MS.

Detailed illustration process is shown in Fig. 7.9.

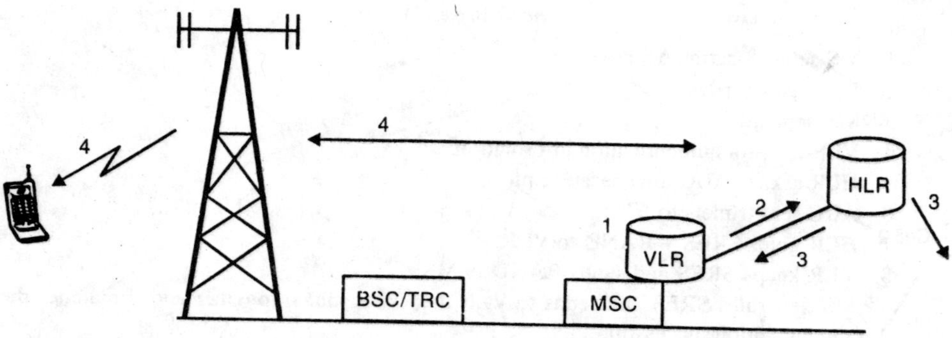

Fig. 7.9. Location updating.

Location Updating (Periodic Type Registration)

Periodic registration is a feature, which forces MSs to send a registration message to the network at predefined intervals. If an MS should miss such a registration, the network will mark the MS as detached. This may occur if an MS is out of the area of coverage and ensures that needless paging is not performed.

Detaching from the Network

IMSI Detach

IMSI detach enables the MS to indicate to the network that it is switched off. At power off, the MS sends an IMSI detach message to the network. On reception, the VLR marks the corresponding IMSI as detached.

The HLR is not informed. No acknowledgment is sent to the MS.

IMPLICIT Detach

If the MS sends an IMSI detach message to the system and the radio link quality is poor, the system might not be able to decode the information. Because no acknowledgment is sent to the MS, no further attempt is made. In this case, the system still regards the MS as attached. If periodic registration is in use, the system will soon determine that the MS is detached. The VLR then performs an implicit detach, marking the MS as detached.

Cell Selection

Cell selection will be performed by MS, when first time power will be ON.

MS will detect signal strength of all BTS which are rounded by MS but MS will detect strongest signal strength transmitting by BTS/cell.

Cell Re-Selection

This term arise when conversation/Data is going on. Actually, MS sends a measurement report to network in every 460 miliseconds which consists of own signal strength and six surrounding signal strength. On the basis of this measurement report network will decide which BTS connection will be allocated to MS.

Paging

Paging is the process to search a mobile subscriber in case of mobile terminated call. Whenever a mobile terminated call for a MS then BSC instruct to page (announce) TMSI to all BTS in location area. This TMSI will listen by all MS

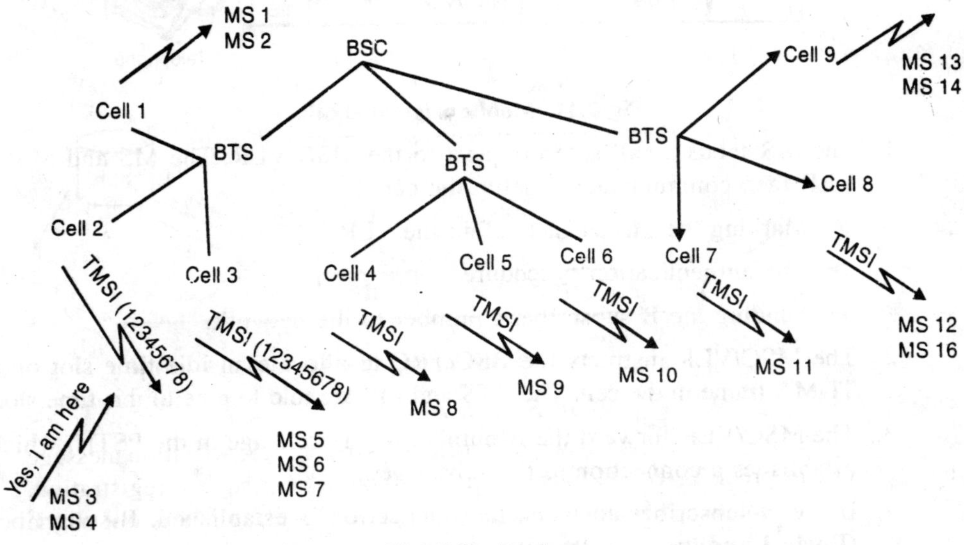

Fig. 7.10. Paging.

present in that location area but response will be given by only that MS which is having this TMSI.

For example in a class if I am announcing a message for you by your name then it will be listened by all students but response will be given by you only because in announcement I have mentioned your name. This is called paging.

7.3.9.3. When MS in Active Mode

An MS is in active mode when there is call (speech, fax or data), or a call setup procedure taking place. When MS is in active mode following process takes place.

1. Call from MS (speech, fax, data, short message)
2. Call to MS (speech, fax, data, SMS, cell broadcast)
3. Handover—inter cell intra BSC
4. Handover—inter BSC, intra MSC
5. Handover—inter MSC.

Call from an MS

In this section, we describe what happens when a mobile subscriber wants to setup a voice call to a subscriber in the PSTN. As illustrated in Fig. 7.11.

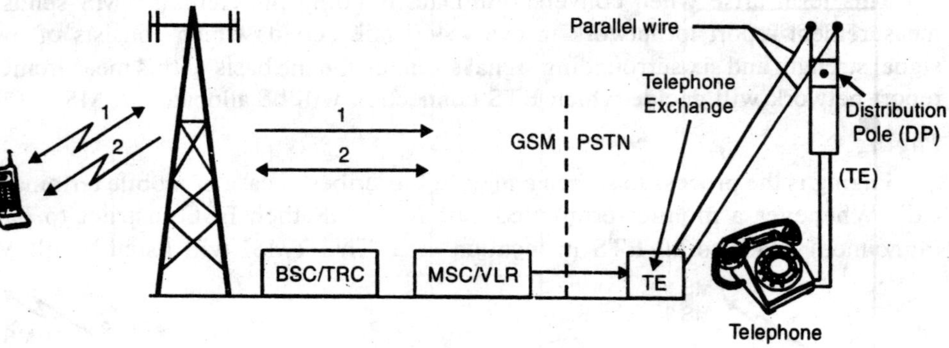

Fig. 7.11. Mobile originated call.

1. The MS sends a call setup request to the MSC/VLR. The MS and MSC/VLR then communicate to setup the call.
   - ➤ Marking the MS as "active" in the VLR.
   - ➤ The authentication procedure
   - ➤ Sending the B-subscriber's number to the network.
2. The MSC/VLR instructs the BSC/TRC to allocate an idle time slot on a TDMA frame in the cell. The BTS and MS are told to tune to this time slot.
3. The MSC/VLR forward the B-number to an exchange in the PSTN, which establishes a connection to the subscriber.
4. If the B-subscriber answers, the connection is established. B-subscriber (**Basic Landline Customer**).

Call to MS from PSTN

The major difference between a call to an MS and a call from an MS is that in a call to an MS the exact location of the mobile subscriber is unknown. Therefore, the MS must be located using paging before a connection can be established.

Below is the description of the call setup procedure for a call from a PSTN subscriber to a mobile subscriber. A call from an MS to a mobile subscriber operates according to the same process, the only difference being that the GMSC is connected by another MSC/VLR instead of by a PSTN node.

1. The PSTN subscriber keys in the MS's telephone number, which is analyzed in the PSTN which identifies that this is a call to mobile network subscriber. A connection is established to the MS's home GMSC.
2. PSTN analyses MSISDN and contacts the Gateway MSC (GMSC).
3. GMSC analyse MSISDN and sends a message to concerned HLR to provide MSRN of mobile.

   Note: It is step where GMSC can interrogate directly HLR and MSRN number to provide exact location of MS.

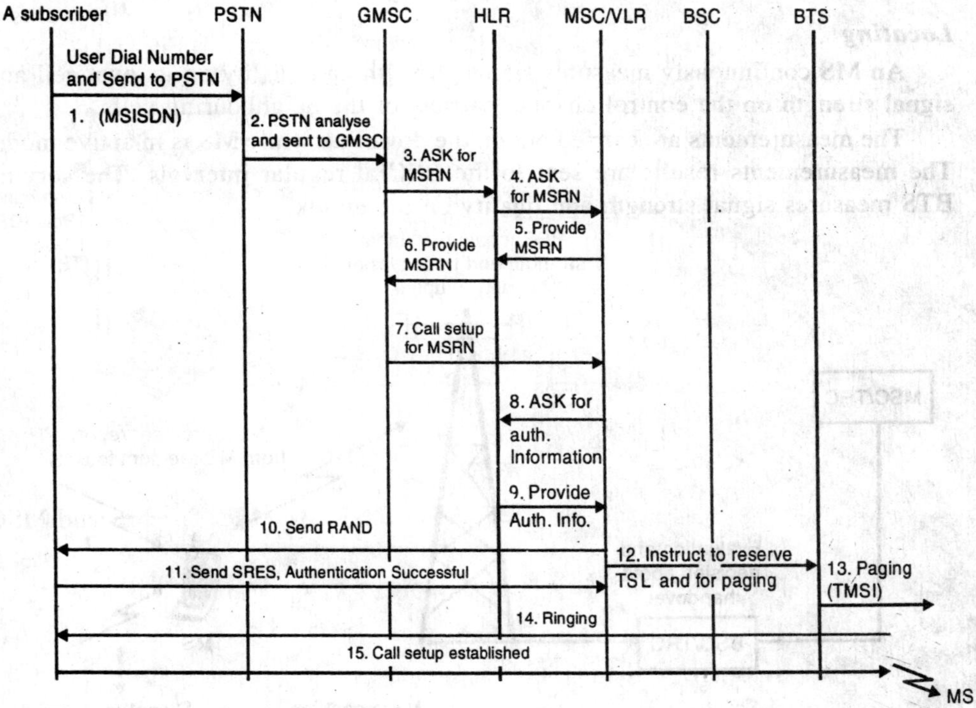

Fig. 7.12. PSTN originated call.

4. HLR analyses MSISDN which is provided by GMSC and by getting IMSI corresponds to MSISDN ask to MSC/VLR to provide MSRN.

   Note: It is step where HLR is required to store information of VLR connected to HLR. Important to note HLR will not send IMSI instead of sending MSISDN. Now in network MSISDN will not be used.

5. VLR stores IMSI, TMSI both in the database hence VLR analyses IMSI and provides MSRN to HLR.
6. HLR sends MSRN to GMSC.
7. Now GMSC analyses MSRN and directly connect to concern MSC.
8. Now MSC asks for authentication information from HLR and asks to MS to complete authentication.

9 to 11. Authentication complete.

12. MSC/VLR instructs to allocate time slot (TSL) for MS and page.
13. BSC reserves TSL and instructs to all BTS for paging using TMSI number.
14. If MS is there ringing will start.
15. If B subscriber responses then call will be setup.

Note: Steps will be same if call will be originated from another operator network. MSISDN number first sends to GMSC then same as above.

7.3.9.4. Handover

The process of changing cells during a call is called handover. To choose the best target cell, measurements are performed by the MS and the RBS.

Locating

An MS continuously measures signal strength and quality on its own cell and signal strength on the control channel carriers of the neighbouring cells.

The measurements are carried out on the downlink while MS is in active mode. The measurements results are sent to the BSC at regular intervals. The serving BTS measures signal strength and quality on the uplink.

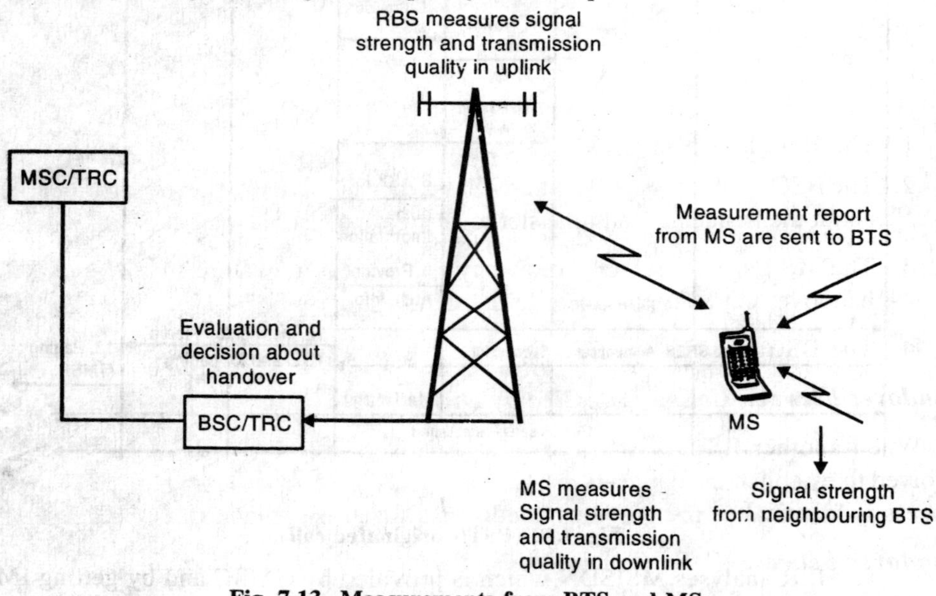

Fig. 7.13. Measurements from BTS and MS.

Based on these measurements, the BSC decides if a handover is necessary and to which cell. This is called locating. As soon as a neighbouring cell is considered to be better than the serving cell, a handover is attempted.

There are several types of handover, including

- Handover between calls controlled by the same BSC.
- Handover between cells controlled by different BSCs, but the same MSC/VLR.
- Handover between cells controlled by different MSC/VLRs.

Handover between Cells Controlled by the Same BSC

When performing a handover between two cells controlled by the same BSC, the MSC/VLR is not involved. However, the MSC/VLR will be informed when a handover has taken place. If the handover involves different LAs, location updating is performed once the call is finished as shown in Fig. 7.12.

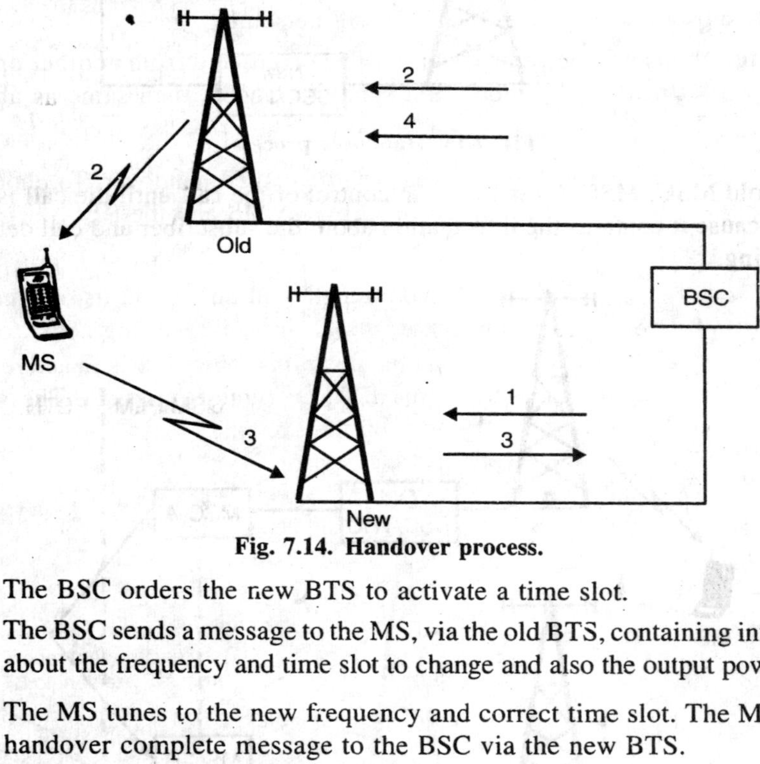

Fig. 7.14. Handover process.

1. The BSC orders the new BTS to activate a time slot.
2. The BSC sends a message to the MS, via the old BTS, containing information about the frequency and time slot to change and also the output power to use.
3. The MS tunes to the new frequency and correct time slot. The MS sends a handover complete message to the BSC via the new BTS.
4. The BSC tells the old BTS to release the old time slot.

Handover between Cells Controlled by Different BSCs but the same MSC/VLR

When another BSC is involved in a handover, the MSC/VLR must also be involved to establish the connection between the two BSCs, otherwise, the handover process is identical to the previous handover traffic case shown in Fig. 7.15.

Handover between Cells Controlled by Different MSC/VLRs

Handover between Cells Controlled by different MSC/VLRs can only be performed within one PLMN and not between two PLMNs. Cells controlled by different MSC/VLRs also means that they are controlled by different BSCs.

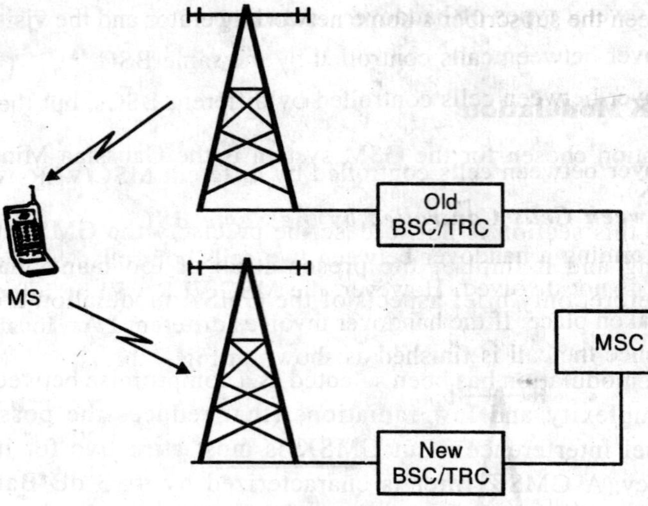

Fig. 7.15. Handover process.

The old MSC, MSC-A retains main control of the call until the call is cleared, this is because it contains the information about the subscriber and call details such as changing.

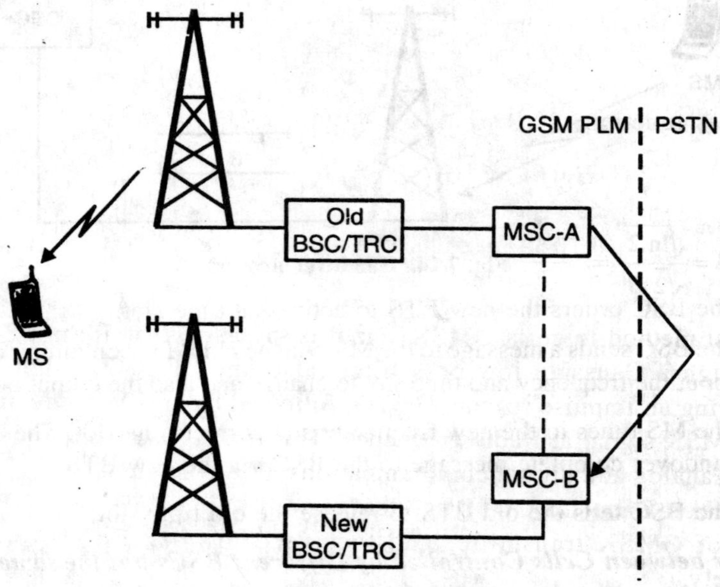

Fig. 7.16. Handover cells controlled by different MSCs.

After call release, the MS must perform locating updating because an LA never belongs to more than one MSC/VLR service area. The HLR is updated by the VLR-B, and will in turn tell VLR-A to delete all information about the mobile subscriber. Detailed illustration shown in Fig. 7.16.

International Traffic Cases

One of the primary features of GSM is the ability to perform international roaming and to handle international call cases. In order for a mobile subscriber to

be able to make calls while roaming in a different GSM network, there must be an agreement between the subscriber's home network operator and the visited network operator.

7.3.10. GMSK Modulation

The modulation chosen for the GSM system is the Gaussian Minimum Shift Keying (GMSK).

The aim of this section is not to describe precisely the GMSK modulation as it is too long and it implies the presentation of too many mathematical concepts. Therefore, only brief aspects of the GMSK modulation are presented in this section.

The GMSK modulation has been selected as a compromise between spectrum efficiency, complexity and low radiations (that reduces the possibilities of adjacent channel interference). But GMSK is most attractive for its excellent power efficiency. A GMSK filter is characterized by its 3 dB Bandwidth-bit duration product (BT).

If BT for a GMSK filter is greater than 0.5 then transmission error will be very less and we get good spectral efficiency and constant envelope.

Impulse response of a GMSK filter is given by

$$h_G(t) = \frac{\sqrt{\pi}}{\beta}\exp\left(-\frac{\pi^2}{\beta^2}t^2\right) \qquad ...(7.1)$$

and transfer function

$$H_G(f) = \exp(-\beta^2 f^2) \qquad ...(7.2)$$

where $\left\{\beta = \frac{\sqrt{\ln 2}}{\sqrt{2}\,B} \Rightarrow \frac{0.58}{B}\right\}$

Easiest method to get a GMSK signal is to pass a Non Return Zero (NRZ) data through a Gaussian low pass filter (also known as Gaussian Base band filter) having an impulse response $h_G(t)$, followed by an FM modulator. Finally we get GMSK signal at Radio frequency. This modulation technique is used in variety of analog and digital communications. For example, used in US Cellular Digital Packet Data (US CDPD) system as well as in GSM system. Basic block diagram of GMSK transmitter and receiver shown in Fig. 7.17(*a*) and (*b*) respectively.

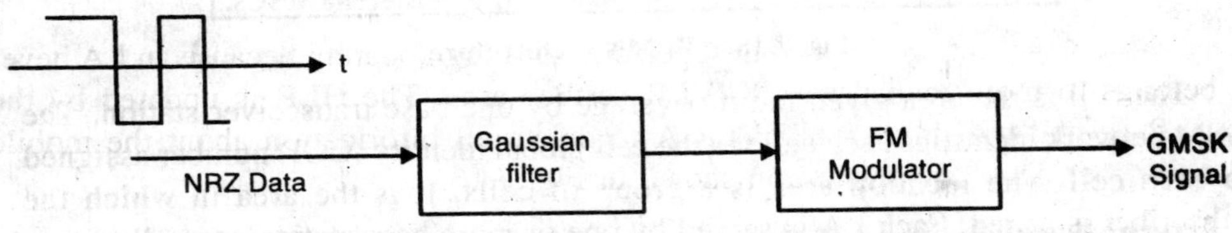

Fig. 7.17(*a*). GMSK transmitter using direct FM generation.

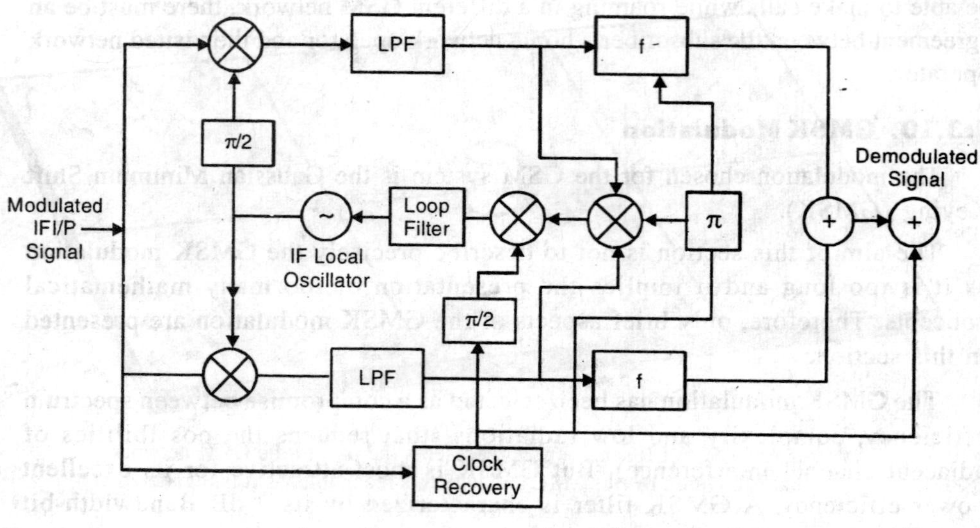

Fig. 7.17(*b*). GMSK receiver.

7.3.11. GSM Network Areas

The GSM network is made up of geographic areas. These areas include cells, location areas (LAs), MSC/VLR service areas, and public land mobile network (PLMN) areas.

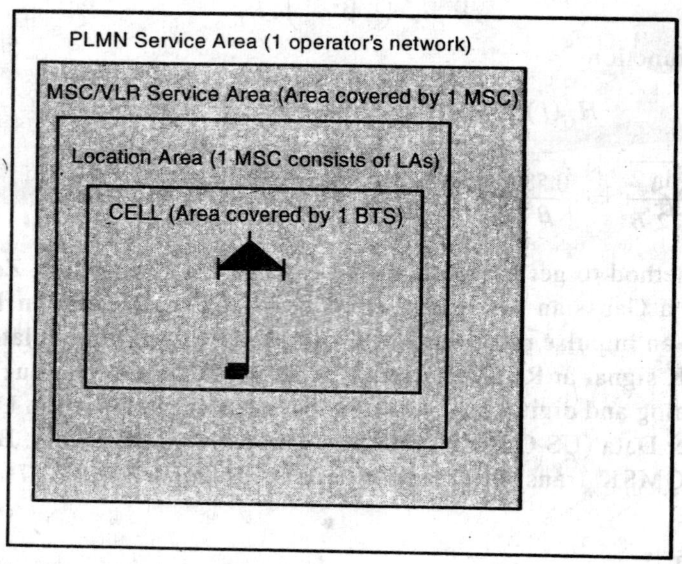

Fig. 7.18. PLMN architecture.

The cell is the area given radio coverage by one base transceiver station. The GSM network identifies each cell via the cell global identity (CGI) number assigned to each cell. The location area is a group of cells. It is the area in which the subscriber is paged. Each LA is served by one or more base station controllers, yet only by a single MSC (Fig. 7.19). Each LA is assigned a location area identity (LAI) number.

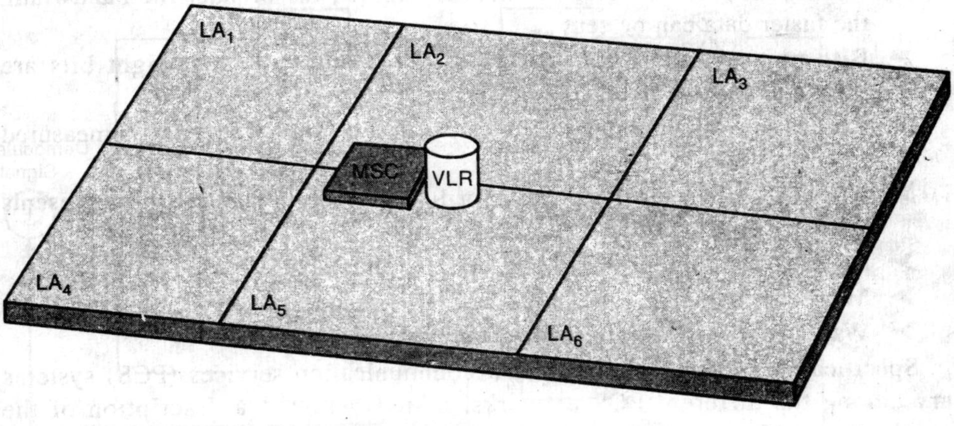

(*a*) An MSC/VLR service area represents the part of the GSM network that is covered by one MSC and which is reachable, as it is registered in the VLR of the MSC.

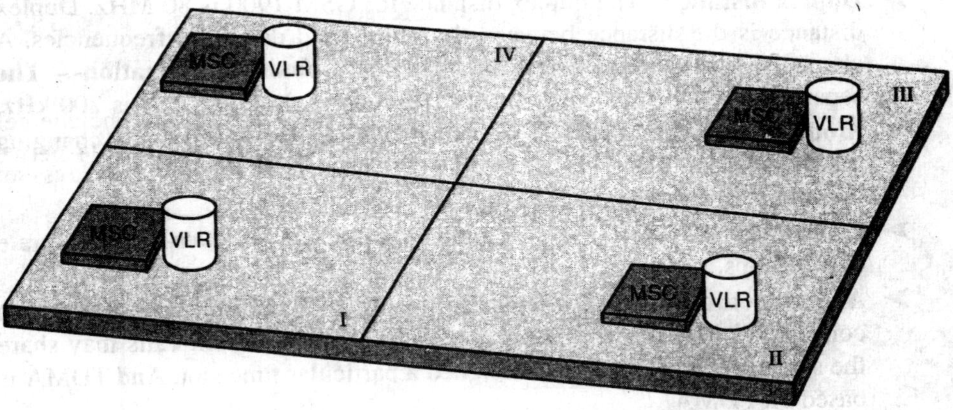

(*b*) The PLMN service area is an area served by one network operator.

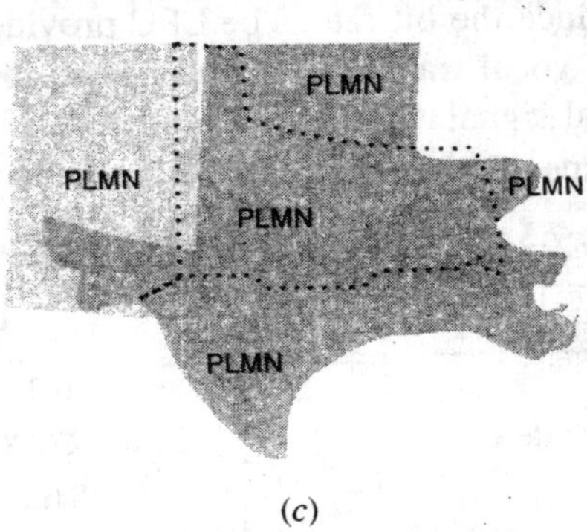

(*c*)

Fig. 7.19. Various network areas.

7.3.12. GSM Air Specifications

Before looking at the GSM specifications, it is important to understand the following basic terms:

- ➤ **Bandwidth**—the range of a channel's limits; the broader the bandwidth, the faster data can be sent
- ➤ **Bits per second (bps)**—a single on-off pulse of data; eight bits are equivalent to one byte
- ➤ **Frequency**—the number of cycles per unit of time; frequency is measured in hertz (Hz)
- ➤ **Kilo (k)**—kilo is the designation for 1,000; the abbreviation kbps represents 1,000 bits per second
- ➤ **Megahertz (MHz)**—1,000,000 hertz (cycles per second)
- ➤ **Milliseconds (ms)**—one-thousandth of a second
- ➤ **Watt (W)**—a measure of power of a transmitter

Specifications for different personal communication services (PCS) systems vary among the different PCS networks. Listed below is a description of the specifications and characteristics for GSM.

- ➤ **Frequency band:** The frequency range specified for GSM 1900 is 1,850 to 1,990 MHz (mobile station to base station).
- ➤ **Duplex distance:** The duplex distance for GSM 1900 is 80 MHz. Duplex distance is the distance between the uplink and downlink frequencies. A channel has two frequencies, 80 MHz apart. **channel separation— The** separation between adjacent carrier frequencies. In GSM, this is 200 kHz.
- ➤ **Modulation:** Modulation is the process of sending a signal by changing the characteristics of a carrier frequency. This is done in GSM via Gaussian minimum shift keying (GMSK).
- ➤ **Transmission rate**: GSM is a digital system with an over-the-air bit rate of 270 kbps.
- ➤ **Access method:** GSM utilizes the time division multiple access (TDMA) concept. TDMA is a technique in which several different calls may share the same carrier. Each call is assigned a particular time slot. And TDMA is based on FDMA.
- ➤ **Speech coder:** GSM uses linear predictive coding (LPC). The purpose of LPC is to reduce the bit rate. The LPC provides parameters for a filter that mimics the vocal tract. The signal passes through this filter, leaving behind a residual signal. Speech is encoded at 13 kbps.

Note: Some more specifications are shown in Table 7.5.

Table 7.5. GSM Air Interface Specifications.

Parameter	Specifications
Modulation	0.3 GMSK
Modulation data rate	270.83 Kbps
User per frame	8 (maximum)
Frame per frame	4.6 m second
Time slot period	577 μ second
Bit period	3.7 μ second
Voice codex bit rate	13.4 KbPs
Forward channel frequency	935-960 MHz
Reverse channel frequency	890-915 MHz

7.3.13. GSM Subscriber Services

There are two basic types of services offered through GSM: telephony (also referred to as teleservices) and data (also referred to as bearer services). Telephony services are mainly voice services that provide subscribers with the complete capability (including necessary terminal equipment) to communicate with other subscribers. Data services provide the capacity necessary to transmit appropriate data signals between two access points creating an interface to the network. In addition to normal telephony and emergency calling, the following subscriber services are supported by GSM:

- **Dual-tone multifrequency (DTMF):** DTMF is a tone signalling scheme often used for various control purposes via the telephone network, such as remote control of an answering machine. GSM supports full-originating DTMF.
- **Facsimile group III:** GSM supports CCITT Group 3 facsimile. As standard fax machines are designed to be connected to a telephone using analog signals, a special fax converter connected to the exchange is used in the GSM system. This enables a GSM-connected fax to communicate with any analog fax in the network.
- **Short message services:** A convenient facility of the GSM network is the short message service. A message consisting of a maximum of 160 alphanumeric characters can be sent to or from a mobile station. This service can be viewed as an advanced form of alphanumeric paging with a number of advantages. If the subscriber's mobile unit is powered off or has left the coverage area, the message is stored and offered back to the subscriber when the mobile is powered on or has reentered the coverage area of the network. This function ensures that the message will be received.
- **Cell broadcast:** A variation of the short message service is the cell broadcast facility. A message of a maximum of 93 characters can be broadcast to all mobile subscribers in a certain geographic area. Typical applications include traffic congestion warnings and reports on accidents.
- **Voice mail:** This service is actually an answering machine within the network, which is controlled by the subscriber. Calls can be forwarded to the subscriber's voice-mail box and the subscriber checks for messages via a personal security code.
- **Fax mail:** With this service, the subscriber can receive fax messages at any fax machine. The messages are stored in a service center from which they can be retrieved by the subscriber via a personal security code to the desired fax number.

GSM supports a comprehensive set of supplementary services that can complement and support both telephony and data services. Supplementary services are defined by GSM and are characterized as revenue-generating features. A partial listing of supplementary services follows:

- **Call forwarding:** This service gives the subscriber the ability to forward incoming calls to another number if the called mobile unit is not reachable, if it is busy, if there is no reply, or if call forwarding is allowed unconditionally.

- **Barring of outgoing calls:** This service makes it possible for a mobile subscriber to prevent all outgoing calls.
- **Barring of incoming calls:** This function allows the subscriber to prevent incoming calls. The following two conditions for incoming call barring exist: barring of all incoming calls and barring of incoming calls when roaming outside the home PLMN.
- **Advice of charge (AoC):** The AoC service provides the mobile subscriber with an estimate of the call charges. There are two types of AoC information: one that provides the subscriber with an estimate of the bill and one that can be used for immediate charging purposes. AoC for data calls is provided on the basis of time measurements.
- **Call hold:** This service enables the subscriber to interrupt an ongoing call and then subsequently reestablish the call. The call hold service is only applicable to normal telephony.
- **Call waiting:** This service enables the mobile subscriber to be notified of an incoming call during a conversation. The subscriber can answer, reject, or ignore the incoming call. Call waiting is applicable to all GSM telecommunications services using a circuit-switched connection.
- **Multiparty service:** The multiparty service enables a mobile subscriber to establish a multiparty conversation—that is, a simultaneous conversation between three and six subscribers. This service is only applicable to normal telephony.
- **Calling line identification presentation/restriction:** These services supply the called party with the integrated services digital network (ISDN) number of the calling party. The restriction service enables the calling party to restrict the presentation. The restriction overrides the presentation.
- **Closed user groups (CUGs):** CUGs are generally comparable to a PBX. They are a group of subscribers who are capable of only calling themselves and certain numbers.

7.3.14. GSM Security

GSM was designed with a moderate level of security. The system was designed to authenticate the subscriber using shared-secret cryptography. Communications between the subscriber and the base station can be encrypted. The development of UMTS introduces an optional USIM, that uses a longer authentication key to give greater security, as well as mutually authenticating the network and the user—whereas GSM only authenticated the user to the network (and not vice versa). The security model, therefore, offers confidentiality and authentication, but limited authorization capabilities, and no non-repudiation.

GSM uses several cryptographic algorithms for security. The A5/1 and A5/2 stream ciphers are used for ensuring over-the-air voice privacy. A5/1 was developed first and is a stronger algorithm used within Europe and the United States; A5/2 is weaker and used in other countries. A large security advantage of GSM over earlier systems is that the Ki, the crypto variable stored on the SIM card that is the key to any GSM ciphering algorithm, is never sent over the air interface. Serious weaknesses have been found in both algorithms, and it is possible to break A5/2 in real-time in a ciphertext-only attack. The system supports multiple algorithms so operators may replace that cipher with a stronger one.

7.4. SETUP CHANNELS

Setup Channel can be classified in following types:

7.4.1. GSM Traffic Channels (TCHs)

GSM traffic channel may carry digitally encodeḋ user speech user data with identical functions on forward and reverse channel both. For flowing of data on channel it may get full rate or half rate.

When data is transmitted as half rate, user data is mapped on to the same time slot, but is sent in alternate frames, It means two users can share the same time slot but would alternately transmit during every other frame. While in full rate, single user occupied the whole time slots. Following are the specifications that are supported by full rate and half rate.

Half Rate TCH:

1. *Half rate speech channels.* The half rate speech channels have been designed to carry the digitized speech or user data sampled at a rate half of the full rate channel. The half rate speech channels carry the data at the rate of 11.4 KbPs.
2. *Half rate data channel for 4800 bPs (TCH/H 4.8).* The half rate traffic data channel carries raw user data which is sent at 4800 bPs. This 4800 bPs data is sent at 11.4 KbPs.

Full rate TCH: Full rate speech channel (TCH/FS). The full rate channel carries user speech in digitized at a raw data rate of 13 KbPs. With GSM channel coding added to the digitized speech. The full rate speech channel carries 22.8 KbPs.

The full rate traffic data channel carries raw user data which is sent at 9600 bPs, 4800 bPs, 2400 bPs at 22.8 KbPs.

7.4.2. GSM Control Channel

GSM control channel can be specified in three parts

(*a*) Broadcast Channel (BCH)
(*b*) Common Control Channel (CCCH)
(*c*) Dedicated Control Channel (DCCH)

7.4.2.1. Broadcast Channel (BCH)

Broadcast channel operates on the forward line of a specified ARFCN (Absolute Radio Frequency Channel Number) - within each cell and transmits data only in the first time slot TS0 of GSM frame. In GSM, the Forward Channel (935-960) base to subscriber and reverse channel (890-915 MHz) subscriber to base transmission is divided into 200 kHz wide channel called ARFCN. The ARFCN denotes a forward and reverse channel pair which is separated in frequency by 45 MHz and each channel is time shared between as many as eight subscribers using TDMA. Each of the 8 subscribers uses the same ARFCN and occupies a unique time slot per Frame. Radio transmission on both the forward and reverse link are made at a channel data rate of 1625 KbPs.

BCH only uses the forward link unlike TCH uses both forward and reverse. Although BCH data is transmitted in TC0, therefore, other seven (7) time slots in GSM Frame for that same ARFCN are available for TCH data. BCH is defined by three separate channels which are given access to TS0 during various frames of the 51 frame sequence as shown in Fig. 7.20. For GSM, TDM frame three types of BCH are described.

Broadcast Control Channel (BCCH)

BCCH is used to broadcast information on forward channel such as cell and network identity and operating characteristic of the cell (like channel availability, channel structure and congestion). The BCCH also used to broadcast list of channels that are currently in use within the cell.

Frequency Correction Channel (FCCH)

FCH is a specific type of data that occupies TSO frame and it is repeated every ten frames within a control channel multiframe. FCCH is also used to synchronize internal frequency standard to base station frequency.

Synchronization Channel (SCH)

SCH is broadcast in TS0 frame, immediately following the FCCH frame used to identify the serving base station while following each mobile to frame synchronize with the base station.

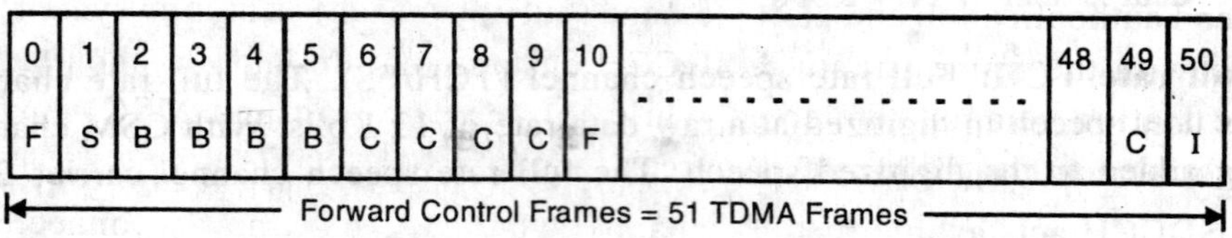

F ⟶ FCCH burst (BCH)

S ⟶ SCH burst (BCH)

B ⟶ BCCH burst (BCH)

C ⟶ CCCH

I ⟶ Idle

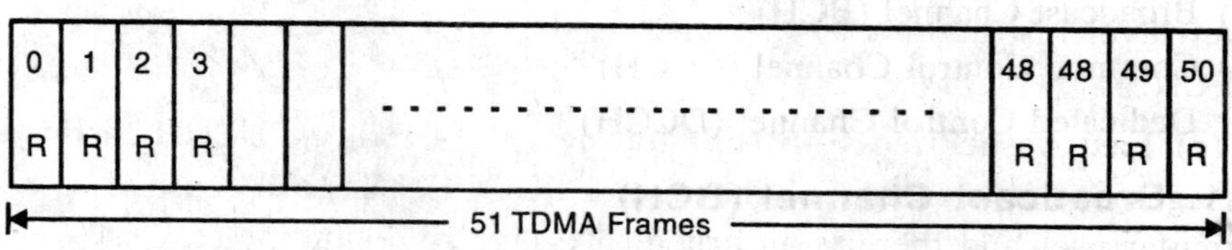

R ⟶ Reverse (CCCH) bursts.

Fig. 7.20.

The frame number (0 to 2715647) is sent with the base station identity code during the SCH burst.

7.4.2.2. Common Control Channel (CCCH)

CCCH consists of three types of channels.

(*a*) Paging Access Channel (PCH). PCH provides paging signals from the base station to all mobile phones in the cell and also notifies that a specific call is originated from PSTN line. PCH may be used to broadcast text message to all subscribers as SMS features in GSM.

(*b*) Random Access Channel (RACH). It is used by reverse channel of a subscriber unit to acknowledge the page from PCH.

Access grant Channel (AGCH). AGCH is used by the base station to provide forward link communication to the mobile, and carries data which instructs the mobile to operate in a particular physical channel with a specific dedicated control channel.

7.4.2.3. Dedicated Control Channel (DCC)

Dedicated control channels are bidirectional in nature and having the same format and function on both the forward and reverse channel.

DCCH may exist in any time slot and on any ARFCN like TCCH except TSO frame. This is classified in three parts.

(*a*) Stand Alone Dedicated Control Channels (SDCCHs). SDCCH carries signalling data following the connection of the mobile with the base station and just before a TCH bursts is broadcast by the base station. The SDCCH is used to send authentication and alert message (but not speech) as mobile synchronizes itself with the frame structure. SDCCH ensures that the mobile phone and the base station remain connected while the base station and MSC verify the subscriber unit and allocate resources for the mobile.

SDCCH acts as intermediate channel which accepts a newly connected call from the BCCH and holds the traffic while waiting for the base section to allocate a TCH channel.

(*b*) Slow Associated Control Channel. SACCH carries general information between the MS and BTS. On the forward channel, the SACCH is used to send slow but regularly changing control information to mobile such as transmit power level instructions and specific timing instructions for each user on the ARFCN.

(*c*) Fast Associated Control Channels (FACCHs). FACCH carries urgent messages and it is assigned whenever a SDCCH has not been dedicated for a particular user and there is an urgent message like handoff request.

A flow chart of control channel is shown below.

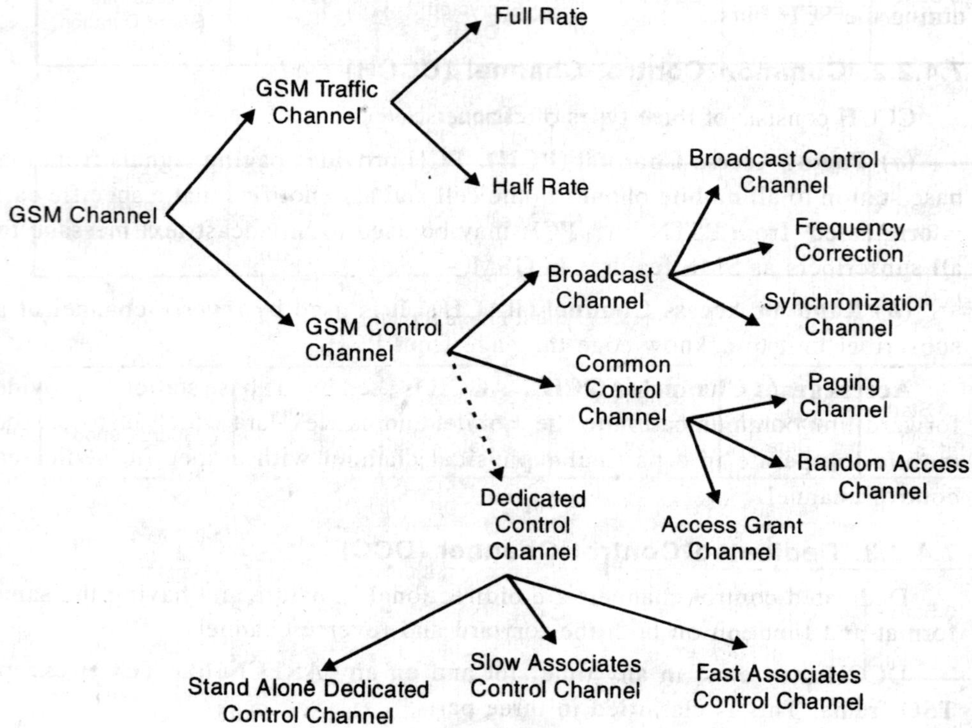

Fig. 7.21. Frame structure.

7.5. FRAME STRUCTURE OF GSM

As we know GSM technology uses the TDM access technique during conversation of a cellphone, user transmit bursts of data, this burst of data before broadcasting is arranged in a specific manner by using various overheads bits. Overhead bits depend upon channel specification. Therefore a Frame structure can be classified in five parts as shown in Fig. 7.22.

Normal burst frame structures are used in DCCH broadcasting on forward and reverse link.

RACH burst is used by all GSM cellular phones to access service from any base station. While FCCH and SCH burst frame structure are used in TSO of specific frames to transmit the time and frequency control bits for synchronization on the forward channel.

Dummy burst frame structure is used as filter information for unused time slots on the forward channel.

Figure 7.23, shows the various timeslots (TS) in normal burst frame structure. It is understood from Fig. 7.20, that normal frame structure consists of 156.25 bits. Out of these bits 148 bits are transmitted at the rate of 270.83 Kbps while 8.25 bits are provided at the end of normal burst frame as guard period.

3 Start Bit	58 Bits of Encrypted Data	26 Training Bits	58 Bits of Encrypted Data	3 Stop Bits	8.25 Bits Guard Duration

(*a*) Normal Frame Structure.

3 Start Bit	142 Fixed Bits of All Zeroes	3 Stop Bits	8.25 Bits Guard Period

(*b*) FCCH Burst Frame Structure.

3 Start Bit	39 Bits of Encrypted Data	64 Bits of Encrypted Data	39 Bits of Encrypted Data	3 Stop Bits	8.25 Bits Guard Period

(*c*) SCH Burst Frame Structure.

3 Start Bit	41 Bits of Synchronization	36 Bits of Encrypted Data	3 Stop Bits	68.25 Bit Extended Guard Period

(*d*) RACH Burst Frame Structure.

3 Start Bit	58 Mixed Bits	26 Training Bits	58 Mixed Bits	8.25 Bits Guard Period

(*e*) Dummy Burst Frame Structure for GSM.

Fig. 7.22. Various GSM Frame Structures.

Out of 148 bits per time slot (TS), 114 are message bits which are broadcast as two 57 bit sequence close to start of burst Frame and end of burst frame. While out of rest 34 bits, 26 bits are training bits. These 26 bits are called midamble bits because these are placed in middle of frame.

These training bits are used to analyse the radio channel characteristics before the decoder at base station or in mobile station.

On either side of training bits 1 bit is used as stealing flags. These stealing flags are used to distinguish whether the time slot (TS) contains voice (TCH) or control (FACH) data.

In GSM communication, during a Frame structure GSM mobile phones use one Time Slot (TS) to transmit, and one TS to receive, and six spare time slots may be used to measure the signal strength on 5 (five) adjacent base stations as well as its own base station.

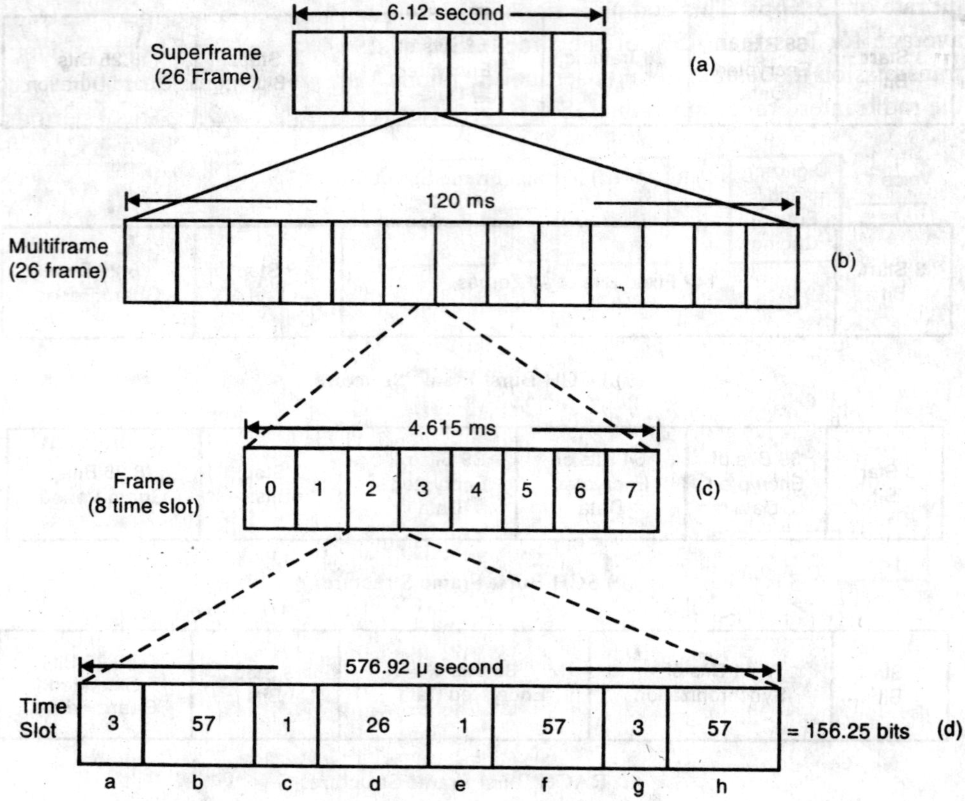

a, g ⟶ Tail bit or start bit
h ⟶ Guard period
b, f ⟶ Coded data
c, e ⟶ Stealing flag
d ⟶ Midamble bit

Fig. 7.23. Various time slots in normal frame structure.

In Fig. 7.23(*c*), consists of one frame, which having 8 time slots and Fig. 7.23(*d*), represent 1 time slot. In one time slot Fig. 7.23(*d*), number of bits is equal to 156.25 hence number of bits in a Frame is equal to

$$156.25 \times 8 = 1250 \text{ bits}$$

And for transmission purpose, these frames are grouped into multiframes and multiframes are grouped into super frames.

7.6. SIGNAL PROCESSING IN GSM

Signal processing is very important phenomena of any communication system. Processing means to convert a signal into such a form so that it can be broadcast and received with less distortion. Signal processing includes coding, interleaving modulation etc. as shown in Fig. 7.24.

7.6.1. Speech Coding

GSM speech coder based on RELP (Residually Excited Linear Predictive Coder). RELP provides 260 bits for each 20 ms blocks of speech, which provides bit rate of 13 Kbps. This coding is done by assuming that each subscriber speaks on average for less than 45% of the time. GSM systems operate in a discontinuous transmission mode, due to that user mobile phone battery life increases and reduces the radio interference since the GSM transmitter is not active during silent periods.

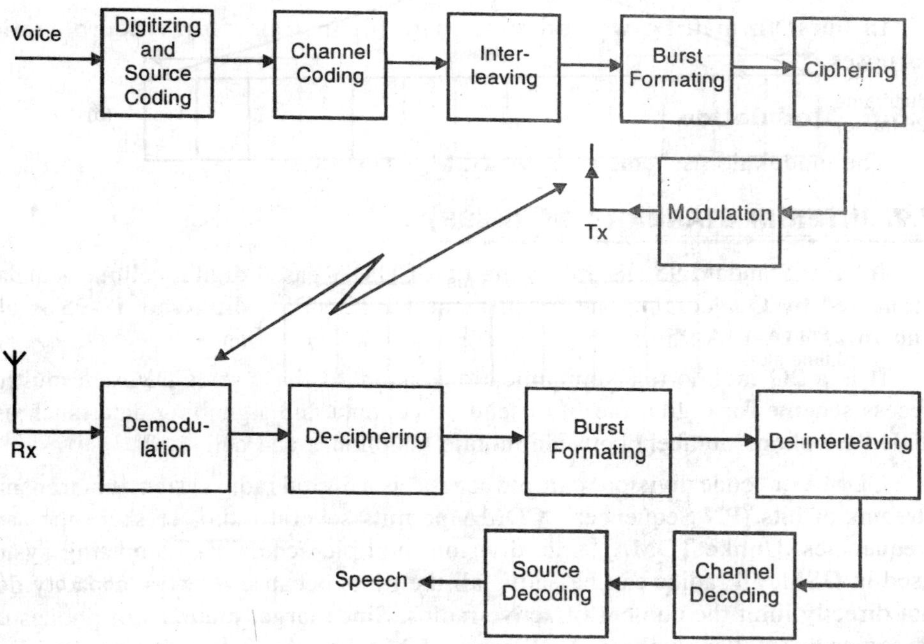

Fig. 7.24. GSM signal processing.

7.6.2. Channel Coding

Output of speech coder is grouped for error protection by means of channel encoding based on their significance contributing in voice quality. Generally CRC (Cyclic Redundancy Check) method is used for error correction.

7.6.3. Interleaving

Interleaving is process to introduce some bits as interleaving bits at specific interval of data bits so that receiver can easily recognize the receiving bits particularly starting and ending of transmitted bit.

Therefore, control message frames are broken into eight 57 bit sub-blocks. These eight sub-blocks which make up a single speech frame are spread over eight consecutive TCH time slots. Due to any reason if a burst is lost due to interference or fading, then channel coding must ensure that enough bits will still be received correctly to allow the error correction to work.

7.6.4. Ciphering

Ciphering is encryption technique. It modifies the received contents after interleaving but encryption technique used for modification of data bit must be known to mobile phone and base station. In GSM communication, two types of ciphering technique used are A_3 and A_5.

A_3 is used to authenticate each mobile by leaving the users passcode within the SIM with the key cryptography key at the MSC. A_5 algorithm provides the scrambling for the 114 coded data bits sent in each time slot.

7.6.5. Burst Formatting

In burst formatting we add some data bit in frame for synchronization purposes.

7.6.6. Modulation

The modulation scheme used for GSM is GMSK.

7.7. INTERIM STANDARD 95 (IS-95)

Interim Standard 95 (IS-95), is the first CDMA-based digital cellular standard pioneered by Qualcomm. The brand name for IS-95 is **cdmaOne.** IS-95 is also known as TIA-EIA-95.

It is a 2G mobile telecommunications standard that uses CDMA, a multiple access scheme for digital radio, to send voice, data and signalling data (such as a dialed telephone number) between mobile telephones and cell sites.

CDMA or "code division multiple access" is a digital radio system that transmits streams of bits (PN Sequences). CDMA permits several radios to share the same frequencies. Unlike TDMA "time division multiple access", a competing system used in GSM, all radios can be active all the time, because network capacity does not directly limit the number of active radios. Since larger numbers of phones can be served by smaller numbers of cell-sites, CDMA-based standards have a significant economic advantage over TDMA-based standards, or the oldest cellular standards that used frequency-division multiplexing.

It is now being supplanted by IS-2000 (CDMA2000), a later CDMA-based standard. It is used in the USA, South Korea, Canada, Mexico, India, Israel, Australia, Venezuela, Brazil and China.

7.7.1. Protocol Revisions

CDMA One's technical history is reflective of both its birth as a Qualcomm internal project, and the world of then unproven competing digital cellular standards under which it was developed. The term IS-95 generically applies to the earlier set of protocol revisions, namely P_REV's one through five.

P_REV=1 was developed under an ANSI standards process with documentation reference *J-STD-008*. J-STD-008, published in 1995, was only defined for the then new North American PCS band (Band Class 1, 1900 MHz). The term *IS-95* properly refers to P_REV=1, developed under the Telecommunications Industry Association (TIA) standards process, for the North American cellular band (Band Class 0, 800 MHz) under roughly the same time frame. IS-95 offered interoperation (including handoff) with the analog cellular network. For digital operation, IS-95 and J-STD-008 have most technical details in common. The immature style and structure of both documents are highly reflective of the "standardizing" of Qualcomm's internal project.

P_REV=2 is termed *Interim Standard 95A (IS-95 A).* IS-95A was developed for Band Class 0 only, as in incremental improvement over IS-95 in the TIA standards process.

P_REV=3 is termed *Technical Services Bulletin 74 (TSB-74).* TSB-74 was the next incremental improvement over IS-95A in the TIA standards process.

P_REV=4 is termed *Interim Standard 95B (IS-95B) Phase I,* and P_REV=5 is termed *Interim Standard 95B (IS-95 B) Phase II.* The IS-95B standards track provided for a merging of the TIA and ANSI standards tracks under the TIA, and was the first document that provided for interoperation of IS-95 mobile handsets in both band classes (dual-band operation). PREV=4 was by far the most popular variant of IS-95, with P_REV=5 only seeing minimal uptake in South Korea.

P_REV=6 and beyond fall under the CDMA2000 umbrella. Besides technical improvements, the IS-2000 documents are much more mature in terms of layout and content. They also provide backwards-compatibility to IS-95.

7.7.1.1. Protocol Details

The IS-95 standards describe an *air interface,* a set of protocols used between mobile units and the network. IS-95 is widely described as a three-layer stack, where L1 corresponds to the physical (PHY) layer, L2 refers to the Media Access Control (MAC) and Link-Access Control (LAC) sublayers, and L3 to the call-processing state machine.

***(a)* Physical Layer:** IS-95 defines the transmission of signals in both the *forward* (network-to-mobile) and *reverse* (mobile-to-network) directions.

In the forward direction, radio signals are transmitted by base stations (BTS's). Every BTS is synchronized with a GPS receiver so transmissions are tightly controlled in time. All forward transmissions are BPSK with a chip rate of 1,228,800 per second. Each signal is spread with a Walsh code of length 64 and a pseudo random noise code (PN code) of length 2^{15} yielding a PN roll-over period of $\frac{80}{3}$ ms.

***(b)* Forward Broadcast Channels:** Every BTS dedicates a significant amount of output power to a *pilot channel,* which is an unmodulated PN sequence (in other words, spread with Walsh code 0). Each BTS sector in the network is assigned a PN offset in steps of 64 chips. There is no data carried on the forward pilot. With its strong autocorrelation function, the forward pilot allows mobiles to determine system timing and distinguish different BTS's for handoff.

When a mobile is "searching", it is attempting to find pilot signals on the network by tuning to particular radio frequencies, and performing an autocorrelation sweep across all possible PN phases. A strong autocorrelation result indicates the proximity of a BTS.

Other forward channels, selected by their Walsh code, carry data from the network to the mobiles. Data consists of network signalling and user traffic. Generally, data to be transmitted is divided into frames of bits. A frame of bits is passed through a convolutional encoder, adding forward error correction redundancy, generating a frame of symbols. These symbols are then spread with the Walsh and PN sequences and transmitted.

BTS's transmit a *sync channel* spread with Walsh code 32. The sync channel frame is $\frac{80}{3}$ ms long, and its frame boundary is aligned to the pilot. The sync channel continually transmits a single message, the *Sync Channel Message,* which has a length and content dependent on the P_REV. The message is transmitted 32 bits per frame, encoded to 128 symbols, yielding a rate of 1200 bit/s. The Sync Channel Message contains information about the network, including the PN offset used by the BTS sector.

Once a mobile has found a strong pilot channel, it listens to the sync channel and decodes a Sync Channel Message to develop a highly-accurate synchronization to system time. At this point the mobile knows whether it is roaming, and that it is "in service".

BTS's transmit at least one, and as many as seven, *paging channels* starting with Walsh code 1. The paging channel frame time is 20 ms, and is time aligned to the IS-95 system (*i.e.,* GPS) 2-second roll-over. There are two possible rates used on the paging channel: 4800 bit/s or 9600 bit/s. Both rates are encoded to 19200 symbols per second.

The paging channel contains signalling messages transmitted from the network to all idle mobiles. A set of messages communicate detailed network overhead to the mobiles, circulating this information while the paging channel is free. The paging channel also carries higher-priority messages dedicated to setting up calls to and from the mobiles.

When a mobile is idle, it is mostly listening to a paging channel. Once a mobile has parsed all the network overhead information, it *registers* with the network, then optionally enters *slotted-mode.* Both of these processes are described in more detail below.

7.7.2. Forward Traffic Channels

The Walsh space not dedicated to broadcast channels on the BTS sector is available for *traffic channels.* These channels carry the individual voice and data calls supported by IS-95. Like the paging channel, traffic channels have a frame time of 20 ms.

Since voice and user data are intermittent, the traffic channels support variable-rate operation. Every 20 ms frame may be transmitted at a different rate, as determined by the service in use (voice or data). P_REV=1 and P_REV=2 supported *rate set 1,* providing a rate of 1200, 2400, 4800, or 9600 bit/s. P_REV=3 and beyond also provided *rate set* 2, yielding rates of 1800, 3600, 7200, or 14400 bit/s.

For voice calls, the traffic channel carries frames of *vocoder* data. A number of different vocoders are defined under IS-95, the earlier of which were limited to rate set 1, and were responsible for some user complaints of poor voice quality. More sophisticated vocoders, taking advantage of modern DSPs and rate set 2, remedied the voice quality situation and are still in wide use in 2005.

The mobile receiving a variable-rate traffic frame does not know the rate at which the frame was transmitted. Typically, the frame is decoded at each possible rate, and using the quality metrics of the Viterbi decoder, the correct result is chosen.

Traffic channels may also carry circuit-switch data calls in IS-95. The variable-rate traffic frames are generated using the IS-95 *Radio Link Protocol (RLP).* RLP provides a mechanism to improve the performance of the wireless link for data. Where voice calls might tolerate the dropping of occasional 20 ms frames, a data call would have unacceptable performance without RLP.

Under IS-95B PREV=5, it was possible for a user to use up to seven supplemental "code" (traffic) channels simultaneously to increase the throughput of a data call. Very few mobiles or networks ever provided this feature, which could in theory offer 115200 bit/s to a user.

7.7.3. Capacity

IS-95 and its use of CDMA techniques, like any other communications system, are limited to their throughput according to Shannon's theorem. Accordingly, capacity improves with SNR and bandwidth. IS-95 has a fixed bandwidth, but fares well in the digital world because it takes active steps to improve SNR.

With CDMA, signals that are not correlated with the channel of interest (such as other PN offsets from adjacent cellular base stations) appear as noise, and signals carried on other Walsh codes (that are properly time aligned) are essentially removed in the despreading process. The variable-rate nature of traffic channels provide lower-rate frames to be transmitted at lower power causing less noise for other signals yet still be correctly received. These factors provide an inherent lower noise level than other cellular technologies allowing the IS-95 network to squeeze more users into the same radio spectrum.

Active (slow) power control is also used on the forward traffic channels, where during a call, the mobile sends signalling messages to the network indicating the quality of the signal. The network will control the transmitted power of the traffic channel to keep the signal quality just good enough, thereby keeping the noise level seen by all other users to a minimum.

The receiver also uses the techniques of the Rake receiver to improve SNR as well as perform Soft handoff.

7.8. DIFFERENCE BETWEEN GSM AND CDMA

In cellular service there are two main competing network technologies: Global System for Mobile Communications (GSM) and Code Division Multiple Access (CDMA).Understanding the difference between GSM and CDMA will allow you to choose a carrier that uses the preferable network technology for your needs.

The GSM Association is an international organization founded in 1987, dedicated to providing, developing, and overseeing the worldwide wireless standard of GSM. CDMA, a proprietary standard designed by Qualcomm in the United States, has been the dominant network standard for North America and parts of Asia. However, GSM networks continue to make inroads in the United States, as CDMA networks make progress in other parts of the world. There are camps on both sides that firmly believe either GSM or CDMA architecture is superior to the other. That said, to the non-invested consumer who simply wants bottom line information to make a choice, the following considerations may be helpful.

Coverage: The most important factor is getting service in the areas you will be using your phone. Upon viewing competitors' coverage maps you may discover that only GSM or CDMA carriers offer cellular service in your area. If so, there is no decision to be made, but most people will find that they do have a choice.

Data Transfer Speed: With the advent of cellular phones doing double and triple duty as streaming video devices, podcast receivers and email devices, speed is important to those who use the phone for more than making calls. CDMA has been traditionally faster than GSM, though both technologies continue to rapidly leapfrog along this path. Both boast "3G" standards, or 3rd generation technologies.

EVDO, also known as CDMA2000, is CDMA's answer to the need for speed with a downstream rate of about 2 megabits per second, though some reports suggest real world speeds are closer to 300-700 kilobits per second (kbps). This is comparable to basic DSL. As of fall 2005, EVDO is in the process of being deployed. It is not available everywhere and requires a phone that is CDMA2000 ready.

GSM's answer is EDGE (Enhanced Data Rates for GSM Evolution), which boasts data rates of up to 384 kbps with real world speeds reported closer to 70-140 kbps. With added technologies still in the works that include UMTS (Universal Mobile Telephone Standard) and HSDPA (High Speed Downlink Packet Access), speeds reportedly increase to about 275-380 kbps. This technology is also known as W-CDMA, but is incompatible with CDMA networks. An EDGE-ready phone is required.

In the case of *EVDO,* theoretical high traffic can degrade speed and performance, while the EDGE network is more susceptible to interference. Both require being within close range of a cell to get the best speeds, while performance decreases with distance.

Subscriber Identity Module (SIM) cards: In the whole world only GSM phones use SIM cards. The removable SIM card allows phones to be instantly activated, interchanged, swapped out and upgraded, all without carrier intervention. The SIM itself is tied to the network, rather than the actual phone. Phones that are card-enabled can be used with any GSM carrier.

The CDMA equivalent, a R-UIM (User Identity Module) card, is only available in parts of Asia but remains on the horizon for the U.S. market. CDMA carriers in the U.S. require proprietary handsets that are linked to one carrier only and are not card-enabled. To upgrade a CDMA phone, the carrier must deactivate the old phone then activate the new one. The old phone becomes useless.

Roaming: For the most part, both networks have fairly concentrated coverage in major cities and along major highways. GSM carriers, however, have roaming contracts with other GSM carriers, allowing wider coverage of more rural areas, generally speaking, often without roaming charges to the customer. CDMA networks may not cover rural areas as well as GSM carriers, and though they may contract with GSM cells for roaming in more rural areas, the charge to the customer will generally be significantly higher.

International Roaming: If you need to make calls to other countries, a GSM carrier can offer international roaming, as GSM networks dominate the world market. If you travel to other countries you can even use your GSM cell phone abroad, providing it is a quad-band phone (850/900/1800/1900 MHz). By purchasing a SIM card with minutes and a local number in the country you are visiting, you can make calls against the card to save yourself international roaming charges from your carrier back home. CDMA phones that are not card-enabled do not have this capability.

According to various International Organization wireless service provider, CDMA networks support over 270 million subscribers worldwide, while GSM.org tallies up their score at over 1 billion. As CDMA phones become R-UIM enabled and roaming contracts between networks improve, integration of the standards might eventually make differences all but transparent to the consumer. The chief GSM carriers in the India are BSNL, AirTel, Hutch, Idea, while Major CDMA carriers are Reliance India Mobile. There are also several smaller cellular companies on both networks.

7.9. PERSONAL COMMUNICATION ACCESS SYSTEM (PACS)

PACS is a third generation system based on commuiization system. PACS is supposed to be support voice, picture, data, moving picture for indoor and microcell. Coverage area of PACS is supposed to be around 500 meters. Main objective of PACS is to integrate all forms of wireless services like Bluetooth, Infrared and cellular based service in one system with all basic features of telephone system.

7.9.1. PACS System Architecture

A simplified PACS system architecture is shown in Fig. 7.25. PACS was developed as a universal wideband wireless access system for the application in

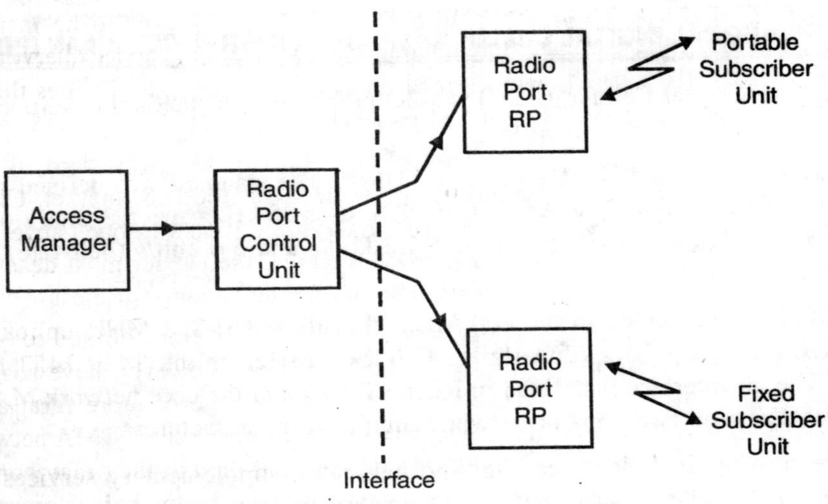

Fig. 7.25. PACS system architecture.

licensed or in unlicensed PACS bands. PACS may be connected to PSTN, EPABX system. PACS is divided and four part, named as subscriber unit, radio parts, and Access manager.

7.9.2. PACS Radio Interface

PACS system is designed for the operation in the USPCS band. In PACS radio interface, number of RF channels may be frequency division multiplexed with 80 MHz separation, or some time it is time division multiplexed. Some of the Radio Interface specification shown in Table 7.7 PACS subscriber unit uses adaptive power control to minimize battery drain during transmission and to minimize the co-channel interference on the reverse path.

Table 7.7. PACS Radio Interface Specifications

Parameter	Specifications
Frequency band	1 - 3 GHz
Modulation	$\frac{\pi}{4}$ DQPSK
Multiple access	TDMA
Channel spacing	300 kHz
Channel coding	CRC
Speech coding	ADPCM (16 bit)
Bit error	10^{-2}
Frame duration	2.5 m second
Time slot per frame	8
Multiple access	TDMA
Duplexing	FDD
Channel spacing	300 kHz
Base station power	800 mW
Subscriber unit power	200 mW

7.10. PERSONAL DIGITAL CELLULAR/PACIFIC DIGITAL CELLULAR (PDC)

Personal Digital Cellular (PDC) is a 2G mobile phone standard developed and used exclusively in Japan.

Like D-AMPS and GSM PDC uses TDMA. The standard was defined by the RCR (later became ARIB) in April 1991. PDC uses 25 kHz carrier, 3 time slots, pi/4-DQPSK modulation and low bit-rate 11.2 kbit/s and 5.6 kbit/s (half-rate) voice codecs.

PDC is implemented in the 800 MHz (downlink 810-888 MHz, uplink 893-958 MHz), and 1.5 GHz (downlink 1477-1501 MHz, uplink 1429-1453 MHz) bands. The air interface is defined in RCR STD-27 and the core network MAP by JJ-70.10. Ericsson is the major network equipment manufacturer.

The services include voice (full and half-rate), supplementary services (call waiting, voice mail, three-way calling, call forwarding, and so on), data service (up to 9.6 kbit/s CSD), and packet-switched wireless data (up to 28.8 kbit/s PDC-P).

Compared to GSM, PDC's weak broadcast strength allows small, portable phones with light batteries at the expense of substandard voice quality and problems maintaining the connection, particularly in enclosed spaces like elevators.

After a peak of nearly 80 million subscriber to PDC, it now has 45.856 million subscribers (December 2005) and is slowly being phased out in favour of 3G technologies like W-CDMA and CDMA2000.

7.10.1. Introduction

The Personal Digital Cellular or Pacific Digital Cellular (PDC) system is a second-generation mobile phone technology introduced in 1991. Although it is only found in Japan, there its use is very widespread and there are a considerable number of users. Accordingly it accounts for over 10% of the world market for second-generation mobile phone users.

PDC was introduced in Japan in 1991 with the move from analogue to digital technology. It uses TDMA technology and it is very similar to the US "TDMA" or IS54/IS136 system but operates in the 800 and 1500 MHz bands. The modulation scheme, voice frame size, TDMA frame duration, and interleaving remain the same. The major difference is that it uses a 25 kHz channel spacing instead of 30 kHz.

7.10.2. Technology

As already mentioned PDC is a TDMA system and it operates by splitting each channel into several time slots and thereby allowing several users to use the same frequency channel. For each channel it is possible to support three users under normal circumstances. However, when traffic levels are high it is possible to use half data rate speech. Although this reduces the speech quality, it enables six calls to be supported by each channel. This compares very favourably to GSM that manages eight within each 200 kHz channel.

Speech encoding is an important factor. PDC uses a different encoder to that used on IS54/IS136. The standard rate is 9.6 kbps along with similar technologies such as GSM, but when half rate encoding is used this falls to 5.6 kbps. Although this gives a significant reduction in voice quality, it is still adequate to maintain intelligibility and enables the network capacity to be increased to accommodate further calls.

Whilst it may appear that the spectral efficiency of PDC would result in a lower level of features, it manages to support many of the expected features including text messaging and caller identification. It also has Intelligent Network (IN) capabilities and these enable it to support features such as prepaid calling, Universal Access Numbers, personal numbers and virtual private networks (VPN). These VPNs are limited access groups that allow people working in different locations to use the mobile phone network as thought it was an internal office phone system.

7.10.2.1. Data

To facilitate data transmission to a data rate of 28.8 kbps the PDC-P (PDC mobile Packet Data Communication System) has been introduced. The normal system uses a circuit switched approach where a complete circuit is switched for a given user. This is fine for voice communications, but where data applications are concerned a packet switched approach is far more efficient. Here individual packets of data are routed to the required user and there is no dedicated channel. This makes far more efficient use of the channels available because data transmissions are notoriously "bursty" in nature. By adopting this approach users can be charged for the data transferred rather than the connection time.

7.10.2.2. Coverage

One very important feature in Japan is that of coverage, and that of coverage within buildings such as shopping malls, offices and subway stations. In fact this is a major sales differentiator between the different operators and as such it is of great importance. In view of this there is a large number of micro and pico cells that have been set up. The base stations for these cells use low power combined with distributed antenna systems and repeaters to ensure the required coverage.

7.10.2.3. Migration

One of the important aspects of any system today is the migration path to 3G. With the move to the greater use of packet data on PDC this provides an important step towards the IMT2000. This will provide a data transfer rate of up to 2 Mbps.

7.11. PERSONAL HANDY-PHONE SYSTEM

The Personal Handy-phone System (PHS), also marketed as the **Personal Access System** (PAS) and known as *Xiaolingtong* in China, is a mobile network system operating in the 1880-1930 MHz frequency band, used mainly in Japan, China, Taiwan and some other Asian countries.

7.11.1. Technology

PHS is, essentially, a cordless telephone like DECT, with the capability to handover from one cell to another. PHS cells are small, with transmission power of base station a maximum of 500 mW and range typically measures in tens or at most hundreds of meters (some of brand-new base stations can range about 2 kilometers at line-of-sight), as opposed to the multi-kilometer ranges of GSM. This makes PHS suitable for dense urban areas, but impractical for rural areas, and the small cell size also makes it difficult if not impossible to make calls from rapidly moving vehicles.

PHS uses TDMA/TDD for its radio channel access method, and 32 kbit/s ADPCM for its voice codec. Modern PHS phone can also support many value-added services such as high speed wireless data / Internet connection (64 kbit/s and higher), WWW access, e-mailing, text messaging and even colour image transfer.

PHS technology is also a popular option for providing a wireless local loop, where it is used to bridge the "last mile" gap between the POTS network and the subscriber's home. Actually, it was developed under the concept that it makes up a wireless front-end of ISDN network. So a base station of PHS has a compatibility with, and is often connected directly to ISDN telephone exchange equipments (aka digital switch).

As for its low-price base station, micro-cellular system and 'Dynamic Cell Assign' system, PHS can afford more number-of-digits frequency use efficiency with lower cost (throughput per area basis), compared with typical 3G cellular telephone systems. It makes possible the flat-rate wireless service such as AIR-EDGE all over Japan.

The speed of AIR-EDGE data connection is accelerated by combining lines, each of which basically is 32 kbit/s. AIR-EDGE 1x or first version introduced in 2001 provide only 32 kbit/s service. In 2002, 128 kbit/s service (AIR-EDGE 4x) started. In 2005, 256 kbit/s (AIR-EDGE 8x) service started. Furthermore, in 2006, the speed of each line was also upgraded to 1.6 times, so the speed of AIR-EDGE 8x is up to 402 kbit/s with the latest instrument. And it may exceed the speeds of popular W-CDMA 3G service like NTTDoCoMo's FQMA in Japan.

7.12. THE INDUSTRIAL, SCIENTIFIC AND MEDICAL (ISM BAND)

The **industrial, scientific and medical (ISM)** radio bands were originally reserved internationally for non-commercial use of RF electromagnetic fields for industrial, scientific and medical purposes.

The **ISM** bands are defined by the ITU-R in 5.138 and 5.150 of the Radio Regulations. Individual countries' use of the bands designated in these sections may differ due to variations in national radio regulations. In the United States of America ISM is governed by Part 18 of the FCC rules and should not be confused with Part 15 rules. Communication is not permitted under Part 18 (ISM) rules.

In recent years they have also been shared with license-free error-tolerant communications applications such as wireless LANs and Bluetooth:

- 900 MHz band (33.3 cm wavelength)
- 1.8 GHz Band (16.7 cm wavelength)
- 2.4 GHz band (12.5 cm wavelength)
- 5.8 GHz band (5.2 cm wavelength)

(using an approximation to the speed of light of 3×10^8 m/s)

IEEE 802.11b/g wireless Ethernet also operates on the 2.4 GHz band, and most microwave ovens use 2.45 GHz. Although these devices share the (Part 18) ISM band, they are not ISM devices in the United States, and fall under the regulation of Part 15.

7.13. WIRELESS PERSONAL AREA NETWORK (WPAN)

A wireless network that is typically limited to a small cell radius. In an office environment, a WPAN would be used to transfer data between a handheld device and desktop machine or printer. For example, a mobile user could download e-mail or Web data into a dual mode smartphone or PDA and then exchange that data with a machine in the office. In the home, WPANs provide cable-free connections for alarms, appliances and entertainment systems. See UWB. 802.15, Bluetooth, HomeRF and wireless LAN.

7.14. NETWORK DATABASE

The network database model conceived as a flexible way of representing objects and their relationships. Its original inventor was Charles Bachman. and it was developed into a standard specification published in 1969 by the CODASYL Consortium. Where the hierarchical model structures data as a tree of records, with each record having one parent record and many children, the network model allows each record to have multiple parent and child records, forming a lattice structure.

The chief argument in favour of the network model, in comparison to the hierarchic model, was that it allowed a more natural modeling of relationships between entities. Although the model was widely implemented and used, it failed to become dominant for two main reasons. Firstly, IBM chose to stick to the hierarchical model in their established products such as IMS and DL/I. Secondly, it was eventually displaced by the relational model, which offered a higher-level, more declarative interface. Until the early 1980s the performance benefits of the low-level navigational interfaces offered by hierarchical and network databases were persuasive for many large-scale applications, but as hardware became faster, the extra productivity and flexibility of the relational model replaced the network model in corporate enterprise usage. .

The navigational interface offered by the network model bears some resemblance to the hyperlink-based models that have become popular with the advent of the Internet and World Wide Web. However, the network model (like the relational model) assumes that the entire database has a centrally-managed schema, and as such it is not well suited to distributed, heterogeneous environments.

7.14.1. History

In 1969, the Conference on Data Systems Languages (CODASYL) established the first specification of the network database model. This was followed by a second publication in 1971 which became the basis for most implementations. Subsequent work continued into the early 1980s, culminating in an ISO specification, but this had little influence on products.

7.15. SPECTRUM ALLOCATION

The **electromagnetic spectrum** is an aspect of the physical world, like land, water, and air. Use of **radio frequency** bands of the electromagnetic spectrum is regulated by governments in most countries, in a process known as **frequency allocation** or **spectrum allocation**. Like weather and internationally traded goods, radio propagation and RF technology do not stop at national boundaries. Giving technical and economic reasons, governments have sought to harmonise spectrum allocation standards.

A number of forums and standards bodies work on standards for frequency allocation, including:

- ITU (International Telecom Union)
- CEPT (Conference of European the Post Telecom)
- ETSI (European Telecom Standard Institute)
- International Special Committee on Radio Interference
- ANSI (American National Standard Institute)

High-demand sections of the electromagnetic spectrum may sometimes be allocated through auctions. In India frequency allocation is done by TRAI — (Telephone Regulatory Authority of India) through auctions to different operators for different purposes in coordination with International authorities. The range of "radio frequencies" is a matter of international convention. At the international radio conference at Atlantic City in 1947, Hertzian (radio) waves were defined as electromagnetic waves of frequencies between 10 Kc/s and 3000000 Mc/s. The lower limit was dropped in subsequent international radio regulations. More recently there have been proposals to raise the upper limit.

As a matter of physics, many objects and actions generate low-level, wide-band radiation. The frequency allocation process traditionally has not been concerned with many types of radiation.

In many countries spectrum allocation is one of the major problem that is faced by the radio communication industry. In setting of spectrum allocation policy it must be kept in mind that minimum bandwidth is given to the operator and with this minimum bandwidth operator must provide high usage with customer satisfaction.

A mobile telephone system would operate within a limited assigned frequency band and would serve an almost unlimited number of users in unlimited areas. Following are the major approaches in spectrum allocation.

1. Single sideband, which divides the allocated frequency band into maximum numbers of channels.
2. Cellular, which reuses the allocated frequency in different geographical areas.
3. Spread spectrum or frequency hoped, which generates many code over a wide frequency band.

7.15.1. Frequency Utilization Measurement Factor

In a cellular phone system, the frequency utilization measurement (X) is defined as the maximum number of customers that could be served by one channel at busy hour.

$$X = \frac{\text{Number of Customers}}{\text{Channel}}$$

7.15.2. Spectrum Efficiency

The frequency spectrum is a limited resource, we should utilize it very effectively.

In order to approach this goal, spectrum efficiency should be clearly defined from either a total system point of view or a fixed point to point link.

For most radio system spectrum efficiency is the same as channel efficiency. But this is true for a point to point system not true where frequency reuse concepts used like mobile cellular system. In cellular system, spectrum efficiency is not channel efficiency. But system capacity is directly related to spectrum efficiency and the parameter for system comparison are voice quality, transmitted power, and cell size. Satisfactory voice quality is generally accepted as governed by the carrier noise ratio $C/N_{FM} = 18$ dB.

This is the level at which voice quality is excellent. For point to point radio links, the wider the channel bandwidth, the lower the required level of transmitted power.

To maintain voice quality when channel bandwidth is reduced, it is necessary to increase the S/N ratio in order to improve the reception. Transmission power is then also increased.

Every time power is increased, interference problems are created. For point to point radio links, these problems are adjustable because no frequency reuse is involved. However, when frequency reuse system like cellular phone frequency is used, the concept is different. We know that a celiular system includes many subscriber and on demand at any given time in a given system, identical channels will be operating simultaneously in different geographic locations.

As the number of mobile phones increases in a given area, interference may appear in one of several forms, co-channel, adjacent channel, multichannel at co-locations. Interference may also result from received power level differences.

In a frequency reuse system, however, co-channel separation is more critical to the system than adjacent channel interference because adjacent channel interference may be eliminated by the use of sharp filters.

Number of frequency reuse cells K:

$$K = \sqrt{\frac{2}{3}\frac{C}{I}}$$

C/I is carrier to interference ratio.

A higher required carrier to interference ratio at the boundary of a cell results in the need for more frequency reuse cells.

Number of channels per cell:

$$m = \frac{B_t}{B_c K} = \frac{B_t}{B_c\sqrt{\frac{2}{3}\frac{C}{I}}}$$

B_t = Total bandwidth (transmitted or received)

B_c = Channel bandwidth.

7.15.3. GSM Spectrum Allocation in India

For any type of radio link, frequency is allocated by TRAI (Telephone Regulatory Authority of India) to the operator. This frequency allocation is different for different services. Like TRAI, each country has a authority, who works with co-ordination of ITU (International Telecom Union) or like other international organisation.

As we know GSM is a 2G standard and operated world wide in name of GSM 900 and GSM 1800. Bandwidth of GSM is as follows.

Table 7.8.

	Uplink frequency range	Bandwidth	Downlink frequency range
GSM 900	890–915 MHz25 MHz	935–960 MHz	
GSM 1800	1710–1785 MHz	75 MHz	1805–1880 MHz

Both GSM 900 and GSM 1800 is divided in bandwidth of 200 kHz and from GSM 900, we get

$$\text{Number of channels} = \frac{25\text{ MHz}}{200\text{ kHz}}$$

$$= 125 \text{ channels or } 125 \text{ carrier frequency}$$

and from GSM 1800, we get

$$= \frac{75\text{ MHz}}{200\text{ kHz}}$$

$$= 375 \text{ channels or carrier frequency}$$

It is important to note that in GSM 900, carrier frequency starts from 890.2 MHz to 915.2 MHz instead of starting from 890.0 MHz to 915.0 MHz and same for GSM 1800. Due to that, we get effectively 124 channels in GSM 900 and 374 channels in GSM 1800 for commercial purpose.

To divide these carrier frequency among network operator, whole India is divided into various telecom circles. For example Delhi (NCR) is a one Telecom Circle and Haryana, U.P. East, U.P. West are different isolated telecom circles. As per Indian Government rule TRAI allows only 4 operators in one telecom circle to provide GSM services. For example in Delhi NCR Telecom Circle (Delhi, Noida, Faridabad, Gaziabad, Gurgaon are combined) and following operators are providing GSM services.

Airtel

MTNL

Vodafone

Idea

Out of these operators Airtel, Vodafone is providing services in GSM 900 and in GSM 1800 bands while MTNL is providing GSM services on GSM 900 and Idea is providing services on GSM 1800 bands. As per various network operator performance which include service quality, number of services, number of subscribers, number of drop calls per month, quality of service etc. TRAI allocates frequencies to operators. In GSM 900 band total number of carriers are 124 and it is divided in all 4 operators. So every operator gets 31 carrier frequency to provide GSM services in whole telecom circle.

This 31 carrier frequency is distributed in various cell areas and used on time sharing basis.

But whole Delhi Telecom Circle has more than 150 cell areas under one operator so it is very difficult to distribute 31 frequency in 150 cell areas.

But this 31 frequency is more than sufficient because it is distributed on frequency re-use concept.

Practically each cell area is having three different frequency and this frequency may be used in some another cell but not in adjacent cell.

It is important to note that cell is geographical area of 500 meters to several kilometers depending upon density of people.

7.16. FREQUENCY HOPPING

A frequency hopping system can be used as a CDMA system (FH-CDMA).

The hopping pattern can be formed as a code sequence. Frequency hopping has been used in the past to overcome enemy jammers in military applications. There are two kinds of hopping, fast frequency hopping and slow frequency hopping.

Frequency hopping can be divided in two parts.

Fast frequency hopping (FFH)

Slow frequency hopping (SFH)

Slow frequency hopping is defined as sending multiple bits on a single hop. Depending upon enemy's and their reaction it can be adjusted.

FFH (Fast Frequency Hopping) is defined as sending a bit on a pseudo random pattern of frequency channels then sending the next bit on a different pseudo random pattern of frequency channels.

The multiple frequency channels form a code for one bit which is sent out simultaneously or sequentially. The bandwidth of the channel depends on the transmission bit rate.

The scheme of simultaneously sending out the same bit on different frequency channels requires a large bandwidth for sending each bit. This is name wideband (WCDMA) system.

FFH CDMA system requires a larger bandwidth than SFH CDMA system.

7.16.1. Frequency Hopping Spread Spectrum

Frequency hopping spread spectrum (FHSS) is a radio transmission process where a message or voice communication is sent on a radio channel that regularly changes frequency (hops) according to a predetermined code. The receiver of the message or voice information must also receive on the same frequencies using the same frequency hopping sequence. Frequency hopping was first used for military electronic counter measures, because radio communication occurs only for brief periods on a radio channel and the frequency hop channel numbers are only known to authorized receivers of the information, transmitted signals that use frequency hopping are difficult to detect and monitor.

The following figure shows a simplified diagram of how the Bluetooth system uses frequency hopping to transfer information (data) from a transmitter to a receiver using 79 communication channels. This diagram shows a transmitter that has a preprogrammed frequency tuning sequence and this frequency sequence occurs by hopping from channel frequency to channel frequency. To receive information from the transmitter, the receiver uses the exact, same hopping sequence is used. When the transmitter and receiver frequency hopping sequences occur exactly at the same time, information can transfer from the transmitter to the receiver. This diagram shows that after the transmitter hops to a new frequency, it transmits a burst of information (packet of data). Because the receiver hops to the same frequency, it can receive the packet of data each time.

7.17. WALSH CODE

The **Walsh code** is used to uniquely define individual communication channels. Walsh codes are "orthogonal" mathematical codes. As such, if two Walsh codes are correlated, the result is intelligible only if these two codes are the same. As a result, a Walsh encoded signal appears as random noise to a CDMA capable mobile terminal, unless that terminal uses the same code as the one used to encode the incoming signal.

The walsh code is calculated by the Walsh function.

Example: Denote $W(k, n)$ as Walsh code k in n-length walsh matrix. It means the kth row of $H(m)$ Hadamard matrix, where $n = 2^m$. Here k could be 0, 1, ..., $n - 1$.

$W(0, 1) = 1$

$W(0, 2) = 1, 1$

$W(1, 2) = 1, -1$

$W(0, 4) = 1, 1, 1, 1$

$W(1, 4) = 1, -1, 1, -1$

$W(2, 4) = 1, 1, -1, -1$

$W(3, 4) = 1, -1, -1, 1$

$W(0, 8) = 1, 1, 1, 1, 1, 1, 1, 1$

$W(1, 8) = 1, -1, 1, -1, 1, -1, 1, -1$

$W(2, 8) = 1, 1, -1, -1, 1, 1, -1, -1$

$W(3, 8) = 1, -1, -1, 1, 1, -1, -1, 1$

$W(4, 8) = 1, 1, 1, 1, -1, -1, -1, -1$

$W(5, 8) = 1, -1, 1, -1, -1, 1, -1, 1$

$W(6, 8) = 1, 1, -1, -1, -1, -1, 1, 1$

$W(7, 8) = 1, -1, -1, 1, -1, 1, 1, -1$

7.18. CDMA (CODE DIVISION MULTIPLE ACCESS)

Code Division Multiple Access is a method for transmitting simultaneous signals over a shared portion of the spectrum. The foremost application of CDMA is the digital cellular phone technology that operates in the 800 MHz band and 1.9 GHz PCS (Personal Communication Service) band. CDMA phones are noted for their call quality. **Code division multiple access (CDMA)** is a form of multiplexing (not a modulation scheme) and a method of multiple access that does not divide the channel by time (as in TDMA), or frequency (as in FDMA), but instead of that encodes data with a special code associated with each channel and uses the constructive interference properties of the special codes to perform the multiplexing. **CDMA** also refers to digital cellular telephony systems that make use of this multiple access scheme, such as those pioneered by Qualcomm and W-CDMA by the International Telecommunication Union or ITU. CDMA has since been used in many communications systems, including the Global Positioning System (GPS) and in the OmniTRACS satellite system for transportation logistics.

CDMA is Efficient

CDMA requires fewer cell sites than the GSM and TDMA digital cellphone systems and provides three to five times the calling capacity. Providing more than 10 times the capacity of the analog cellphone system (AMPS), CDMA has become widely used in North America and is also expected to become the third-generation (3G) technology for GSM. For example, in the U.S., the Verizon and Sprint cellphone services are based on CDMA and in India Reliance Communications is based on CDMA.

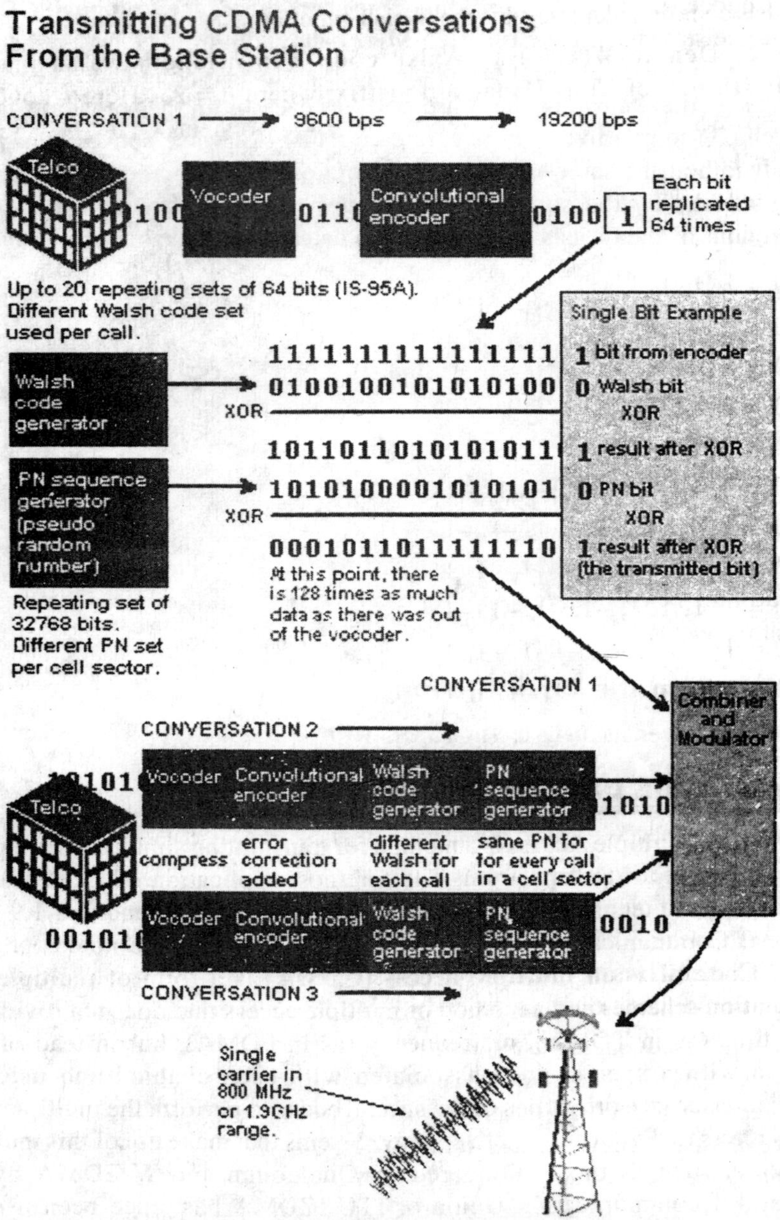

Fig. 7.26. (Transmitting) Receiving at the Cellphone.

Spread Spectrum

Unlike the other digital systems that divide the spectrum into different time slots, CDMA's spread spectrum technique overlaps every transmission on the same carrier frequency by assigning a unique code to each conversation. The often-used analogy for this is your ability to detect your own language in a room full of people speaking other languages.

A Different Kind of Chip

Speech Coddec (Coding and Decoding chip) converts voice to digital, CDMA spreads the voice stream over the full 1.25 MHz bandwidth of the CDMA channel, coding each stream separately so it can be decoded at the receiving end. The rate of the spreading signal is known as the "chip rate", as each bit in the spreading signal is called a "chip" (no relation to an integrated circuit). All voice conversations use the full bandwidth at the same time. One bit from each conversation is multiplied into 128 coded bits by the spreading techniques, giving the receiving side an enormous amount of data it can average just to determine the value of one bit.

More Secure

CDMA transmission has been used by the military for secure phone calls. Unlike FDMA and TDMA, CDMA's wide spreading signal makes it difficult to detect and jam. For more information, contact the CDMA Development Group (CDG) at www.cdg.org.

How the Technology Works

The following illustration, which was created with the assistance of Klein Gilhousen, co-inventor of CDMA, shows how bits are encoded at the base station and decoded in the cellphone. A single bit example is used to take you through the Boolean math.

Transmitting from the Base Station

Each voice conversation is compressed with a vocoder (Voice Coder). The output is doubled by a convolutional encoder that adds redundancy for error checking. Each bit from the encoder is replicated 64 times and exclusive OR'd with a Walsh code that is used to identify that call from the rest.

The output of the Walsh code is exclusive OR'd with the next string of bits (PN-pseudorandom number sequence) from a pseudorandom number generator, which is used to identify all the calls in a particular cell's sector. At this point, there is 128 times as many bits as there were from the vocoder's output. All the calls are combined and modulated onto a carrier frequency in the 800 MHz range.

The received frequencies are quantized into bits ("chips") by the analog-to-digital converter (ADC). The output is run through the Walsh code and PN sequence correlation receiver to recover the transmitted bits of the original signal. When 20 ms of voice data is received, a Viterbi decoder corrects the errors using the convolutional code. The Viterbi output goes to the vocoder and digital-to-analog converter (DAC), which decompresses the bits and turns them back into waveforms (sound).

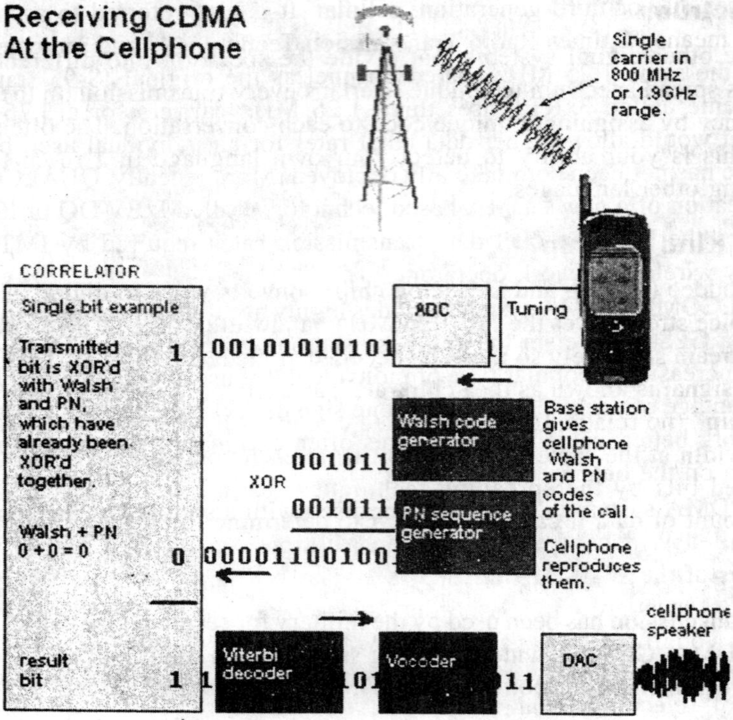

Exclusive OR

IN	IN	OUT
0	0	0
0	1	1
1	0	1
1	1	0

Fig. 7.27. (Receiving) Follow the Single Bit Example.

This exclusive OR truth table shows you the Boolean algebraic rules to prove the single bit example in the illustrations above. The example bit is a 1, and the Walsh and PN codes are 0.

7.18.1. Usage in Mobile Telephony

A number of different terms are used to refer to CDMA implementations. The original U.S. standard defined by QUALCOMM was known as IS-95, the IS referring to an Interim Standard of the Telecommunications Industry Association (TIA). IS-95 is often referred to as 2G or second generation cellular. The QUALCOMM brand name cdmaOne may also be used to refer to the 2G CDMA standard. Whereas the Global System for Mobile Communications (GSM) standard is a specification of an entire network infrastructure, the CDMA interface relates only to the air interface—the radio part of the technology. For example GSM specifies an infrastructure based on internationally approved standard while CDMA allows each operator to provide the network features as it finds suited.

After a couple of revisions, IS-95 was superseded by the 1S-2000 standard. This standard was introduced to meet some of the criteria laid out in the IMT-2000 specification for 3G, or third generation, cellular. It is also referred to as 1xRTT which simply means "1 times Radio Transmission Technology" and indicates that IS-2000 uses the same 1.25 MHz shared channel as the original IS-95 standard. A related scheme called 3xRTT uses three 1.25 MHz carriers for a 3.75 MHz bandwidth that would allow higher data burst rates for an individual user, but the 3xRTT scheme has not been commercially deployed. More recently, QUALCOMM has led the creation of a new CDMA-based technology called IxEV-DQ or IS-856, which provides the higher packet data transmission rates required by IMT-2000 and desired by wireless network operators.

The QUALCOMM CDMA system includes highly accurate time signals (usually referenced to a GPS receiver in the cell base station), so cellphone CDMA-based clocks are an increasingly popular type of radio clock for use in computer networks. The main advantage of using CDMA cellphone signals for reference clock purposes is that they work better inside buildings, thus often eliminating the need to mount a GPS antenna on the outside of a building.

The US CDMA system is frequently confused with a similar but incompatible technology called Wideband Code Division Multiple Access (W-CDMA) which forms the basis of the W-CDMA air interface. The W-CDMA air interface is used in the global 3G standard UMTS and the Japanese 3G standard FOMA, by NTT DoCoMo and Vodafone: however, the CDMA family of US national standards (including cdmaOne and CDMA2000) are not compatible with the W-CDMA family of International Telecommunication Union (ITU) standards.

Another important application of CDMA—predating and entirely distinct from CDMA cellular—is the Global Positioning System, GPS.

7.18.2. Coverage and Applications

The size of a given cell depends on the power of the signal transmitted by the handset and the radio frequency being used. Various algorithms can reduce the noise introduced by variations in cell area, but require extra information be sent to validate the transfer. Hence, the radio frequency and power of the handset effectively determine the cell size. Long wavelengths need less energy to travel a given distance vs. short wavelengths, so lower frequencies generally result in greater coverage while higher frequencies result in shorter coverage. These characteristics are used by mobile network planners in determining the size and placement of the cells in the network. In cities, many small cells are needed; the use of high frequencies allows sites to be placed more closely together, with more subscribers provided service. In rural areas with a lower density of subscribers, use of lower frequencies allows each site to provide broader coverage.

Various companies use different variants of CDMA to provide fixed-line networks using Wireless local loop (WLL) technology. Since they can plan with a specific number of subscribers per cell in mind, and these are all stationary, this application of CDMA can be found in most parts of the world.

CDMA is suited for data transfer with bursty behaviour and where delays can be accepted. It is, therefore, used in Wireless LAN applications; the cell size here is 500 feet because of the high frequency (2.4 GHz) and low power. The suitability for data transfer is the reason for why W-CDMA seems to be "winning technology" for the data portion of third-generation (3G) mobile cellular networks.

7.18.3. Technical Details

7.18.3.1. Code Division Multiplexing (Synchronous CDMA)

Synchronous CDMA, also known as Code Division Multiplexing (CDM), exploits at its core mathematical properties of orthogonality. Suppose we represent data signals as vectors. For example, the binary string "1011" would be represented by the vector (1, 0, 1, 1). We may wish to give a vector a name, we may do so by using boldface letters, *e.g., a.* We also use an operation on vectors, known as the dot product, to "multiply" vectors, by summing the product of the components. The operation is denoted with a dot between the vectors. For example, the dot product of $a = (1, 0, 1, 1)$ and $b = (1, -1, -1, 0)$, written as $a \,.\, b$, would be $(1) \times (1) + (0) \times (-1) + (1) \times (-1) + (1) \times (0) = 1 + (-1) = 0$. For the special case when the dot product of two vectors is identically 0, the two vectors are said to be *orthogonal* to each other.

The dot product has a number of properties, which will aid in understanding how CDM works. For vectors *a, b, c*:

$$a \,.\, (b + c) = a \,.\, b + a \,.\, c,$$

and $a \,.\, kb = k(a \,.\, b)$, where k is an arbitrary constant, and not a vector.

The square root of $a \,.\, a$ is a real number, and is called the magnitude of the vector *a*. It is denoted as

$$\| a \| = \sqrt{a \,.\, a}$$

$$a \,.\, (a + b) = \| a \|^2 \text{ since } a \,.\, a + a \,.\, b = \| a \|^2 + 0,$$

$$a \,.\, (-a + b) = -\| a \|^2 \text{ since } -a \,.\, a + a \,.\, b = -\| a \|^2 + 0,$$

$$b \,.\, (a + b) = \| b \|^2 \text{ since } b \,.\, a + b \,.\, b = 0 + \| b \|^2,$$

$$b \,.\, (a - b) = -\| b \|^2 \text{ since } b \,.\, a - b \,.\, b = 0 - \| b \|^2.$$

Example : (Fig. 7.28)

Suppose now we have a set of vectors that are mutually orthogonal to each other. Usually these vectors are specially constructed for ease of decoding—they are columns or rows from Walsh matrices that are constructed from Walsh functions—but strictly mathematically the only restriction on these vectors is that they are orthogonal. An example of orthogonal functions is shown in the picture on the right. Now, associate with one sender a vector from this set, say v, which is called the *chip code.* Associate a zero digit with the vector $-v$, and a digit 1 with the vector v. For example, if $v = (1, -1)$, then the binary vector (1, 0, 1, 1) would correspond to (1, –1, –1, 1, 1, –1, 1, –1). For the purposes of this article, we call this constructed vector the *transmitted vector.*

Each sender has a different, unique vector chosen from that set, but the construction of the transmitted vector is identical.

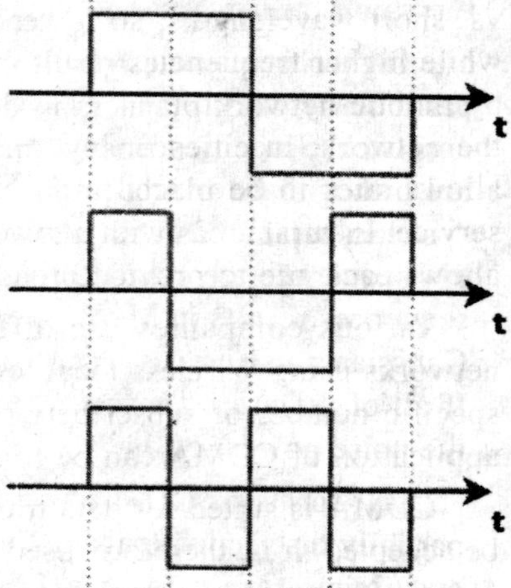

Fig. 7.28. An example of 4 orthogonal digital signals.

Now, the physical properties of interference say that if two signals at a point are in phase, they will "add up" to give twice the amplitude of each signal, but if they are out of phase, they will "subtract" and give a signal that is the difference of the amplitudes. Digitally, this behaviour can be modelled simply by the addition of the transmission vectors, component by component. So, if we have two senders, both sending simultaneously, one with the chip code (1, –1) and data vector (1, 0, 1, 1), and another with the chip code (1, 1), and data vector (0, 0, 1, 1), the raw signal received would be the sum of the transmission vectors (1, –1, –1, 1, 1, –1, 1, –1) + (–1, –1, –1, –1, 1, 1, 1, 1) = (0, –2, –2, 0, 2, 0, 2, 0).

Suppose a receiver gets such a signal, and wants to detect what the transmitter with chip code (1, –1) is sending. The receiver will make use of the property described in the above foundation section, and take the dot product to the received vector in parts. Take the first two components of the received vector, that is, (0, –2). Now, (0, –2).(1, –1) = (0)(1) + (–2)(–1) = 2. Since this is positive, we can deduce that a one digit was sent. Taking the next two components, (–2, 0), (–2, 0) . (1, –1) = (–2)(1) + (0)(–1) = –2. Since this is negative, we can deduce that a zero digit was sent. Continuing in this fashion, we can successfully decode what the transmitter with chip code (1, –1) was sending: (1, 0, 1, 1).

Likewise, applying the same process with chip code (1, 1): (1, 1) . (0, –2) = –2 gives digit 0, (1, 1) . (–2, 0) = (1) (–2) + (1)(0) = –2 gives digit 0, and so on, to give us the data vector sent by the transmitter with chip code (1, 1): (0, 0, 1, 1).

7.18.3.2. Asynchronous CDMA

The previous example of orthogonal Walsh sequences describes how 2 users can be multiplexed together in a synchronous system, a technique that is commonly referred to as Code Division Multiplexing (CDM). The set of 4 Walsh sequences shown in the figure will afford up to 4 users, and in general, an $N \times N$ Walsh matrix can be used to multiplex N users. Multiplexing requires all of the users to be coordinated so that each transmits their assigned sequence v (or the complement, $-v$) starting at exactly the same time. Thus, this technique finds use in base-to-mobile links, where all of the transmissions originate from the same transmitter and can be perfectly coordinated.

On the other hand, the mobile-to-base links cannot be precisely coordinated, particularly due to the mobility of the handsets, and require a somewhat different approach. Since it is not mathematically possible to create signature sequences that are orthogonal for arbitrarily random starting points, unique "pseudo-random" or "pseudo-noise" (PN) sequences are used in *Asynchronous* CDMA systems. These PN sequences are statistically uncorrelated, and the sum of a large number of PN sequences results in Multiple Access Interference (MAI) that is approximated by a Gaussian noise process (via the theorem of the "law of large numbers" in statistics). If all of the users are received with the same power level, then the variance (*e.g.*, the noise power) of the MAI increases in direct proportion to the number of users.

All forms of CDMA use spread spectrum process gain to allow receivers to partially discriminate against unwanted signals. Signals with the desired chip code and timing are received, while signals with different chip codes (or the same spreading code but a different timing offset) appear as wideband noise reduced by the process gain.

Since each user generates MAI, controlling the signal strength is an important issue with CDMA transmitters. A CDM (Synchronous CDMA), TDMA or FDMA receiver can in theory completely reject arbitrarily strong signals using different codes, time slots or frequency channels due to the orthogonality of these systems. This is not true for Asynchronous CDMA; rejection of unwanted signals is only partial. If any or all of the unwanted signals are much stronger than the desired signal, they will overwhelm it. This leads to a general requirement in any Asynchronous CDMA system to approximately match the various signal power levels as seen at the receiver. In CDMA cellular, the base station uses a fast closed-loop power control scheme to tightly control each mobile's transmit power.

7.18.3.3. Advantages of Asynchronous CDMA Over Other Techniques

Asynchronous CDMA's main advantage over COM *(Synchronous* CDMA), TDMA and FDMA is that it can use the spectrum more efficiently in mobile telephony applications. TDMA systems must carefully synchronize the transmission times of all the users to ensure that they are received in the correct timeslot and do not cause interference. Since this cannot be perfectly controlled in a mobile environment, each timeslot must have a guard-time, which reduces the probability that users will interfere, but decreases the spectral efficiency. Similarly, FDMA systems must use a guard-band between adjacent channels, due to the random doppler shift of the signal spectrum which occurs due to the user's mobility. The guard-bands will reduce the probability that adjacent channels will interfere, but decrease the utilization of the spectrum.

Most importantly, Asynchronous CDMA offers a key advantage in the flexible allocation of resources. There are a fixed number of orthogonal codes, timeslots or frequency bands that can be allocated for CDM, TDMA and FDMA systems, which remain underutilized due to the bursty nature of telephony and packetized data transmissions. There is no strict limit to the number of users that can be supported in an *Asynchronous* CDMA system, only a practical limit governed by the desired bit error probability, since the SIR (Signal to Interference Ratio) varies inversely with the number of users. In a bursty traffic environment like mobile telephony, the advantage afforded by Asynchronous CDMA is that the performance (bit error rate) is allowed to fluctuate randomly, with an average value determined by the number of users times the percentage of utilization. Suppose there are $2N$ users that only talk half of the time, then $2N$ users can be accommodated with the same *average* bit error probability as N users that talk all of the time. The key difference here is that the bit error probability for N users talking all of the time is constant, whereas it is a *random* quantity (with the same mean) for $2N$ users talking half of the time.

In other words, Asynchronous CDMA is ideally suited to a mobile network where large numbers of transmitters each generate a relatively small amount of traffic at irregular intervals. CDM *(Synchronous* CDMA), TDMA and FDMA systems cannot recover the underutilized resources inherent to bursty traffic due to the fixed number of orthogonal codes, time slots or frequency channels that can be assigned to individual transmitters. For instance, if there are N time slots in a TDMA system and $2N$ users that talk half of the time, then half of the time there will be more than N users needing to use more than N timeslots. Furthermore, it would require significant overhead to continually allocate and deallocate the orthogonal code, time slot or frequency channel resources. By comparison, Asynchronous CDMA transmitters simply send when they have something to say, and go off the air when they don't, keeping the same PN signature sequence as long as they are connected to the system.

7.18.4. Soft Handoff

Soft handoff (or soft handover) is an innovation in mobility. It refers to the technique of adding additional base stations (in IS-95 as many as 5) to a connection to be certain that the next base is ready as you move through the terrain. However, it can also be used to move a call from one base station that is approaching congestion to another with better capacity. As a result, signal quality and handoff-robustness is improved compared to TDMA systems.

In TDMA and analog systems, each cell transmits on its own frequency, different from those of its neighbouring cells. If a mobile device reaches the edge of the cell currently serving its call, it is told to break its radio link and quickly tune to the frequency of one of the neighbouring cells where the call has been moved by the network due to the mobile's movement. If the mobile is unable to tune to the new frequency in time the call is dropped.

In CDMA, a set of neighbouring cells all use the same frequency for transmission and distinguish cells (or base stations) by means of a number called the "PN offset", a time offset from the beginning of the well-known pseudo random noise sequence that is used to spread the signal from the base station. Because all of the cells are on the same frequency, listening to different base stations is now an exercise in digital signal processing based on offsets from the PN sequence, not RF transmission and reception based on separate frequencies.

As the CDMA phone roams through the network, it detects the PN offsets of the neighbouring cells and reports the strength of each signal back to the reference cell of the call (usually the strongest cell). If the signal from a neighbouring cell is strong enough, the mobile will be directed to "add a leg" to its call and start transmitting and receiving to and from the new cell in addition to the cell (or cells) already hosting the call. Likewise, if a cell's signal becomes too weak the mobile is directed to drop that leg. In this way, the mobile can move from cell to cell and add and drop legs as necessary in order to keep the call up without ever dropping the link.

When there are frequency boundaries between different carriers or sub-networks, a CDMA phone behaves in the same way as TDMA or analog and performs a hard handoff in which it breaks the existing connection and tries to pick up on the new frequency where it left off.

7.18.5. CDMA Features

- Narrowband message signal multiplied by wideband spreading signal or pseudonoise code.
- Each user has his own pseudonoise (PN) code
- Soft capacity limit: system performance degrades for all users as number of users increases

- Cell frequency reuse: no frequency planning needed
- Soft handoff increases capacity
- Near-far problem
- Interference limited: power control is required
- Wide bandwidth induces diversity: rake receiver is used

7.18.6. Comparison of Mobile Phone Standards

GSM and CDMA are the two most prevalent mobile communication technologies. Both technologies have the same goal: to divide the finite RF spectrum between multiple users.

TDMA (Time Division Multiple Access - underlying technology used in GSM) does it by chopping up the channel into sequential time slices. Each user of the channel takes turns to transmit and receive signals. In reality, only one person is actually using the channel at a specific moment. This is analogous to time-sharing on a large computer server. CDMA on the other hand, uses special digital modulation **Spread Spectrum** which spreads the voice over very wide channel in *pseudo random fashion.* The receiver undoes the randomization to collect the bits together and produce the sound.

As a trivial comparison imagine a UN party where couples from different countries are invited. In GSM, each couple takes turns talking. They talk for a short time and then stop to let another couple talk. As there is never more than one person speaking in the room, no one has to worry about being heard over the background noise. In CDMA, each couple talks at the same time; however, they all use different languages. Because (we assume) that none of the listeners understand any other language than that of their partner, the background noise does not create any problem.

Advantages of GSM

- GSM is mature, this maturity means a more stable network with robust features.
- Less signal deterioration inside buildings.
- Ability to use repeaters
- Talktime is generally higher in GSM phones due to pulse nature of transmission.
- The availability of Subscriber Identity Modules allows to users to switch networks and handsets at will.
- GSM covers virtually all parts of world so international roaming is not a problem.

Disadvantages of GSM

- Pulse nature of transmission interferes with some electronics, especially certain audio amplifiers.
- Intellectual property is concentrated among a few industry participants, creating barriers to entry for new entrants and limiting competition among phone manufacturers.
- GSM has a fixed maximum cell site range of 35 km, which is imposed by technical limitations.

Advantages of CDMA

- Capacity is CDMA's biggest asset. It can accommodate more users per MHz of bandwidth than any other technology.
- CDMA has no built-in limit to the number of concurrent users.
- CDMA uses precise clocks that do not limit the distance a tower can cover.
- CDMA consumes less power and covers large areas so cell size in CDMA is larger.
- CDMA is able to produce a reasonable call with lower signal (cellphone reception) levels.
- CDMA uses Soft Handoff, reducing the likelihood of dropped calls.
- CDMA's variable rate voice coders reduce the rate being transmitted when speaker is not talking, which allows the channel to be packed more efficiently.
- Has a well-defined path to higher data rates.

Disadvantages of CDMA

- Most technologies are patented and must be licensed from Qualcomm.
- *Breathing* of base stations, where coverage area shrinks under load. As the number of subscribers using a particular site goes up, the range of that site goes down. Because CDMA towers interfere with themselves, they are normally installed on much shorter towers. Because of this, CDMA may not perform well in hilly region.

7.18.7. CDMA 2000

CDMA 2000 is a family of third-generation (3G) mobile telecommunications standards that use CDMA, a multiple access scheme for digital radio, to send voice, data, and signalling data (such as a dialed telephone number) between mobile phones and cell sites. It is the second generation of CDMA digital cellular.

CDMA (code division multiple access) is a mobile digital radio technology that transmits streams of bits and whose channels are divided using codes (PN sequences). CDMA permits many radios to share the same frequency channel. Unlike TDMA (time division multiple access), a different technique used in GSM and D-AMPS, all radios can be active all the time, because network capacity does not directly limit the number of active radios. Since larger numbers of phones can be served by smaller numbers of cell sites, CDMA-based standards have a significant economic advantage over TDMA-based standards, or the oldest cellular standards that used frequency division multiple access (FDMA).

CDMA2000 has a relatively long technical history, and remains compatible with the older CDMA telephony methods (such as cdmaOne) first developed by Qualco.mm, a commercial company, and holder of several key international patents on the technology.

The CDMA2000 standards **CDMA2000 Ix, CDMA2000 IxEV-DO,** and **CDMA2000 IxEV-DV** are approved radio interfaces for the ITU's IMT-2000 standard and a direct successor to 2G CDMA, IS-95 (cdmaOne). CDMA2000 is standardized by 3GPP2.

CDMA2000 is a registered trademark of the Telecommunications Industry Association (TIA-USA) in the United States, not a generic term like CDMA. (This is similar to how TIA has branded their 2G CDMA standard, IS-95, as cdmaOne.)

CDMA2000 is an incompatible competitor of the other major 3G standard UMTS. It is defined to operate at 400 MHz, 800 MHz, 900 MHz, 1700 MHz, 1800 MHz, 1900 MHz, and 2100 MHz.

Below are the different types of CDMA2000, in order of increasing complexity.

7.18.7.1. CDMA2000 1x

CDMA2000 1x, the core CDMA2000 wireless air interface standard, is known by many terms: 1x, 1xRTT, IS-2000, CDMA2000 1X, IX, and cdma2000 (lowercase). The designation "1×RTT" (1 times Radio Transmission Technology) is used to identify the version of CDMA2000 radio technology that operates in a pair of 1.25 MHz radio channels (one times 1.25 MHz, as opposed to three times 1.25 MHz in 3xRTT). 1xRTT almost doubles voice capacity over IS-95 networks. Although capable of higher data rates, most deployments have limited the peak data rate to 144 kbit/s. While 1xRTT officially qualifies as 3G technology, 1xRTT is considered by some to be a 2.5G (or sometimes 2.75G) technology. This has allowed it to be deployed in 2G spectrum in some countries which limit 3G systems to certain bands.

The main differences between IS-95 and IS-2000 signalling are: 64 more traffic channels on the forward link that are orthogonal to the original set. Some changes were also made to the data link layer to accommodate the greater use of data services—IS-2000 has media and link access control protocols and QoS control. In IS-95, none of these were present, and the data link layer basically consisted of a "best effort delivery" RLP—this arrangement is still used for voice.

In the United States, Verizon Wireless, Sprint PCS, Alltel, and U.S. Cellular use 1x, also it is in Canada by Bell Mobility and TELUS Mobility, and in Mexico by lusacell and Unefon.

In India, BSNL, Reliance and Tata Teleservices are major wireless services providers on CDMA 2000 1x.

In Indonesia, Mobile-8 is the major mobile wireless service provider on CDMA 2000 Ix (with EVDO in Western Java). The other CDMA providers are fixed wireless (such as Bakrie Telkom, Telkom Flexi, and Indosat's Starone).

7.18.7.2. CDMA2000 3x

CDMA2000 3x utilizes a pair of 3.75 MHz radio channels (*i.e.*, 3 × 1.25 MHz) to achieve higher data rates. The 3x version of CDMA2000 is sometimes referred to as Multi-Carrier or MC. The 3x version of CDMA2000 has not been deployed and is not under development at present.

7.18.7.3. CDMA2000 1xEV-DO

CDMA2000 1xEV-DO (1x Evolution-Data Optimized, originally Ix Evolution-Data Only), also referred to as IxEV-DO, EV-DO, EVDO, or just DO, is an evolution of CDMA2000 Ix with High Data Rate (HDR) capability added and where the forward link is time-division multiplexed. This 3G air interface standard is denoted as IS-856.

CDMA2000 1xEV-DO in its latest revision, Rev. A, supports downlink (forward link) data rates up to 3.1 Mbit/s and uplink (reverse link) data rates up to 1.8 Mbit/s in a radio channel dedicated to carrying high-speed packet data. IxEV-DO Rev. A was first deployed in Japan and will be deployed in North America in 2006. The Rev. 0 that is currently deployed in North America has a peak downlink data rate of 2.5 Mbit/s and a peak uplink data rate of 154 kbit/s.

Verizon Wireless, Sprint Nextel Corporation, Bell Canada, and TELUS are in the midst of nationwide deployment of IxEV-DO Rev. A in North America, and Alaska Communications Systems (ACS) is deploying IxEV-DO in the main population centers of Alaska. Verizon Wireless currently uses its 1xEV-DO Rev. 0 network for its VCAST services, while Sprint Nextel Corporation currently uses its 1xEV-DO Rev. 0 network for its Power Vision services.

Japanese operator KDDI uses the brand "CDMA 1X WIN" for their CDMA2000 1xEV-DO network, but this is only in reference to its building on past marketing promotions.

South Korean Operator SK Telecom currently uses its 1xEV-DO Rev.0 network for its "June" brand, while KTF currently uses its 1xEV-DO Rev.0 network for its "fimm" brand services.

7.18.7.4. CDMA2000 1xEV-DV

CDMA2000 IxEV-DV (Ix Evolution-Data/Voice), supports downlink (forward link) data rates up to 3.1 Mbit/s and uplink (reverse link) data rates of up to 1.8 Mbit/s. 1xEV-DV can also support concurrent operation of legacy 1x voice users, 1x data users, and high speed 1xEV-DV data users within the same radio channel.

In 2005, Qualcomm put the development of EV-DV on an indefinite halt, due to lack of carrier interest, mostly because both Verizon Wireless and Sprint are using EV-DO.

7.19. DESIGN CONCEPT OF CDMA

In previous chapters, in different sections I have already introduced to the students with CDMA, that implementing a CDMA is just driving a sophisticated car, as we know for proper tuning of car and for safely driving, we need a sophisticated computer aided system. In same way, for better efficient parameter CDMA needs a sophisticated computer aided system.

As we know in Digital and Analog system based on TDMA and FDMA, as capacity increases the interference decreases from the desired signal.

While in a CDMA system, the capacity increases is based on, how much interference desired signal can tolerate. The signal level of a desired signal is always below the interference level.

Also all the customer have to share the same radio channel.

If one user takes more power than it needs, then the others will suffer and system capacity will be reduced. This situation is same as that if you are watching T.V. at your home with other family member. But suddenly some member starts a conversation then definitely you will be interrupt and you cannot enjoy the T.V. programme.

But if other member is talking at very low level of volume than T.V. then you will be less interfered. Mean if one group is talking at high level conversation then you have to either increase the T.V. volume or request to that person to decrease the volume of conversation. This is the concept of CDMA, that all the voice channels are sharing one big radio channel, if other member watching T.V. start to raise their voices, then you have to either leave or raise their voices too. Mean your reception of signal based on other users.

7.19.1. Design of a CDMA System

Direct-Sequence CDMA. Let us start with the equation

$$m = \frac{M}{N} \qquad \text{...(7.1)}$$

where N is the number of cells in frequency reuse

M is the total number of frequency channels

In a FDMA system, $M = \frac{B_t}{B_c}$ is fixed.

In a TDMA system, M is the total number of slot channels and is fixed. In CDMA, M, total number of available channels is a variable and depends on the interference situation. Interference comes from the home cell and the adjacent cells and the value of $\frac{C}{I}$ is expressed as:

$$\frac{C}{I} = \frac{E_b}{I_0} \times \frac{R}{W} = \frac{E_b / I_0}{G_P} \qquad \text{...(7.2)}$$

where E_b = energy per bit

I_0 = interference per hertz ($I_0 >> N_0$, where N_0 is thermal noise per hertz)

R = information rate

W = bandwidth per channel

$$G_P = \text{processing gain} = 10 \log \left(\frac{W}{R}\right) \qquad \text{...(7.3)}$$

In DS-CDMA, since G_P is greater than one, C is always smaller than I in Eq. (7.2) even if there is only one active user. The processing gain is used to overcome I and determine the number of traffic channels that can be created.

From two sets of given values, $\frac{E_b}{I_0}$ and R/W, we can find the $\frac{C}{I}$ from Eq. (7.2). Then, $\frac{C}{I}$ at the mobile location can be used to find the number M which is considered a worst case. Assume that the interference level is much higher than the thermal noise level; then

$$\frac{C}{N+1} \simeq \frac{C}{I}$$

Assume that in a DS-CDMA system, W = 1.23 MHz and R = 9.6 kbps. Then G_P = 1.23 MHz/9.6 kbps or 21 dB.

In this system, a voice quality is accepted at a frame error rate FER = 10^{-2}, which typically corresponds to $\frac{E_b}{I_0}$ = 7 dB. Knowing the values of G_P and $\frac{E_b}{I_0}$, $\frac{C}{I}$ of Eqn. (7.2) can be obtained:

$$\frac{C}{I} = 7 - 21 = -14 \text{ dB or } 0.03981 \qquad ...(7.4)$$

Find the number of traffic channels m_i in each cell from $\frac{C}{I}$ by:

$$\frac{C}{I} = \frac{\alpha_1 (R)^{-4}}{\underbrace{\alpha_1 (m_1 - 1)R^{-4}}_{\text{own cell}} + \underbrace{(\alpha_2 m_2 + \alpha_3 m_3)(R)^{-4}}_{\text{2 adjacent cells}} + \underbrace{\beta.(2R)^{-4}}_{\text{own cell}} + \underbrace{\lambda(2.63R)^{-4}}_{\text{6 distant cells}}} \qquad ...(7.5)$$

$$= 0.03981 = \frac{1}{25.1}$$

where $$\beta = \alpha_4 m_4 + \alpha_5 m_5 + \alpha_6 m_6 \qquad ...(7.6)$$

$$\lambda = \sum_{i=7}^{12} \alpha_i m_i \qquad ...(7.7)$$

and m_i and α_i are the number of traffic channels and the power level, respectively, in each of the i cells. Solving Eqn. (7.7), we obtain

$$m_1 = 26.1 - \left[\frac{\alpha_2 m_2 + \alpha_3 m_3}{\alpha_1}\right] - \frac{\beta}{\alpha_1}(2)^{-4} - \frac{\lambda}{\alpha_1}(2.63)^{-4} \qquad ...(7.8)$$

Case *A*: Single cell case ($\alpha_i = 0$ for $1 \neq 1$). From Eqn. (7.8)

$$m_1 = \frac{1}{C} + 1 = 25.1 + 1 = 26.1 \text{ traffic channels/cell}$$

$$m_1 = \frac{I}{C} + 1 = 25.1 + 1 = 26.1 \text{ traffic channels/cell}$$

Case *B*: Identical-cell case. All the cells have the same power and the same number of traffic channels: $m_i = m_i$ and $\alpha_i = \alpha_i$ we may substitute $m_1 = m$ and $\alpha_i = \alpha$ into Eqn. (7.7) or Eqn. (7.8) and solve for m as follows:

$$26.1 = m[3 + 3.(2)^4 + 6.(2.633)^{-4}]$$

$$m = 7.85 \text{ traffic channels/cell}$$

Both traffic channels appearing in Case *A* and Case *B* include the overhead channels for sync and set-up but do not take into consideration the voice activity cycle or sector-reuse factor as used in real commercial systems.

7.20. DIRECT SEQUENCE CDMA (DS-CDMA)

In CDMA, the broad band frequency channel can be re-used in every adjacent cell so that N is close to 1.

As we know $$N = \frac{1}{3} \cdot \left(\frac{D}{R}\right)^2$$

Ideally $D = 2R$

So $N = 1.33$

In DS-CDMA, N is fixed, but M (the total number of available channels) is a variable and depends on the interference situation.

Let us consider the Fig. 7.29, one mobile station is at position A that is boundary of many cells at A from all the adjacent cell he will experience interference.

Let us calculate that value C/I can be expressed as

$$\frac{C}{I} = \frac{E_b}{I_C} \times \frac{R_b}{B} \Rightarrow \frac{E_b/I_0}{P_G}$$

E_b ⇒ Energy per bit

I_o ⇒ Interference per Hz ($I_0 >> N_0$, where N_0 is thermal noise per Hz)

R_b ⇒ Information Rate

B ⇒ Bandwidth per channel

P_G ⇒ Processing gain ⇒ 10 log (B/R_b)

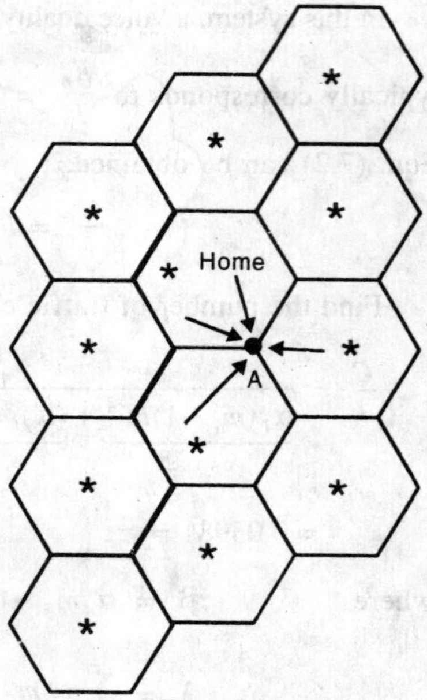

Fig. 7.29.

In (DS-CDMA), since P_G is greater than one, therefore, C is always smaller than 1 which determines interference is very high, and C/I location A can be used to find the number M which is considered a worst case here.

7.21. 3G (WCDMA) ARCHITECTURE

3G (WCDMA) Architecture shown in Fig. 7.30. There are various improvement in 3G as compared to 2G (GSM) architecture. Some of them as follows :

Changes with Respect to BSS

1. BSS (Base Station Subsystem) in 2G named as RAN in 3G (WCDMA).
2. It is important to note that WCDMA is abbreviated as Wide band code division multiple access. It is almost similar to existing CDMA 2000 technique. But existing CDMA 2000 is having bandwidth of 1.25 MHz while WCDMA is having the bandwidth of 5 MHz.
3. WCDMA implemented over existing 2G (GSM) system architecture.
4. Main aim of WCDMA implementation was to provide more bandwidth more speed.
5. CDMA 2000 is having all the feature of 3G (WCDMA) but both implemented separately.
6. BTS of 2G replaced to node B in 3G.
7. BSC of 2G replaced by RNC in 3G.
8. New interface for added between RNC to RNC.
9. WCDMA commonly known as 3G with respect to 2G in GSM.
10. It is used on CMDA + FDMA while 2G is based on TDMA + FDMA.

Changes with Respect to Core Network (NSS)

1. NSS named as switch core network (SCN).

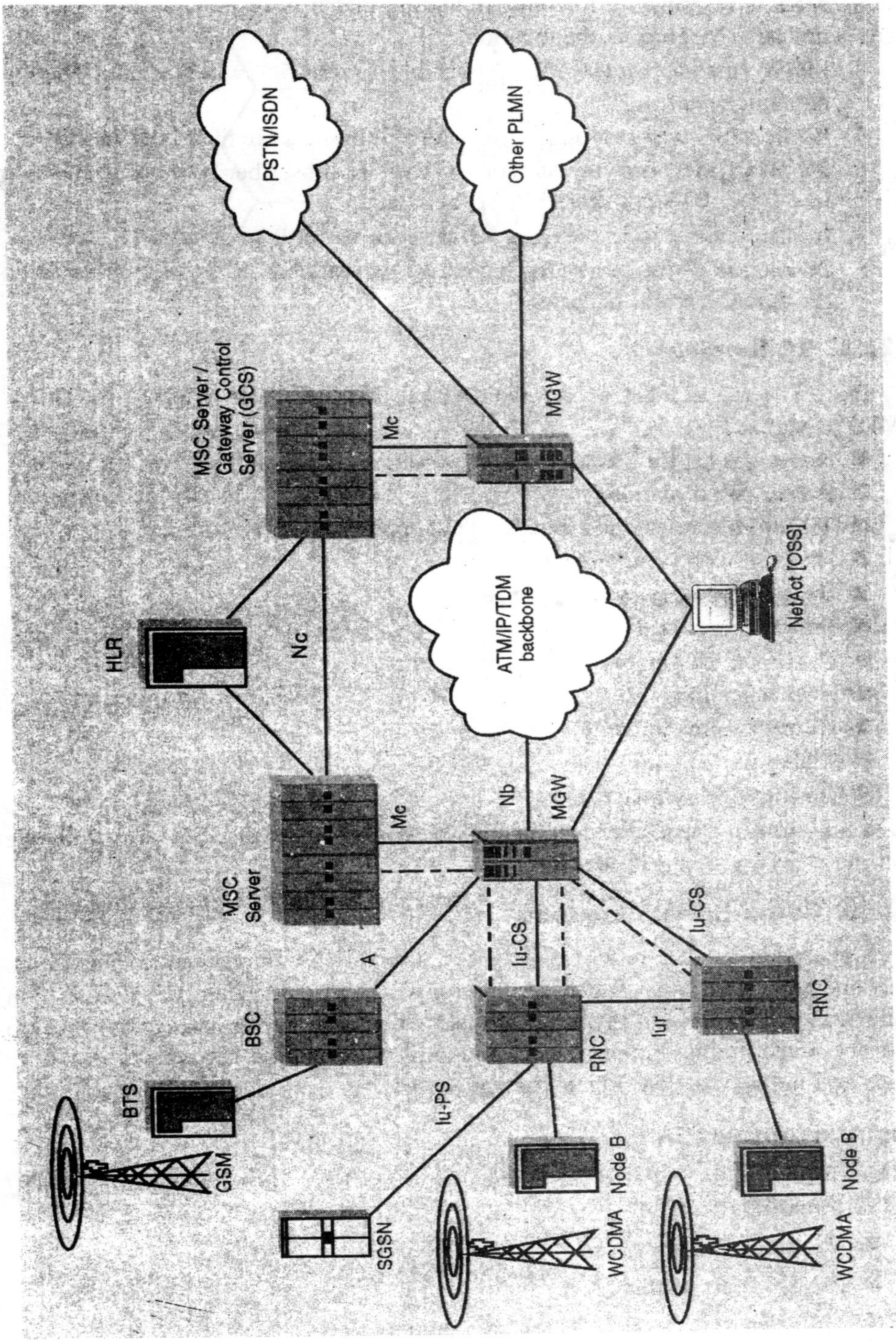

Fig. 7.30. 3G (WCDMA) Architecture.

2. MSC replaced by MSS + MGW.
3. MSS is responsible for only signalling and it is responsible for location update, charging, authentication.
4. MGW (media gateway) is responsible for traffic switching and playing announcement.
5. While MSC was responsible for signalling and switching both in 2G.
6. 3G (WCDMA) network support 2G and 3G subscriber both but if 2G node like BSC, BTS connected in 3G network.
7. It is flexibility to implement 3G network standalone or with 2G network.
8. 2G mobile phone cannot latch over 3G network but 3G mobile phone (UE) can latch over 2G network.

7.21.1. 3G Services

The following is a list of the applications that have been planned for GSM/ UMTS to realize:

- News and traffic flashes
- Public video phoning
- Ticketing services and interactive shopping
- Desktop video conferencing
- Voice recognition and response
- Interactive virtual school
- Universal SIM with credit card function
- Virtual banking
- Currency downloading
- Video on demand
- On-line library and books

In addition to these, the supplementary services used in GSM are available from the very beginning of the 3G.

7.21.2. Potential Applications

Applications are the "end user services". They are no longer standardized. It is up to operators and value added service providers to determine the need for an application and implement them. GSM/UMTS offer the bearer and call control to exchange content and content related signalling information between the mobile device and the application driven content server.

Potential Applications

The following is a list of the applications that have been planned for GSM/ UMTS to realize:

- News and traffic flashes
- Public video phoning
- Ticketing services and interactive shopping
- Desktop video conferencing
- Voice recognition and response
- Interactive and virtual school

- Universal SIM with credit card function
- Virtual banking
- Currency downloading
- Video on demand
- On-line library and books

In addition to these, the supplementary services used in GSM are available from the very beginning of the 3G.

7.21.3. Applications Categorisation From the Business Area Point of View

The different potential applications can be categorized into five distinct groups:

- Person-to-Person Multimedia Communications
- Mobile Internet
- Business Solutions
- Mobile Commerce
- Location Based Services

7.21.4. 3G—Services and Required Data Rates

Different services have different requirements regarding the appropriate data rate. Only a few kbit/s are required for conventional voice transmission with the use of efficient speech data compression functions. Data rates to the order of several 10 kbit/s are helpful and meaningful for convenient e-mail transfers. Greater bandwidth ranging from several kbit/s to more than 100 kbit/s is required for efficient general data transmissions, Internet access, mobile banking, shopping, etc.

Even greater data rates from several 10 kbit/s to several 100 kbit/s are necessary for high-quality image transmission and video telephony. The highest requirements for data rates from 100 kbit/s to more than 1 Mbit/s are demanded by video conferences and video-on-demand applications, in addition to different multimedia applications.

UMTS will be able to dynamically and flexibly provide these data rates ranging from 8 kbit/s to a maximum of 2 Mbit/s.

The following diagram illustrates the services provided by the 3G:

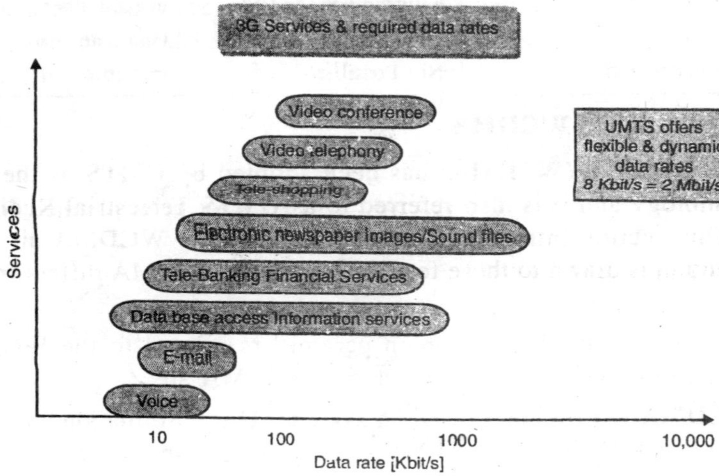

Fig. 7.31. The Services Provided by the 3G.

7.21.5. Comparison between 2G and 3G Technologies

Table 7.9. Comparison between 2G and 3G

Parameter	2G	3G
Band	GSM 900, 1800,1900	GSM 2 100
Access technique	TDMA+FDMA	CDMA + FDMA or CDMA+TDMA
Channel Bandwidth	200 kHz	5 MHz
Maximum theoretical speed	9.6 to 236 kbps	1 .3 to 2 mbps
Frequency per cell	2 to 6	1
Cell size	500 meter to 25 Kilometer	500 meter to 5 Kilometer
Mobile Phone	Simple to design	Very Complex
Security	Only network authenticate to mobile	Mobile & Network both authenticate to each other
Video Telephony	Not Possible	Possible
Interference Level	Very high	Very less
Handover	Hard handover	Only soft handover
Online Game	Not Possible	Possible
Inter System handover	Not Possible (2Gto 3G) 2G mobile can not latch over 3G Network.	Possible (3G to 2G) 3G mobile can latch over 2G networks.
Technology name	GSM	WCDMA
Network node for radio access	BTS & BSC	Node B & RNC
Network Node for Core Network	MSC	MSS + MOW
Radio Interfaces	Um, Abis, Ater	Uu, lu-cs, lu-r
Core Interfaces	C,D,E	C,D,Mc, Nc , Nb,
Physical Media Support	Only TDM	It support TDM, IP, ATM
Radio network connection	There is no connection between BSC to BSC	RNC & RNC connected to each other.
Quality of service	No Guarantee	Guarantee
Bandwidth on demand	Not Possible	Possible

7.21.6. Features of WCDMA

Wideband-CDMA (WCDMA) has been adopted by UMTS as the multiple access technology and it is also referred to as UMTS Terrestrial Radio Access (UTRA). This section introduces the principles of the WCDMA air interface. Special attention is drawn to those features by which WCDMA differs from GSM and IS-95.

Table above summarises the main parameters related to the WCDMA air interface. Some of the items that characterise WCDMA are:

(*i*) **WCDMA is a wideband CDMA system:** User information bits are spread over a wide bandwidth (5 MHz) by multiplying with spreading codes before transmission and are recovered by decoding in the receiver.

(*ii*) The **chip rate of 3.84 Mchip/s** used leads to a carrier bandwidth of approximately 5 MHz. In GSM, carrier bandwidth is only 200 kHz. Even in narrowband CDMA systems, such as IS-95, the carrier bandwidth is only 1.25 MHz. The inherently wide carrier bandwidth of WCDMA supports high user data rates and also has certain performance benefits, such as increased multipath diversity [TROO].

(*iii*) **WCDMA supports highly variable user data rates;** in other words the concept of obtaining Bandwidth on Demand (BoD) is well supported. Each user is allocated frames of 10 ms duration, during which the user data rate is kept constant. However, the data capacity among the users can change from frame to frame.

(*iv*) **WCDMA supports two basic modes of operation:** Frequency Division Duplex (FDD) and Time Division Duplex (TDD). In FDD mode, separate 5 MHz carriers are used for the uplink and downlink respectively, whereas in TDD only one 5 MHz is time-shared between uplink and downlink.

(*v*) **WCDMA supports the operation of asynchronous base stations:** Unlike the synchronous IS-95 system, there is no need for a global time reference, such as a OPS, so making deployment of indoor and micro base stations easier.

(*vi*) **WCDMA employs coherent detection on uplink and downlink** based on the use of pilot symbols or common pilot. In IS-95 coherent detection is only used on the downlink. The use of coherent detection on uplink will result in an overall increase of coverage and capacity on the uplink. This makes the downlink more likely to be the bottleneck of the whole system.

(*vii*) The WCDMA air interface has been crafted in such a way that advanced CDMA receiver concepts, such as multiuser detection (MUD) and smart adaptive antennas, can be deployed by the network operator as a system option to increase capacity and/or coverage. In most second generation systems no provision has been made for such concepts.

(*viii*) WCDMA is designed to be deployed in conjunction with GSM. Therefore, handovers between GSM and WCDMA are supported.

Multiple access method	DS-CDMA
Duplexing method	FDD/TDD
Base station synchronisation	Asynchronous operation
Chip rate	3.84 Mcps
Frame length	10 ms
Service multiplexing	Multiple services with different quality of service requirements multiplexed on one connection
Multirate concept	Variable spreading factor and multicode
Detection	Coherent using pilot symbols or common pilet
Multiuser detection, smnart antennas	Supported by the standard, optional in the implementation

7.22. OVERVIEW OF HANDOVER IN MOBILE NETWORKS

Mobile networks allow users to access services while on the move so giving end users "freedom" in terms of mobility. However, this freedom does bring uncertainties to mobile systems. The mobility of the end users causes dynamic variations both in the link quality and the interference level, sometimes requiring that a particular user change its serving base station. This process is known as handover (HO).

Handover is the essential component for dealing with the mobility of end users. It guarantees the continuity of the wireless services when the mobile user moves across cellular boundaries.

In first-generation cellular systems like AMPS, handovers were relatively simple. Second-generation systems like GSM and PACS are superior to 1G systems in many ways, including the handover algorithms used. More sophisticated signal processing and handover decision procedures have been incorporated in these systems and the handover decision delay has been substantially reduced. Since the introduction of CDMA technology, another idea that has been proposed for improving the handover process is soft handover and this is the focus of the work in this thesis.

7.22.1. Types of Handover in 3G WCDMA Systems

There are four different types of handovers in WCDMA mobile networks. They are:

Intra-system HO

Intra-system HO occurs within one system. It can be further divided into Intra-frequency HO and Inter-frequency HO. Intra-frequency occurs between cells belonging to the same WCDMA carrier, while Inter-frequency occurs between cells operate on different WCDMA carriers.

Inter-system HO

Inter-system HO takes places between cells belonging to two different Radio Access Technologies (RAT) or different Radio Access Modes (RAM). The most frequent case for the first type is expected between WCDMA and GSM/EDGE systems. Handover between two different CDMA systems also belongs to this type. An example of inter-RAM HO is between UTRA FDD and UTRA TDD modes.

Hard Handover (HHO)

HHO is a category of HO procedures in which all the old radio links of a mobile are released before the new radio links are established. For real-time bearers it means a short disconnection of the bearer; for non-real-time bearers HHO is lossless. Hard handover can take place as intra or inter-frequency handover.

Soft Handover (SHO) and Softer HO

During soft handover, a mobile simultaneously communicates with two (2-way SHO) or more cells belonging to different BSs of the same RNC (intra-RNC) or different RNCs (inter-RNC). In the downlink (DL), the mobile receives both signals for maximal ratio combining: in the uplink (UL), the mobile code channel is detected by both BSs (2-way SHO), and is routed to the RNC for selection combining. Two active power control loops participate in soft handover: one for each BS. In the softer handover situation, a mobile is controlled by at least two sectors under one BS, the RNC is not involved and there is only one active power control loop. SHO and softer HO are only possible within one carrier frequency and therefore, they are intra-frequency handover processes.

Figure 7.32 shows some scenarios of different types of HO.

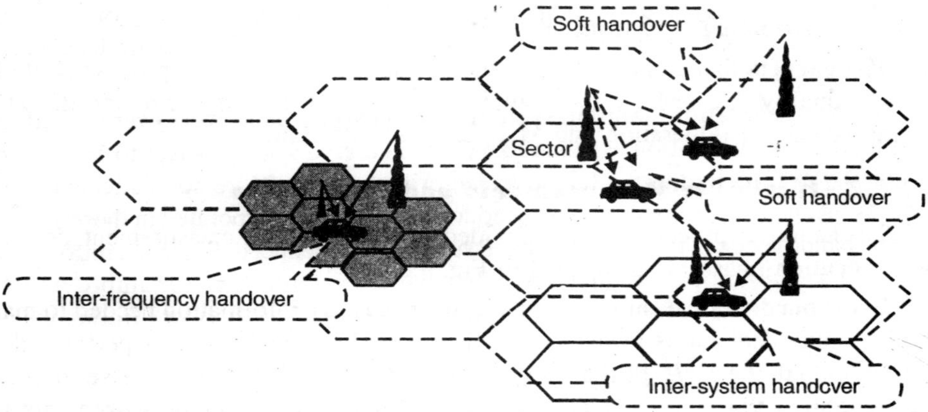

Fig. 7.32. Scenarios of Different Types of Handover, Macro Cells of WCDMA Network, Pico Cells of WCDMA Network, Micro Cells of GSM Network.

7.22.2. Objectives of Handover

Handover can be initiated in three different ways: mobile initiated, network initiated and mobile assisted.

(*i*) **Mobile Initiated:** The Mobile makes quality measurements, picks the best BS, and switches, with the network's cooperation. This type of handover is generally triggered by the poor link quality measured by the mobile.

(*ii*) **Network Initiated:** The BS makes the measurements and reports to the RNC, which makes the decision whether to handover or not. Network initiated handover is executed for reasons other than radio link control, *e.g.*, to control traffic distribution between cells. An example of this is the BS-controlled Traffic Reason Handover (TRHO). TRHO is a load-based algorithm that changes the handover threshold for one or more outgoing adjacencies for a given source cell depending on its load. If the load of the source cell exceeds a given level, and the load in a neighbouring cell is below another given level, then the source cell will shrink its coverage, handing over some traffic to the neighbouring cell. Therefore, the overall blocking rate can be reduced, leading to a greater utilisation of the cell resource.

(*iii*) **Mobile Assisted:** Here the network and the mobile both make measurements. The mobile reports the measurement results from nearby BSs and the network makes the decision of handing over or not.

The objectives of handover can be summarised as follows:

(*i*) Guaranteeing the continuity of wireless services when the mobile user moves across the cellular boundaries

(*ii*) Keep required QoS

(*iii*) Minimising interference level of the whole system by keeping the mobile linked to the strongest BS or BSs.

(*iv*) Roaming between different networks

(*v*) Distributing load from hot spot areas (load balancing)

The triggers that can be used for the initiation of a handover process could be the link quality (UL or DL), the changing of service, the changing of speed, traffic reasons or O&M (Operation and Maintenance) intervention.

7.22.3. Handover Measurements and Procedures

The handover procedure can be divided into three phases: measurement, decision and execution phases as illustrated in Fig. 7.33.

In the handover measurement phase, the necessary information needed to make the handover decision is measured. Typical downlink measurements performed by the mobile are the E_c/I_0^2 of the Common Pilot Channel (CPICH) of its serving cell and neighbouring cells. For certain types of handover, other measurements are needed as well. For example, in an asynchronous network like UTRA FDD (WCDMA), the relative timing information between the cells needs to be measured in order to adjust the transmission timing in soft handover to allow coherent combining in the Rake receiver. Otherwise, the transmissions from the different BSs would be difficult to combine and especially the power control operation in soft handover would suffer additional delay.

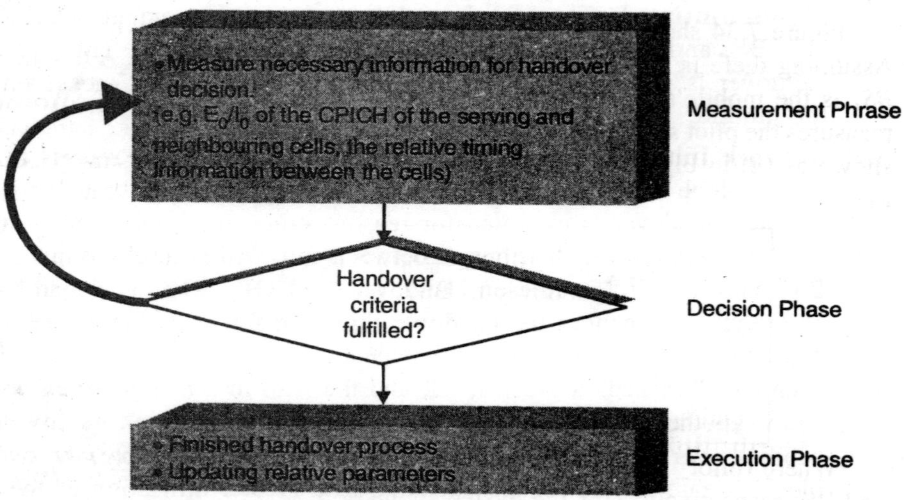

Fig. 7.33. Handover procedures.

In the handover decision phase, the measurement results are compared against the predefined thresholds and then it is decided whether to initiate the handover or not. Different handover algorithms have different trigger conditions.

In the execution phase, the handover process is completed and the relative parameters are changed according to the different types of handover. For example, in the execution phase of the soft handover, the mobile enters or leaves the soft handover state, a new BS is added or released, the active set is updated and the power of each channel involved in soft handover is adjusted.

7.22.4. Soft Handover (SHO)

Soft handover was introduced by CDMA technology. Compared to the conventional hard handover, soft handover has quite a few inherent advantages. However, it also has the disadvantages of complexity and extra resource consumption. Planning of soft handover overhead is one of the fundamental components of the radio network planning and optimisation [HT00]. In this section, the basic principles of soft handover are presented.

7.22.5. Principles of Soft Handover

Soft handover is different from the traditional hard handover process. With hard handover, a definite decision is made on whether to handover or not and the mobile only communicates with one BS at a time. With soft handover, a conditional decision is made on whether to handover or not. Depending on the changes in pilot signal strength from the two or more BSs involved, a hard decision will eventually be made to communicate with only one. This normally happens after it is clear that the signal coming from one BS is considerably stronger than those come from the others. In the interim period of soft handover, the mobile communicates simultaneously with all the BSs in the active set. The difference between hard and soft handover is like the difference between a swimming relay and track-and-field relay events. Hard handover happens on a time point; while, soft handover lasts for a period of time.

Figure 7.34 shows the basic process of hard and soft handover (2-way case). Assuming there is a mobile terminal inside the car moving from cell 1 to cell 2, BS_1 is the mobile's original serving *BS*. While moving, the mobile continuously measures the pilot signal strength received from the nearby BSs. With hard handover shown as *(a)* in Figure 7.34, the trigger of the handover can be simply described as :

```
If (pilot_E0/I0)2 - (pilot_E0/I0)1 > D and BS1 is the serving
        BS Handover to BS2;
else
do not handover;
end
```

Where $(\text{pilot\_}\ E_c/I_0)_l$ and $(\text{pilot\_}E_c/I_0)$ 2 are the received pilot E_c/I_0 from BS_1 and BS_2 respectively; *D* is the hysteresis margin.

The reason for introducing the hysteresis margin in the hard handover algorithm is to avoid a "ping-pong effect", the phenomenon that when a mobile moves in and out of cell's boundary, frequent hard handover occurs. Apart from the mobility of the mobile, fading effects of the radio channel can also make the "ping-pong" effect more serious. By introducing the hysteresis margin, the "ping-pong" effect is mitigated because the mobile does not handover immediately to the better *BS*. The bigger the margin, the less the "ping-pong" effect. However, a big margin, means more delay. Moreover, the mobile causes extra interference to neighbouring cells due to the poor quality link during the delay. Therefore, to hard handover, the value of the hysteresis margin is fairly important. When hard handover occurs, the original traffic link with BS_1 is dropped before the setting up of the new link with BS_2 so hard handover is a process of "break before make".

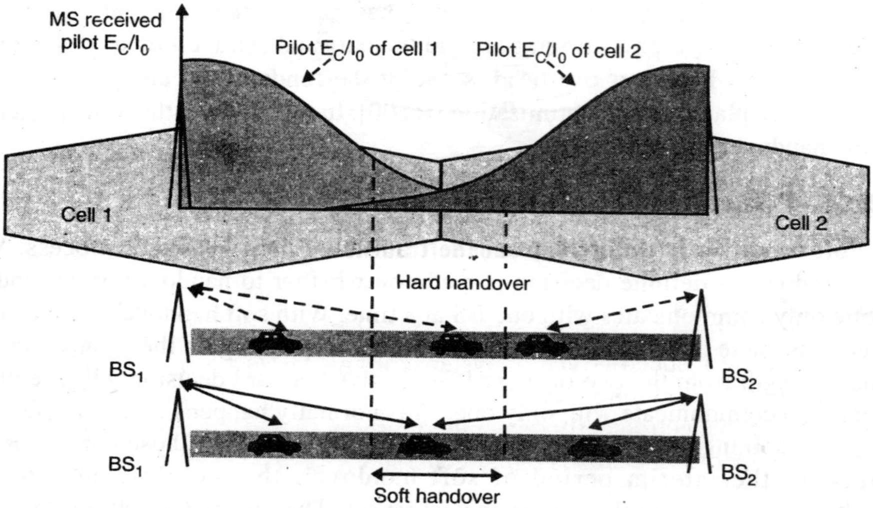

Fig. 7.34. Comparison Between Hard and Soft Handover.

In the case of soft handover, shown as (*b*) in Figure 7.34, before $(pilot\_E_c/I_0)_2$ goes beyond $(pilot\_E_c/I_0)_1$, as long as the soft handover trigger condition is fulfilled, the mobile enters the soft handover state and a new link is set up. Before BS_2 is dropped (handover dropping condition is fulfilled), the mobile communicates with both BS_1 and BS_2 simultaneously. Therefore, unlike hard handover, soft handover is a process of "make before break". So far, several algorithms have been proposed to support soft handover and different criteria are used in different algorithms.

The soft handover process is not the same in the different transmission directions. Figure 7.35 illustrates this. In the uplink, the mobile transmits the signals to the air through its omnidirectional antenna. The two BSs in the active set can receive the signals simultaneously because of the frequency reuse factor of one in CDMA systems. Then, the signals are passed forward to the RNC for selection combining. The better frame is selected and the other is discarded. Therefore, in the uplink, there is no extra channel needed to support soft handover.

In the downlink, the same signals are transmitted through both BSs and the mobile can coherently combine the signals from different BSs since it sees them as just additional multipath components. Normally maximum ratio combining strategy is used, which provides an additional benefit called macrodiversity. However, to support soft handover in the downlink, at least one extra downlink channel (2-way SHO) is needed. This extra downlink channel acts to other users acts like additional interference in the air interface. Thus, to support soft handover in the downlink, more resource is required. As a result, in the downlink direction, the performance of the soft handover depends on the trade-off between the macrodiversity gain and the extra resource consumption.

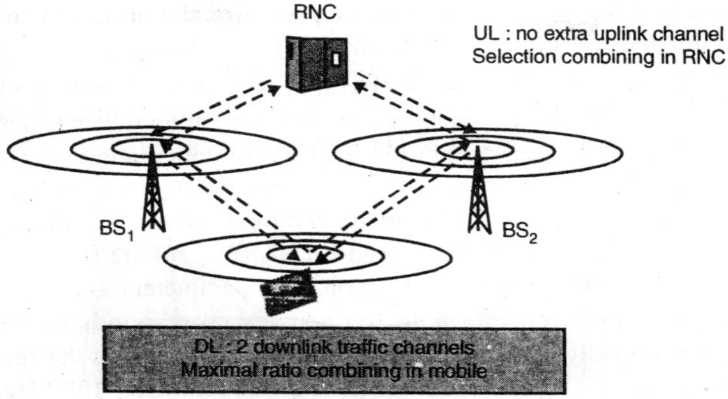

Fig. 7.35. Principles of Soft Handover (2-Way Case)

7.23. VOCODERS

Vocoders are a specific type of system for analyzing the speech, voice signals at the transmitter in mobile communication system. All vocoder analyzes speech, transforms it into electronically transmitted information, and recreates it. The Voder (from Voice Operating Demonstrator) generates synthesized speech by means of a console with fifteen touch sensitive keys and a pedal. It was originally developed as a speech coder for telecommunications applications in the 1930s, the idea being to code speech for transmission. Its primary use in this fashion is for secure radio communication, where voice has to be encrypted and then transmitted. The advantage of this method of "encryption" is that no signal is sent, but rather envelopes of the band pass filter. The receiving unit needs to be set up in the same channel configuration to resynthesize a version of the original signal spectrum. The vocoder as both hardware and software has also been used extensively as an electronic musical instrument.

7.23.1. Vocoder Theory

The human voice consists of sound generated by the opening and closing of the glottis by the vocal cards, which produces a periodic waveform with many harmonics. This basic sound is then filtered by the nose and throat to produce differences in harmonic content in a controlled way. Creating the wide variety of sounds used in speech. There is another set of sound, known as the unvoiced and plosive sounds which are created or modified by the mouth in different fashions. The vocoder examines speech by measuring how it spectral.

Characteristics change over time. This results in a series of numbers representing these modified frequencies at any particular time as the user speaks. In simple terms, the signal is split into a number of frequency bands and the level of signal present at each frequency band gives the instantaneous representation of the spectral energy content. Thus, the vocoder dramatically reduces the amount of information needed to share speech, from a complete recording to a series of numbers. To recreate speech, the vocoder simply reverses the process, processing a broadband noise source by passing it through a stage that filters the frequency content based on the originally recorded series of numbers.

Information about the instantaneous frequency of the original voice signal is discarded.

Following Fig. 7.36 is showing the traditional speech generation model that it is the basis of all vocoding systems. The second generating mechanisms forms the source and is binary separated from the intelligence modulating vocal tract filter which forms the system. The speech signal is assumed to be of two types : voice and unvoiced. Voice sound (*m, n, v* pronunciations) are a result of quasiperiodic vibrations of the vocal chord and unvoiced sounds (*f, s, sh*) are fricatives produced by turbulent air flow through a construction. The parameters associated with this model are the voice pitch, the pole frequencies corresponds to the resonant frequencies of the vocal tract and are often called the formants of the speech signal. For adult speakers, the formants are centered around 500 Hz, 1500 Hz, 2500 Hz, 3500 Hz.

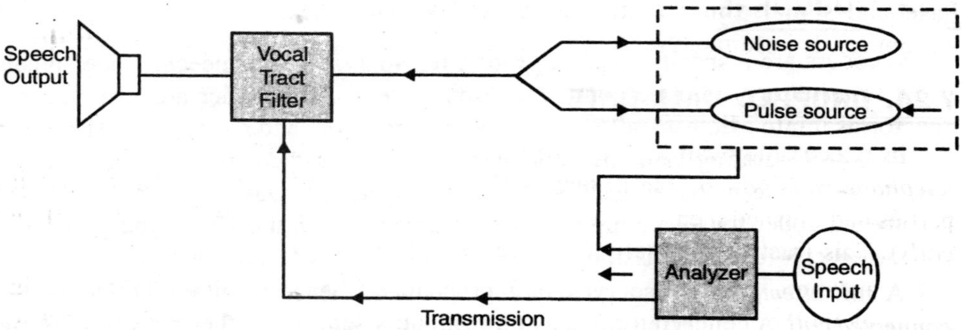

Fig. 7.36. Speech Generation Model

7.23.2. Channel Vocoders

Channel vocoder was the first among the analysis synthesis systems of speech demonstrated practically. Channel vocoders are frequency domain vocoders that determine the envelope of the speech signal for a number of frequency bands and then sample, encode and multiplex these samples with the encoded outputs of the other filters. The sampling is done synchronously every 10 ms to 30 ms. Along with the energy information about each band, the voiced/unvoiced decision and the pitch frequency for voiced speech are also transmitted.

7.23.3. Cepstrum Vocoders

The cepstrum vocoder separates the excitation and vocal tract spectrum by method called inverse fourier transforming of the log magnitude spectrum to produce the cepstrum of the signal. The low frequency coefficients in the cepstrum corresponds to the vocal tract spectral envelope, with the high frequency excitation coefficients forming a periodic pulse train at multiples of the sampling period. Linear filtering is performed to separate the vocal tract cepstral coefficients from the excitation coefficients. IN the receviersredeivers, the vocal tract cepstral coefficients are fourier transformed to produce the vocal tract impulse response. By convolving this impulse response with a synthetic excitation signal, the original speech is reconstructed.

7.23.4. Formant Vocoders

Formant vocoders can operate a lower bit rates than the channel vocoder because it uses fewer control signals. Instead of sending samples of the power spectrum envelope, the formant vocoder attempts to transmit the positions of the peaps of the spectral envelops. Typically, a formant vocoder must be able to identify at least three formants for representing the speech sounds, and it must also control the intensities of the formants. It can reproduce speech at bit rate lower than 1200 bits/second.

7.23.5. Voice Excited Vocoder

It eliminate the need for pitch extraction and voicing detection operations. It uses a hybrid combination of PCM transmission for the low frequency band of speech, combined with channel vocoding of higher frequency bands. A pitch signal is generated at the synthesizer by rectifying, bandpass filtering, and clipping the baseband signal, thus creating a spectrally that signal with energy and pitch harmonics. It is suitable for 7200 bits/sec to 9600 bits/sec operation.

7.24. DROPPED CALL RATE

In telecommunications, the **dropped-call rate (DCR)** is the fraction of the *telephone calls* which, due to technical reasons, were cut off before the speaking parties had finished their conversation and before one of them had hung up (*dropped calls*). This fraction is usually measured as a percentage of all calls.

A *call attempt* invokes a *call setup* procedure, which, if successful, results in a *connected call*. A connected call may be terminated (disconnected) due to a technical reason before the parties making the call would wish to do so (in ordinary phone calls this would mean before either of the parties has hung up). Such calls are classified as dropped calls. In many practical cases this definition needs to be further expanded with a number of detailed specifications describing which calls exactly are counted as dropped, at what stage of the call setup procedure a call is counted as connected, etc. In modern telecommunication systems, such as *cellular* (*mobile*) *networks*, the call setup procedure may be very complex and the point at which a call is considered successfully connected may be defined in a number of ways, thus influencing the way the dropped-call rate is calculated.

The dropped-call rate in conventional (so-called land-line) networks is extremely low and is significantly less than 0.01%. In mobile communication systems using radio channels the dropped-call rate is higher and may range for commercial networks between 0.1% and a few percent. The main reasons for dropped calls in mobile networks are lack of radio coverage (either in the *downlink* or the *uplink*), *radio interference* between different subscribers, imperfections in the functioning of the network (such as failed *handover* or *cell-resellection* attempts), overload of the different elements of the network (such as cells), etc.

The dropped-call rate is one of the key performance indicators (*KPI*) used by the network operators to assess the performance of their networks. It is assumed to have direct influence on the customer satisfaction with the service provided by the network and its operator. The dropped-call rate is usually included, together with other technical parameters of the network, in a key performance indicator known as *call retainability*.

The operators of telecommunication networks aim at reducing the call dropped rate as much as practical and affordable. In mobile networks this is achieved by improving radio coverage, expanding the capacity of the network and optimising the performance of its elements, all of which may require considerable effort and significant investments on the part of the network operator.

7.24.1. Dropped Call Rate Type

There are two types of call drop:

1. TCH call drop means TCH channel is released abnormally after it is occupied successfully.
2. SDCCH call drop means SDCCH channel is released abnormally after it is occupied successfully.

7.24.1.1. Causes of TCH call Drop

(*a*) Radio interface failure

(*b*) Operation and Maintenance

(*c*) Equipment failure

(*d*) Protocol error

Formula of TCH (Traffic Channel) call drop rate

$$\text{TCH call drop rate} = \frac{\text{TCH call drops}}{\text{Successful TCH seizures (all)}}$$

Measurement Point of TCH Call Drop

1. In Intra-BSC handover procedure, the source cell, excluding the source cell for directed retry, measures the item when the timer for the source cell to wait for Inter Clear Request with cause value HANDOVER COMPLETE from the target cell expires.
2. In Intra-BSC handover procedure, when the target cell AM/CM net-drive fails (due to timeout or negative acknowledgement), the target cell sends Inter Clear Request to source cell, in this case, this item is measured in the source cell.
3. In the case that MS reverses to the original channel after intra BSC handover fails, the source cell first releases the terrestrial connection but the AM/CM re-net-drive fails (due to timeout or negative acknowledgement). In this case this item is measured in the source cell.
4. The resource of the lower-priority call will be preempted by the higher-priority call if MSC and BSC both supports preemption, which will cause call drop.
5. This item is measured when the RSL link of the TRX that the call is using disconnects, which will cause call drop.

7.24.1.2. Formula of SDCCH (Stand alone Dedicated Control Channel)

$$\text{SDCCH call drop rate} = \frac{\text{SDCCH call drops}}{\text{Successful SDCCH seizures}} \times 100\%$$

(a) Reason for Call Drops

According to the definition of call drop measurement point, call drop is usually caused by the following:

1. Radio link fault. During the communication, messages can not be received correctly.
2. Abis link (Interface between BSC and BTS) broken during conversation.
3. Call drop during handover.
4. Other system faults.
5. Timers that may cause call drops (BSC timer):
   - starting from sending HANDOVER CMD (Command) and ending at receiving HANDOVER CMP (Complete). Time out of the timer will cause call drop.
   - starting from sending CHAN REL (Channel Release) and ending at receiving REL IND (Release End).

The causes of radio link fault can be:

Interference:

- Internal interference
- External interference
- Equipment interference

Poor coverage:

- Coverage hole
- Isolated island
- Uplink/downlink imbalance

Improper parameter setting:

- Radio link timeout, SACCH multi-frames
- Handover parameters
- Power control parameters

Equipment problem (Antenna, feeder, combiner, TRX)

Clock problem

Transmission problem

(b) Main parameters that may affect the call drop rate:

- Handover related parameters.
- Power control related parameters.
- Frequency planning parameters

1. Handover Problem: In inter-cell handover measurement function, it occurs frequently that the handover fails and the reversion also fails.

In inter-cell handover measurement function, handover failures with successful reversions occur many times. In undefined adjacent cell measurement function, observe the receiving level of the undefined adjacent cells and number of the measurement reports.

In outgoing inter-cell handover measurement function, the outgoing inter-cell handover success rate is low (for a certain cell). Find the adjacent cell to which the handover success rate is low and find the cause.

Incoming inter-cell handover success rate is low. The handover judgment parameter setting of the target cell may be improper. In TCH measurement function, handover times are not in normal proportion to successful TCH seizures for call. (handovers/calls>3)

2. Equipment Problem: Call drop arising from equipment problem, Hardware problem, Transmission problem, Antenna and feeder fault, Other causes.

SUMMARY

1. AMPS is a first generation technology based system.
2. IS-54 is also named as USDC.
3. IS-136 is improvement version of IS-54.
4. IS-95 is first step towards CDMA.
5. ISM band is licence free band in 2.4 GHz band used for Industry, School, Medical category.
6. GSM is 2G based system.
7. Wireless cable T.V. uses microwave communication.
8. Pacific Digital Cellular is also named as Personal Digital Cellular.

REVIEW QUESTIONS

1. How are different channels within the forward link in IS-95 system identified ?
2. Which cellular system uses the most bandwidth efficient modulation ?
3. How many full rate physical channels per cell can a GSM system accommodate?
4. Describe the operation of GSM system.
5. What are the purpose of ISM band and what are their frequency range ?
6. Explain Cordless system.
7. Compare GSM and CDMA with respect to various parameters.
8. Compare the AMPS and Digital Cellular System with respect to various parameters.
9. Describe the personal handyphone system.
10. Describe call flow for mobile originated call.
11. How many types of location update as there ?
12. Why location update is required ?
13. Write short note on "Vocoders".

❑❑❑

APPENDIX—A

Table 1. Analog Cellular Systems (An Overview Worldwide).

Standard	Mobile Tx/ Base Tx (MHz)	Channel Spacing (kHz)	No. of channel	Region
1. AMPS	824-849/869-894	30	832	USA
2. TACS	890-915/935-960	25	1000	Europe
3. ETACS	872-905/917-950	25	1240	U.K.
4. NMT 450	453-457.5/463-467.5	25	180	Europe
5. NMT 900	890-915/935-960	12.5	1999	Europe
6. C-450	450-455.74/460-465.74	10	573	Germany
7. RTMS	450-455/460-465	25	200	Italy
8. Radiocom 2000	192.5-199.5/200.5-207.5 215.5-233.5/207.5-215.5 165.2-168.4/169.8-173 414.8-418/424.8-428	12.5	560 640 256 256	France
9. NTT	925-940/870-885 915-918.5/860-863.5 922-925/867-870	25/6.25 6.25 6.25	600/2400 560 480	Japan
10. JTACS/NTACS	915-925/860-870 898-901/843-846 918.5-922/863.5-867	25/12.5 25/12.5 12.5	400/800 120/240 280	Japan

Table 2. Digital Cordless Air Interface Parameters Summary.

Region	CT2 Europe	CT2+ Canada	DECT Europe	PHS Japan	PACS US
Duplexing	TDD		TDD	TDD	FDD or TDD
Frequency band (MHz)	864-868	944-948	1880-1900	1895-1918	1850-1910/ 1930-1990 or 1920-1930
Carrier spacing (kHz)	100		1728	300	300/300
Number of carriers	40		10	77	400 or 32
Channels/carrier	1		12	4	8 or 4
Channel bit rate kbps	72		1152	384	384
Modulation	GFSK		GFSK	π/4 DQPSK	π/4 QPSK
Speech Coding	32 kb/s		32 kb/s	32 kb/s	32 kb/s
Average handset TX power (mW)	5		10	10	25
Peak handset TX power (mW)	10		250	80	100
Frame duration (ms)	2		10	5	2.5 or 2.0

Table 3. Unlicensed/Short Range PCS Standards.

	PACS-UB	DCS 1800	PHS
Year of Introduction	1992	1993	1993
Frequencies	1920-1930 MHz	1710-1785 MHz (R) 1805-1880 MHz (F)	1895-1907 MHz
Multiple Access	TDMA/FDMA/TDD	TDMA/FDM/FDD	TDMA/FDMA/TDD
Modulation	π/4 QPSK	GMSK	π/4 QPSK
Carrier Separation	300 kHz	200 kHz	300 kHz
Data rate	384 kbps	270.833 kbps	384 kbps
Voice Ch./R.F. Ch.	4	8	4
Speech Coding	ADPCM @ 32 kbps	RELP-LTP @ 13 kbps	ADPCM @ 32 kbps
Channel coding	CRC	Conv. r = 1/2	CRC
Receiver	Coherent	Coherent	Coherent
Portable Tx. Power Max./Avg.	200 mW/20 mW	1 W/125 mW	80 mW/10 mW

Table 4. GSM Air Interface Specifications Summary.

Parameter	Specifications
Reverse Channel Frequency	980-915 MHz
Forward Channel Frequency	935-960 MHz
ARFCN Number	0 to 124 and 975 to 1023
Tx/Rx Frequency Spacing	45 MHz
Tx/Rx Time Slot Spacing	3 Time slots
Modulation Data Rate	270.833333 kbps
Frame period	4.615 ms
Users per frame (Full rate)	8
Time slot period	576.9 μs
Bit Period	3.692 μs
Modulation	0.3 GMSK
ARFCN Channel Spacing	200 kHz
Interleaving (max. delay)	40 ms
Voice Coder Bit Rate	13.4 kbps

Table 5. Second Generation Digital Cellular Standards Summary.

	GSM	IS-136	PDC
Year of Introduction	1990	1991	1993
Frequencies	890-915 MHz (R) 935-960 MHz (F)	824-849 MHz (R) 869-894 MHz (F)	810-830 and 1429-1453 MHz (R) 940-960 and 1477-1501 MHz (F)
Multiple Access	TDMA/FDMA/FDD	TDMA/FDMA/FDD	TDMA/FDMA/FDD
Modulation	GMSK (BT = 0.3)	$\pi/4$ DQPSK	$\pi/4$ DQPSK
Carrier Separation	200 kHz	30kHz	25 kHz
Channel Data Rate	270.833 kbps	48.6 kbps	42 kbps
Number of Voice Channels	1000	2500	3000
Spectrum Efficiency	1.35 bps/Hz	1.62 bps/Hz	1.68 bps/Hz
Speech Coding	RELP-LTP @ 13 kbps	VSELP @7.95 kbps	VSELP @ 6.7 kbps
Channel coding	CRC with $r = 1/2, L = 5$ conv.	7 bit CRC with $r = 1/2; L = 6$ conv.	CRC with Conv.
Equalizers	Adaptive	Adaptive	Adaptive
Portable Tx. Power Max./Avg.	1 W/125 mW	600 mW/200 mW	

Table 6. DECT Radio Specifications Summary.

Parameter	Specifications
Frequency band	1880-1900 MHz
Number of Carriers	10
RF Channel Bandwidth	1.728 MHz
Multiplexing	FDMA/TDMA 24 slots per frame
Duplex	TDD
Spectral Efficiency	500 Erlangs/km^2/MHz
Speech Coder	32 kbps ADPCM
Average Transmitted Power	10 mW
Frame Length	10 ms
Channel Bit Rate	1152 kbps
Data Rate	32 kbps Traffic Channel 6.2 kbps Control Channel
Channel Coding	CRC 16
Dynamic Channel Allocation	Yes
Modulation	GFSK (BT = 0.3)
Speech Channel/RF Channel	12

Table 7. IEEE 802.11b Channels for both DS-SS and FH-ss WLAN Standards.

Country	Frequency range available	DSSS channels available	FHSS channels available
United states	2.4 to 2.4835 Gz	1 through 11	2 through 80
Canada	2.4 to 2.4835 GHz	1 through 11	1 through 80
Japan	2.4 to 2.497 GHz	1 through 14	2 through 95
France	2.4465 to 2.4835 GHz	10 through 13	48 through 82
Spain	2.445 to 2.4835 GHz	10 through 11	47 through 73
Remainder of Europe	2.4 to 2.4835	1 through 13	2 through 80

Table 8. Comparison of Mobile Communication Systems—Mobile Station.

Service	Coverage range	Required infrastructure	Complexity	Hardware cost	Carrier frequency	Functionality
TV remote control	Low	Low	Low	Low	Infrared	Transmitter
Garage door opener	Low	Low	Low	Low	<100 MHz	Transmitter
Paging system	High	High	Low	Low	<1 GHz	Receiver
Cordless phone	Low	Low	Moderate	Low	1-3 GHz	Transceiver
Cellular phone	High	High	High	Moderate	<2 GHz	Transceiver

Table 9. Comparison of Mobile Communication Systems—Base Station.

Service	Coverage range	Required infrastructure	Complexity	Hardware cost	Carrier frequency	Functionality
TV remote control	Low	Low	Low	Low	Infrared	Receiver
Garage door opener	Low	Low	Low	Low	<100 MHz	Transmitter
Paging system	High	High	High	High	<1 GHz	Transmitter
Cordless phone	Low	Low	Low	Moderate	1-3 GHz	Transceiver
Cellular phone	High	High	High	High	<2 GHz	Transceiver

Table 10. Microwave Bands.

New Designation	Frequency (GHz)	Old Designation	Frequency (GHz)
C	0.5-1.0	VHF	0.5-1
D	1.0-2.0	L	1-2
E	2-3	S	2-4
F	3-4	C	4-8
G	5-6	X	8-12.4
H	6-8	Ku	12.4-18
I	8-10	K	18-26.5
J	10-20	Ka	26.5-40
K	20-40	v	
L	40-60		40-75
M	60-100		

Table 11. Antenna Gains and Effective Areas.

Type of Antenna	Effective Area $Ae(m^2)$	Power Gain (relative to isotropic)
Isotropic	$\frac{\lambda^2}{4\pi}$	1
Infinitesimal dipole or loop	$\frac{1.5\lambda^2}{4\pi}$	1.5
Half wave dipole	$\frac{1.64\lambda^2}{4\pi}$	1.64
Horn/mouth area A	0.81 A	$\frac{10\,A}{\lambda^2}$
Parabolic, face area	0.56 A	$\frac{2A}{\lambda^2}$
Turnstile (two crossed, perpendicular dipoles)	$\frac{1.15\lambda^2}{4\pi}$	1.15
G(Antenna gain) = $\frac{4\pi f^2 Ae}{c^2}$ $f \rightarrow$ Carrier frequency C $\rightarrow$ Speed of light $A_e \rightarrow$ Effective area		

Table 12. Antenna Beamwidth for Various Diameter for Parabolic Reflective Antenna at f = 12 GHz.

Antenna diameter	Beamwidth (Degrees)
0.5	3.5
0.75	2.33
1.0	1.75
1.5	1.166
2.0	0.875
2.5	0.7

Table 13. Frequency band and their application

Band	Frequency range	Free space wavelength range	Propagation character	Typical use
ELF extremely low	30-300 Hz	10000 to 1000 km	GW (Ground wave)	Power line frequencies; used by home control system.
VF Voice frequency	300-3000 Hz	1000 to 100 km	GW	Used by Telephone system (Analog subscriber line)
VLF Very low	3 to 30 kHz	100 to 10 km	GW; low attenuation in day and night High atm. noise level	Low range navigation; submarrine Mcomm.
LF Low frequency	30-300 kHz	10 to 1 km	GW, slightly less than VLF, absorption in day time	Long range navigation, marriage, radio, beacons
MF Medium freq.	300-3000 kHz	1000 to 100 m	GW and night SW (sort wave) attenuation low in night, high in day, atm. noise	Direction finding, AM broad casting
HF High freq.	3-30 MHz	100 to -10 m	SW, quality varies with time of day, season and frequency.	International broadcasting, millitary comm. ship, aircraft
VHF Very High Frequency	30-300 MHz	10 to 1 metre	LOS, scattering because of temp--erature inversion, cosmic noise	VHF T.V., FM broadcast, two-way radio, AM aircraft com.
UHF Ultra High	300 to 3000 MHz	100 to 10 cm	LOS, cosmic noise	UHF T.V., cellular telephone, radar microwave link, personal comm. system.
SHF super high frequency	3 to 30 GHz	10 to 10 m	LOS, rain fall attenuation above 10 GHz atm. atten. due to oxygen, water vapour	Satellite comm. Radar, Terrestion comm. WLL
EHF extremely high freq.	30-300 GHz	10 to 1 mm	LOS, Atm-noise due to O_2, H_2O water	Experimental; WLL
Infra-red	300 GHz-400 THz	1 mm-770 nm	LOS	Infrared LAN, consumer electronic application
Visible Light	400 THz to 900 THz	770 nm-330 nm	LOS	Optical communication.

LOS → Line of Sight communication

Comparison between 2.5G and 3G

2.5G :

1. GPRS is a service comes under 2.5G.
2. GPRS is based on GSM architecture.
3. 3.5G is upgradation of 2G.
4. GPRS is a service which provides TDMA + FDMA access in radio access part but switching side it provide packet access.
5. GPRS designed in GSM900 and GSM1800 band.
6. GPRS is a service which enable to Internet access, Picture messaging.
7. Theoretical speed is 53 kbps in GPRS but practically it will be around 30 kbps because it depend upon your mobile class and coding technique as well as time when are you accessing service.
8. Channel bandwidth is 200 kHz in GPRS.
9. If so many subscribers are trying to access Internet via GPRS then every subscriber will get different speed and quality of reception and transmission will be not guaranteed 3rd Generation (3G).
10. In GPRS network authentication will perform by network only.
11. 2G mobile cannot be used over 3G network.
12. GPRS feature governed under ITU recommendation.

3G :

1. 3G is new technology named as WCDMA.
2. WCDMA is implementing over existing 2G architecture not by replacing.
3. It is new technology not up gradation of GSM architecture.
4. It support CDMA + FDMA access in radio part while in core network it support packet switching and circuit switching both.
5. It supports MMS, internet browsing, Video conferencing etc.
6. Theoretical speed is 2 Mbps while practical speed is around 1.3 Mbps.
7. Channel bandwidth is 5 MHz for every subscriber.
8. Quality of service is guaranteed.
9. If you get connected with 3G network then quality of service will be guaranteed until unless call get drop.
10. Authentication will be performing by mobile phone as well as network.
11. 2G mobile phone cannot be used over 3G network but 3G network canbe use over 2G networks. But 2G mobile subscriber can make call to 3G and 2G can make call to 3G.
12. 3G feature governed under 3GPP recommendation.

❑❑❑

APPENDIX—B

ACRONYMS

A

AC	Access Channel
ACA	Adaptive Channel Allocation
ACF	Autocorrelation function
ACI	Adjacent Channel interference
ADM	Adaptive Delta Modulation
ADPCM	Adaptive Digital Pulse Code Modulation
ADSL	Asynchronous Digital Subscriber Line
AGCH	Access Grant Channel
AIN	Advanced Intelligent Network
AM	Amplitude Modulation
AMPS	Advanced Mobile Phone System
ANSI	American National Standards Institute
APC	Adaptive Predictive Coding
ARDIS	Advance Radio Data Information Systems
ARFCN	Absolute Radio Frequency Channel Numbers
ATM	Asynchronous Transfer Mode
AUC	Authentication Center
AWGN	Additive White Gaussian Noise

B

BER	Bit Error Rate
BERSIM	Bit Error Rate Simulator
BCH	Bose-Chaudhuri-Hocquenghem, *also* Broadcast Channel
BCCH	Broadcast Control Channel
BFSK	Binary Frequency Shift Keying
BISDN	Broadband Integrated Services Digital Network
BIU	Base Station Interface Unit
BOC	Bell Operating Company
BPSK	Binary Phase Shift Keying
BRAN	Broadband Radio Access Networks
BRI	Basic Rate Interface
BSC	Base Station Controller
BSIC	Base Station Identity Code
BSS	Base Station Subsystem

C

CAI	Common Air Interface
CATT	China Academy of Telecommunication Technology
CB	Citizens Band
CCCH	Common Control Channel

CCH	Control Channel
CCI	Co-channel Interference
CCIR	Consultative Committee for International Radiocommunications
CCITT	International Telgraph and Telephone Consultative Committee
CCS	Common Channel Signaling
CDMA	Code Division Multiple Access
CDPD	Cellular Digital Packet Data
CFP	Cordless Fixed Part
C/I	Carrier-to-Interference Ratio
CIC	Circuit Identification Code
CIR	Carrier-to-Interference Ratio
CIU	Cellular Controller Interface Unit
CNR	Carrier-to-Noise Ratio
CO	Central Office
Codec	Coder/decoder
CPFSK	Continuous Phase Frequency Shift Keying
CRC	Cyclic Redundancy Code
CSMA	Carrier Sense Multiple Access
CW	Continuous Wave

D

DAS	Distributed Antenna System
dBi	dB gain with respect to an isotropic antenna
dBd	dB gain with respect to a half-wave dipole
DCA	Dynamic Channel Allocation
DCCH	Dedicated Control Channel
DCS	Digital Communication System
DCS1800	Digital Communication System—1800
DECT	Digital European Cordless Telephone
DM	Delta Modulation
DPCM	Differential Pulse Code Modulation
DQPSK	Differential Quadrature Phase Shift Keying
DS	Direct Sequence
DSAT	Digital Supervisory Audio Tone
DSE	Data Switching Exchange
DS/FHMA	Hybrid Direct Sequence/Frequency Hopped Multiple Access
DSL	Digital Subscriber Line
DSP	Digital Signal Processing
DS-SS	Direct Sequence Spread Spectrum
DST	Digital Signaling Tone
DTE	Data Terminal Equipment
DTMF	Dual Tone Multiple Frequency
DTX	Discontinuous Transmission Mode

E

EDGE	Enhanced Data Rates for GSM Evolution
EIA	Electronic Industry Association
EIR	Equipment Identity Register
EIRP	Effective Isotropic Radiated Power
E_b/N_0	Bit Energy-to-noise Density
erf	Error Function
erfc	Complementary Error Function
E-SMR	Extended — Specialized Mobile Radio
ETACS	Extended Total Access Communication System *also* European Total Access Cellular System
ETSI	European Telecommunications Standard Institute

F

FACCH	Fast Associated Control Channel
FCC	Federal Communications Commission, Inc., *also* Forward Control Channel
FCCH	Frequency Correction Channel
FCDMA	Hybrid FDMA/CDMA
FDD	Frequency Division Duplex
FDMA	Frequency Division Multiple Access
FDTC	Forward Data Traffic Channel
FEC	Forward Error Correction
FH	Frequency Hopping
FHMA	Frequency Hopped Multiple Access
FH-SS	Frequency Hopped Spread Spectrum *also* Frequency Hopping Spread Spectrum
FLEX	4-level FSK-based paging standard developed by Motorola
FM	Frequency Modulation
FN	Frame Number
FPLMTS	Future Public Land Mobile Telephone System
FSK	Frequency Shift Keying
FVC	Forward Voice Channel

G

GIS	Graphical Information System
GIU	Gateway Interface Unit
GMSK	Gaussian Minimum Shift Keying
GOS	Grade of Service
GPRS	General Packet Radio Service
GSC	Golay Sequential Coding (a 600 bps paging standard)
GSM	Global System for Mobile Communication *also* Global System Mobile

H

HSCSD	High Speed Circuit Switched Data
HDR	High Data Rate
HIPERLAN	High Performance Radio Local Area Network
HLR	Home Location Register

I

IEEE	Institute of Electrical and Electronics Engineers
IGA	Improved Gaussian Approximation
IM	Intermodulation
IMSI	International Mobile Subscriber Identity
IMT-2000	International Mobile Telecommuncation 2000
IMTS	Improved Mobile Telephone Service
IP	Internet Protocol
IS-54	EIA Interim Standard for U.S. Digital Cellular with Analog Control Channels
IS-95	EIA Interim Standard for U.S. Code Division Multiple Access
IS-136	EIA Interim Standard 136—USDC with Digital Control Channels
ISDN	Integrated Services Digital Network
ISUP	ISDN User Part
ITFS	Instructional Television Fixed Service
ITU	International Telecommunications Union
ITU-R	ITU's Radiocommunications Secto
IXC	Interexchange Carrier

J

JDC	Japanese Digital Cellular (later called Pacific Digital Cellular)
JRC	Joint Radio Committee
JTACS	Japanese Total Access Communication System
JTC	Joint Technical Committee

L

LAN	Local Area Network
LCC	Lost Call Cleared
LCD	Lost Call Delayed
LCR	Level Crossing Rate
LEC	Local Exchange Carrier
LOS	Line-of-sight
LSSB	Lower Single Side Band
LTE	Linear Transversal Equalizer
LTP	Long Term Prediction

M

MAC	Medium Access Control
MAHO	Mobile Assisted Handoff
MAI	Multiple Access Interference

MAN	Metropolitan Area Network
M-ary	Multiple Level Modulation
MC	Multicarrier
MCS	Multiple Modulation and Coding Schemes
MDBS	Mobile Database Stations
MDLP	Mobile Data Link Protocol
MDS	Multipoint Distribution Service
MDR	Medium Data Rate
MFSK	Minimum Frequency Shift Keying
MIN	Mobile Identification Number
MIRS	Motorola Integrated Radio System (for SMR use)
MoU	Memorandum of Understanding
MPSK	Minimum Phase Shift Keying
MS	Mobile Station
MSB	Most Significant Bit
MSC	Mobile Switching Center
MSCID	MSC Identification
MSK	Minimum Shift Keying
MSU	Message Signal Unit
MTP	Message Transfer Part
MTSO	Mobile Telephone Switching Office
MUX	Multiplexer

N

NADC	North American Digital Cellular
NAMPS	Narrowband Advanced Mobile Phone System
NBFM	Narrowband Frequency Modulation
N-ISDN	Narrowband Integrated Service Digital Network
NMT-450	Nordic Mobile Telephone—450
NSS	Network and Switching Subsystem
NTACS	Narrowband Total Access Communication System

O

OFDM	Orthogonal Frequency Division Multiplexing
OMAP	Operations Maintenance and Administration Part
OMC	Operation Maintenance Center
OQPSK	Offset Quadrature Phase Shift Keying
OSI	Open System Interconnect
OSS	Operation Support Subsystem

P

PABX	Private Automatic Branch Exchange
PACS	Personal Access Communication System
PAD	Packet Assembler Disassembler
PBX	Private Branch Exchange
PCH	Paging Channel

PCM	Pulse Code Modulation
PCN	Personal Communication Network
PCS	Personal Communication System
PDC	Pacific Digital Cellular
pdf	probability density function
PH	Portable Handset
PHP	Personal Handyphone
PHS	Personal Handyphone System
PLMR	Public Land Mobile Radio
PR	Packet Radio
PRI	Primary Rate Interface
PSD	Power Spectral Density
PSK	Phase Shift Keying
PSTN	Public Switched Telephone Network

Q

QAM	Quadrature Amplitude Modulation
QCELP	Qualcomm Code Excited Linear Predictive Coder
QMF	Quadrature Mirror Filter
QPSK	Quadrature Phase Shift Keying

R

RACE	Research on Advanced Communications in Europe
RACH	Random Access Channel
RCC	Reverse Control Channel
RF	Radio Frequency
RFP	Radio Fixed Part
RLC	Resistor Inductor Capacitor
RLS	Recursive Least Square
RMD	RAM Mobile Data
RPCU	Radio Port Control Unit
RPE-LTP	Regular Pulse Excited Long-Term Prediction
RRMS	Radio Resource Management Protocol
RVC	Reverse Voice Channel
Rx	Receiver

S

SACCH	Slow Associated Control Channel
SCH	Synchronization Channel
SCM	Station Class Mark
SCP	Service Control Point
SDCCH	Stand-alone Dedicated Control Channel
SDMA	Space Division Multiple Access
SEP	Switching End Points
SIM	Subscriber Identity Module
SIR	Signal-to-Interference Ratio
SMS	Short Messaging Service *also* Service Management System
SOHO	Small Office/Home Office

SONET	Synchronous Optical Network
SP	Signaling Point
SQNR	Signal-to-Quantization Noise Ratio
SS	Spread Spectrum
SSMA	Spread Spectrum Multiple Access
ST	Signaling Tone
SYN	Synchronization channel

T

TACS	Total Access Communications System
TCAP	Transaction Capabilities Application Part
TCDMA	Time Division CDMA
TCH	Traffic Channel
TCM	Trellis Coded Modulation
TDD	Time Division Duplex
TDFH	Time Division Frequency Hopping
TDMA	Time Division Multiple
TDN	Temporary Directory Number
TD-SCDMA	Time Division-Synchronous Code Division Multiple Access
Tx	Transmitter

U

UMTS	Universal Mobile Telecommunications System
USGS	United States Geological Survey
USSB	Upper Single Side Band
UTRA	UMTS Terrestrial Radio Access

V

VAD	Voice Activity Detector
VCI	Virtual Circuit Identifier
VCO	Voltage Controlled Oscillator
VIU	Visitor Interface Unit
VLR	Visitor Location Register
VLSI	Very Large-Scale Integration
VMAC	Voice Mobile Attenuation Code
VoIP	Voice over Internet Protocol

W

WACS	Wireless Access Communication System (later called PACS)
WAN	Wide Area Network
WAP	Wireless Applications Protocol
WARC	World Administrative Radio Conference
W-CDMA	Wideband CDMA
WLAN	Wireless Local Area Network
WLL	Wireless Local Loop
WRC-2000	ITU World Radio Conference
WUPE	Wireless User Premises Equipment

NUMERICS

1 Base S	The IEEE 802.3 standard for low data rate ethernet using twisted pair cable and daisy chained connections.
10 Base T	IEEE 802.3 standard for twisted pair ethernet.
10 Base 2	IEEE 802.3 standard for ethernet using thin co-axial cable.
10 Base 5	IEEE 802.3 standard for ethernet using thick co-axial cable.
100 base FX	A version of IEEE 802.3 standard for Fast ethernet using 2 optical fibers.
100 base T	A version of IEEE 802.3 standard for Fast ethernet using 2 twisted pair cable.
56 K Modem	A modern technology using two different data rates, one for uploading and one for downloading from the internet.
1XDV	3G extension of IS-95B (shared data and voice)
1XD0	3G extension of IS-95B data only
1XEV	3G extension of IS-95B data with circuit switched voice
1XRTT	3G extension of IS-95B one RF channel
3GPP	3G partnership project for wideband CDMA standards base on backward compatibility with GSM and IS-136/PDC
3GPP2	3G partnership project for CDMA 2000 standards based on backward compatibility with IS-95.
3XRTT	3G extention of IS-95B three RF channels.

❑❑❑

APPENDIX—C

GLOSSARY

Some of the definitions in this glossary are from The New IEEE Standard Dictionary of Electrical and Electronics Terms, IEEE Std. 100-1992, 1993. These are marked with an asterisk.

Amplitude modulation. A form of modulation in which the amplitude of a carrier wave is varied in accordance with some characteristic of the modulating signal.

Amplitude-shift keying. Modulation in which the two binary values are represented by two different amplitudes of the carrier frequency.

Analog data. Data represented by a physical quantity that is considered to be continuously variable and whose magnitude is made directly proportional to the data or to a suitable function of the data.

Analog signal. A continuously varying electromagnetic wave that may be propagated over a variety of media.

Analog transmission. The transmission of analog signals without regard to content. The signal may be amplified, but there is no intermediate attempt to recover the data from the signal.

Angle modulation. Modulation in which the angle of a sine wave carrier is varied. Phase and frequency modulation are particular forms of angle modulation.

Antenna. That part of a transmitting or receiving system that is designed to radiate or to receive electromagnetic waves.

Attenuation. The reduction of strength of a signal as a function of distance traveled.

Aperiodic. A function or signal that is not periodic.

Application layer. Layer 7 of the OSI model. This layer determines the interface of the system with the user.

Asynchronous transfer mode (ATM). A form of packet transmission using fixed-size packets, called cells.

Atmospheric absorption. The loss of energy in transmission of radio waves, due to dissipation in the atmosphere.

Automatic repeat request. A feature that automatically initiates a request for retransmission when an error in transmission is detected.

Bandwidth. The difference in Hertz between the limiting (upper and lower) frequencies of a spectrum.

Bit error rate. The probability that a transmitted bit is received in error.

Block error correction code. A technique in which a *k-bit* block of data is mapped into an n-bit block ($n > K$) called a **codeword,** using an FEC (forward error correction) encoder

Broadband. In data communications generally refers to systems that provide user data rates of greater than 2 Mbps, up to 100s of Mbps.

Burst error. A burst error of length B is a contiguous sequence of B bits in which the first and last bits and any number of intermediate bits are received in error.

Carrier frequency. A continuous frequency capable of being modulated or impressed with a second (information-carrying) signal.

Cell relay. The packet-switching mechanism used for the fixed-size packets called cells. ATM is based on cell relay technology.

Cellular network. A wireless communications network in which fixed antennas are arranged in a hexagonal pattern and mobile stations communicate through nearby fixed antennas.

Channel. A single path for transmitting electric signals. *Note:* The word *path* is to be interpreted in a broad sense to include separation by frequency division or time division. The term *channel* may signify either a one-way or a two-way path.

Channel capacity. The maximum possible information rate through a channel subject to the constraints of that channel.

Checksum. An error-detecting code based on a summation operation performed on the bits to be checked.

Circuit switching. A method of communicating in which a dedicated communications path is established between two devices through one or more intermediate switching nodes. Unlike packet switching, digital data are sent as a continuous stream of bits. Bandwidth is guaranteed, and delay is essentially limited to propagation time. The telephone system uses circuit switching.

Crosstalk. Undesired energy appearing in one signal path as a result of coupling from other signal paths.

Cyclic redundancy check. An error detecting code in which the code is the remainder resulting from dividing the bits to be checked by a predetermined binary number.

Datagram. In packet switching, a packet, independent of other packets, that carries information sufficient for routing from the originating data terminal equipment (DTE) to the destination DTE without the necessity of establishing a connection between the DTEs and the network.

Decibel. A measure of the relative strength of two signals. The number of decibels is 10 times the log of the ratio of the power of two signals, or 20 times the log of the ratio of the voltage of two signals.

Diffraction. Deviation of part of a beam, determined by the wave nature of radiation, and occurring when the radiation passes the edge of an opaque obstacle.

Digital data. Data consisting of a sequence of discrete elements.

Digital signal. A discrete or discontinuous signal, such as voltage pulses.

Digital transmission. The transmission of digital data, using either an analog or digital signal, in which the digital data are recovered and repeated at intermediate points to reduce the effects of noise.

Dipole antenna. Any one of a class of antennas producing a radiation pattern approximating that of an elementary electric dipole.

Direct sequence spread spectrum. A form of spread spectrum in which each bit in the original signal is represented by multiple bits in the transmitted signal, using a spreading code.

Doppler shift. The effective change of frequency of a received signal due to the relative velocity of a transmitter with respect to the receiver.

Downlink. The communications link from satellite to earth station.

Electric dipole. An elementary radiator consisting of a pair of equal and opposite oscillating electric charges an infinitesimal distance apart.

Encapsulation. The addition of control information by a protocol entity to data obtained from a protocol user.

Error correction code. A code in which each expression conforms to specific rules of construction, so that departures from this construction can be automatically detected, and permits the automatic correction of some or all of the errors.

Error-detecting code. A code in which each expression conforms to specific rules of construction, so that if certain errors occur in an expression, the resulting expression will not conform to the rules of construction and thus the presence of the errors is detected.

Fading. The time variation of received signal power caused by changes in the transmission medium or path(s).

Flat fading. That type of fading in which all frequency components of the received signal fluctuate in the same proportions simultaneously.

Flow control. The function performed by a receiving entity to limit the amount or rate of data that is sent by a transmitting entity.

Forward channel. In a cellular or cordless network, the communications link from the base station to the mobile unit.

Forward error correction. Procedures whereby a receiver, using only information contained in the incoming digital transmission, a receiver corrects bit errors in the data.

Frame check sequence. An error-detecting code inserted as a field in a block of data to be transmitted. The code serves to check for errors upon reception of the data.

Packet switching. A method of transmitting messages through a communications network, in which long messages are subdivided into short packets. The packets are then transmitted as in message switching. Usually, packet switching is more efficient and rapid than message switching.

Parity bit. A binary digit appended to an array of binary digits to make the sum of all the binary digits, including the parity bit, always odd (odd parity) or always even (even parity).

Peak amplitude. The maximum value or strength of a signal over time; typically, this value is measured in volts.

Period. The absolute value of the minimum interval after which the same characteristics of a periodic waveform recur.

Periodic waveform. A waveform $f(t)$ that satisfies $f(t) = f(t + nk)$ for all integers n, with k being a constant.

Phase. For a periodic signal $f(t)$, the fractional part t/P of the period P through which t has advanced relative to an arbitrary origin. The origin is usually taken at the last previous passage through zero from the negative to the positive direction.

Phase modulation. Modulation in which the phase angle of a carrier is the characteristic varied.

Phase-shift keying. Modulation in which the phase of the carrier signal is shifted to represent digital data.

Physical layer. Layer 1 of the OSI model. Concerned with the electrical, mechanical, and timing aspects of signal transmission over a medium.

Presentation layer. Layer 6 of the OSI model. Provides for the selection of a common syntax for representing data and for transformation of application data into and from the common syntax.

Protocol. A set of rules governing the exchange of data between two entities.

Protocol data unit. A set of data specified in a protocol of a given layer and consisting of protocol control information of that layer, and possibly user data of that layer.

Pseudonoise. A sequence of numbers that are apparently statistically random.

Pulse code modulation. A process in which a signal is sampled, and the magnitude of each sample with respect to a fixed reference is quantized and converted by coding to a digital signal.

Radiation pattern. A graphical representation of the radiation properties of an antenna as a function of space coordinates. Typically, the distance from the antenna to each point on the radiation pattern is proportional to the power radiated from the antenna in that direction.

Reflection. Occurs when an electromagnetic signal encounters a surface that is large relative to the wavelength of the signal; the angle of incidence equals the angle of reflection.

Refraction. The bending of a beam in transmission through an interface between two dissimilar media or in a medium whose refractive index is a continuous function of position.

Reverse channel. In a cellular or cordless network, the communications link from the mobile unit to the base station.

Router. A device used to link two or more networks. The router makes use of an internet protocol, which is a connectionless protocol operating at layer 3 of the OSI model.

Scattering. The production of waves of changed direction, frequency, or polarization when radio waves encounter matter.

Selective fading. Fading that affects unequally the different spectral components of a radio signal.

Service access point. A means of identifying a user of the services of a protocol entity. A protocol entity provides one or more SAPs for use by higher-level entities.

Session layer. Layer 5 of the OSI model. Manages a logical connection (session) between two communicating processes or applications.

Sky wave. A radio wave propagated obliquely toward, and returned from, the ionosphere.

Spectrum. Refers to an absolute range of frequencies. For example, the spectrum of infrared is 3×10^{11} to 4×10^{14} Hz.

Spread spectrum. A technique in which the information in a signal is spread over a wider bandwidth using a spreading code.

Spreading code. A sequence of digits used to spread bandwidth in a spread spectrum system. Also called a spreading sequence or a chipping code.

Subscriber loop. Synonymous with local loop.

Synchronous time division multiplexing. A method of TDM in which time slots on a shared transmission line are assigned to I/O channels on a fixed, predetermined basis.

Thermal noise. Statistically uniform noise due to the temperature of the transmission medium.

Time division duplex. A link transmission technique in which data are transmitted in one direction at a time, with transmission alternating between the two directions.

Time division multiplexing. The division of a transmission facility into two or more channels by allotting the facility to several different information channels, one at a time.

Time domain. A characterization of a function or signal in terms of value as a function of time.

Transmission medium. The physical path between transmitters and receivers in a communications system.

Transport layer. Layer 4 of the OSI model. Provides reliable, transparent transfer of data between end points.

Troposphere. That portion of the earth's atmosphere in which temperature generally decreases with altitude, clouds form, and convection is active. The troposphere occupies the space from the earth's surface up to a height ranging from about 6 km at the poles to about 18 km at the equator.

Unguided medium. A transmission medium, such as air, vacuum, or water, in which an antenna is employed for transmission and reception.

Uplink. The communications link from earth station to satellite.

Virtual circuit. A packet-switching service in which a connection (virtual circuit) is established between two stations at the start of transmission. All packets follow the same route, need not carry a complete address and arrive in sequence.

Wavelength. The distance between two points in a periodic wave that have the same phase.

Wireless. Refers to transmission through air, vacuum, or water by means of an antenna.

❑❑❑

APPENDIX—D

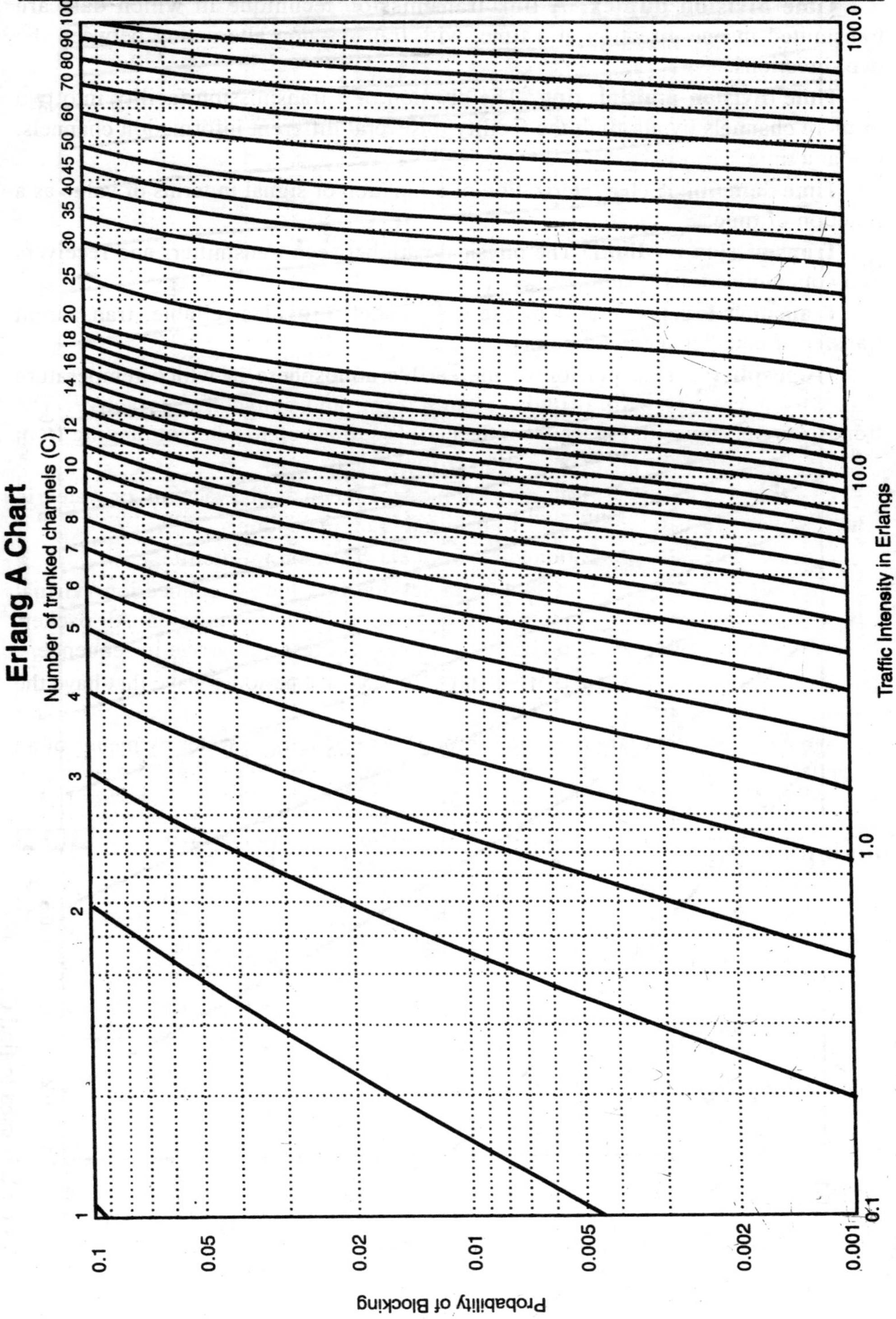
Erlang A Chart
Number of trunked channels (C)
1
2
3
4
5
6
7
8
9
10
12
14
16
18
20
25
30
35
40
45
50
60
70
80
90
100
Traffic Intensity in Erlangs
0.1
1.0
10.0
100.0
Probability of Blocking
0.1
0.05
0.02
0.01
0.005
0.002
0.001

Erlang C Chart

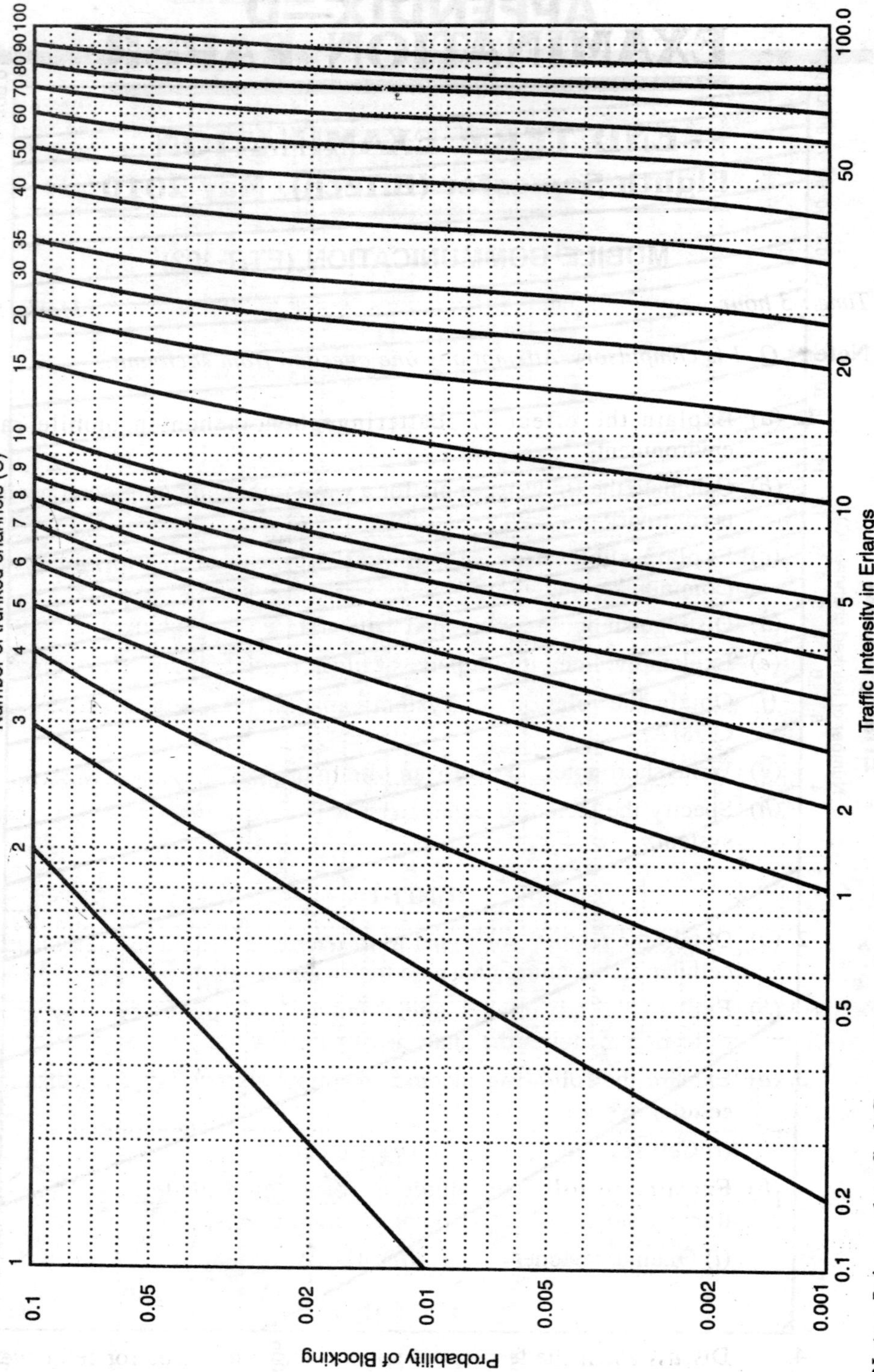

Note : It is used to find C.

❑❑❑

PREVIOUS YEARS EXAMINATION PAPER

END TERM EXAMINATION

Eighth Semester (B.Tech), May 2010

MOBILE COMMUNICATION (ETIT-402)

Time : 3 hour *Marks : 75*

Note : *Q. 1 is compulsory. Attempt any one question from each unit.*

1. (*a*) Explain the effect of scattering phenomenon in mobile radio environment. **(3)**

(*b*) Calculate the Brewster angle for a wave impinging on ground having a permittivity of $\varepsilon_r = 4$. **(4)**

(*c*) Explain subtracting scheme with respect to handoff in Mobile Communication. **(3)**

(*d*) Distinguish between Drapped calls and terminated calls. **(3)**

(*e*) Explain the linearity property of block codes. **(3)**

(*f*) Obtain the relation for system capacity in cellular systems using CDMA. **(3)**

(*g*) Write short notes on Gaussian Minimum Shift Keying (GMSK). **(3)**

(*h*) Specify the technical parameters of D-AMPS mobile communication system. **(3)**

UNIT-I

2. (*a*) Obtain the relation for signal to interference ratio of a 7-cell cluster in cellular architecture. (Assume the shape of cell is hexagon). **(7.5)**

(*b*) Explain the advantages and disadvantages of the two-ray ground reflection model in the analysis of path loss. **(5)**

3. (*a*) Explain the following methods to improve coverage and capacity in cellular systems : **(8.5)**

(*i*) Cell splitting (*ii*) Cell sectoring.

(*b*) Explain the following outdoor propagation models used for radio transmission in Mobile Communication System :

(*i*) Okumura Model (*ii*) HATA model **(4)**

UNIT-II

4. (*a*) Discuss about the features of antenna structures used for cell coverage in cellular architecture based systems. **(6)**

(*b*) Explain various non-fixed channel assignment schemes to manage the call flow in a typical mobile communication environment. **(6.5)**

5. (*a*) Explain the following handoff schemes:

(*i*) Mobile Controlled Handoff.

(*ii*) Network Controlled Handoff.

(*iii*) Mobile Assisted Handoff.

(*b*) What is Fraunhofer distance ? How it influences to find the coverage area of long distance propagation ? **(6.5)**

UNIT-III

6. (*a*) What is Hamming distance and hamming weight ? Explain from first principles how a (15, 12) Hamming code can be used to detect and correct single random error. **(8.5)**

(*b*) Write a short note on concatenated Turbo Codes. **(4)**

7. (*a*) Explain how Orthogonality and Autocorrelation properties are helpful in designing the spreading sequences without interference in CDMA systems. **(7.5)**

(*b*) Write a short note on convolutional codes and explain how these codes differ from block codes.

UNIT-IV

8. (*a*) Explain the mobility management in GSM (Global System for Mobile) network. What are the logical channels used for handling mobility ? **(7.5)**

(*b*) Write a short note on Vocoders. **(5)**

9. (*a*) What is IS95CDMA Network ? How it is different from its competing TDMA based GSM technology ? **(8)**

(*b*) What is soft handoff ? Explain the handoff performance in CDMA systems. **(4.5)**

END TERM EXAMINATION

Eighth Semester (B.Tech), May-June, 2011

MOBILE COMMUNICATION (ETIT-402)

Time : 3 hour *Marks : 75*

Note : *Attempt five questions. Q. 1 is compulsory, choice indicated.*

1. (*a*) What is the 'frequency reuse distance' ? What are the parameters on which it depends. Show that frequency reuse factor for a cellular system is given by 'K/S', where K is average number of channels per cell and S is the total numbers of channels available to service provider. **(5)**

(*b*) Explain different performance criterion of a cellular mobile system with special emphasis on grade of service. **(5)**

(*c*) Explain all fading techniques in wireless communication. How Rayleigh fading can be minimized in mobile radio system. **(5)**

(*d*) Give all differences between fixed channel assignment and nonfixed channel assignment. **(5)**

(*e*) If $P_t = 10$ W, $G_t = 0$ dB, $G_r = 0$ db, $f_c = 900$ MHz.

Find P_6 in watts at a free space distance of 1 Km.

2. (*a*) Why hexagonal shape is preferred over other cell shapes in cellular mobile radio. **(2.5)**

(*b*) Describe all the techniques to improve coverage and capacity in cellular systems. **(5)**

(*c*) If 25 MHz of total spectrum is allocated for a duplex wireless cellular system and each simplex channel has 25 kHz RF bandwidth, find

(*i*) No. of duplex channels

(*ii*) Total number of channels per site, if $N = 4$ cell reuse is used.

3. (*a*) Explain co-channel interference. How it can be minimized. **(2.5)**

(*b*) Prove that for a hexagonal geometry, the co-channel reuse distance is

$Q = \sqrt{3N}$ where $N = i^2 + ij + j^2$. **(5)**

(*c*) Suppose 5 cells originally cover a congested service area with 80 channels per cell. Find its channel capacity. Now, if cell splitting is done with new radius $R_1 = R/2$, find its new channel capacity. Assume path loss exponent $n = 4$. **(5)**

4. (*a*) What is the effect of antenna height on cell area coverage ? Explain in detail. **(6.5)**

(*b*) What are the propagation effects in following areas: **(5.5)**

(*i*) Over flat terrain (*ii*) Over water (*iii*) In an urban area.

OR

5. (*a*) Explain all setup channels in GSM architecture. **(7)**

(*b*) What are type of cell site antennal. Compare them according to geographical area of a cell. **(5.5)**

6. (*a*) Explain three propagation mechanisms : **(6)**

(*i*) Reflection (*ii*) Diffraction (*iii*) Refraction

(*b*) Determine depth loss for a signal with frequency 1 GHz, at a distance of 10 Km. What will be the far field distance for a base station antenna with largest antenna dimension $D = 0.5$ m, frequency of operation $f_1 = 900$ MHz. If frequency of operation is changed to $f_2 = 1800$ MHz, then find far field distance. **(6.5)**

OR

7. (*a*) Explain three types of multiple access techniques. Why CDMA technique is more secure over. **(7)**

(*b*) What are different digital modulation techniques compare QPSK and BPSK in terms of bandwidth and data rate. **(5.5)**

8. (*a*) Draw and explain in detail architecture of a cellular (GSM) system. Explain in detail. How a call is matured. **(5)**

(*b*) What are different handoff strategies. (5)

(*c*) What are training bits and why they are used in GSM burst architecture. **(2.5)**

OR

9. (*a*) Write short notes on **any two** of the following: **(2.5 × 3 = 7.5)**

(*i*) OFDM (*ii*) Turbo coding (*iii*) IS - 95

(*b*) What is the difference between convolution code and linear block codes ? Make block diagram of a convolution encoder. **(5)**

FIRST TERM EXAMINATION
Eighth Semester (B.Tech), February, 2012

MOBILE COMMUNICATION (ETIT-404) (REAPPEAR)

Time : 1½ hour *Marks : 30*

Instructions:

(1) Question 1 is compulsory and attempt any ***Two*** from the other ***Three***.

(2) Write your roll number on the question paper.

(3) Marks are indicating along with the questions.

1. Each part 2 marks:

(*a*) Differentiate between adjacent channel and co-channel interference.

(*b*) What is the relationship between antenna gain and directivity ?

(*c*) Why hexagonal shape is preferred over other cell shapes in cellular N/W ?

(*d*) What is frequency reuse distance ?

(*e*) Explain long distance propagation.

2. (*a*) What are the various stages in wireless communication history and when ? **(5)**

(*b*) Explain the operation of cellular system. **(5)**

3. (*a*) Obtain the relation for signal to interference ratio for a 6 cell cluster in cellular system. **(5)**

(*b*) What are the different factors responsible for interference ? **(2)**

(*c*) What are various antenna parameters ? **(3)**

4. (*a*) Explain mobile to mobile propagation. **(5)**

(*b*) In a mobile radio environment the average cell site antenna height is about 60 m, the mobile antenna height is about 4 m, and the communication path length is 4 km. Calculate:

(*i*) Incident angle

(*ii*) Elevation angle at the antenna of mobile unit

(*iii*) Elevation angle at the location of mobile unit.

SECOND TERM EXAMINATION
Eighth Semester (B.Tech), April, 2012

MOBILE COMMUNICATION (ETIT-404)

Time : 1½ hour *Marks : 30*

Instructions:

(1) Question 1 is compulsory and attempt any ***Two*** from the other ***Three***.

(2) Write your roll number on the question paper.

(3) Marks are indicating along with the questions.

1. Each part 2 marks:

(*a*) How the grouping of channels into subsets is done in frequency management ?

(*b*) What are turbo codes ?

(*c*) Differentiate between soft handoff and hard handoff ?

(*d*) What are the advantages of digital modulation techniques over analog modulation techniques ?

(*e*) Explain mobile high gain antennas.

2. (*a*) How omnidirectional and directional antennas are arranged in system configuration to reduce the interference ? **(5)**

(*b*) Explain the formula of dropped call rate in detail. **(5)**

3. (*a*) What is the role of BSC and MSC in handoff ? **(3)**

(*b*) Distinguish between dropped call and terminated call ? **(2)**

(*c*) Explain QPSK modulation technique in detail. **(5)**

4. (*a*) Why do we prefer MSK over QPSK ? **(2)**

(*b*) Differentiate between block codes and convolutional codes. **(3)**

(*c*) Consider (7, 4) block code generated by

$$G = \underbrace{\begin{bmatrix} 1 & 0 & 0 & 0 \\ 0 & 1 & 0 & 0 \\ 0 & 0 & 1 & 0 \\ 0 & 0 & 0 & 1 \end{bmatrix.}}_{I_4} \begin{matrix} \vdots \\ \vdots \\ \vdots \\ \vdots \end{matrix} \underbrace{\left.\begin{matrix} 1 & 1 & 0 \\ 0 & 1 & 1 \\ 1 & 0 & 1 \\ 1 & 1 & 1 \end{matrix}\right]}_{P}$$

Explain how the error syndrome S helps in correcting a single error.

SECOND TERM EXAMINATION

Eighth Semester (B.Tech), April, 2012 (Reappear)

MOBILE COMMUNICATION (ETIT-404)

Time : 1½ hour *Marks : 30*

Instructions:

(1) Question 1 is compulsory and attempt any ***Two*** from the other ***Three***.

(2) Write your roll number on the question paper.

(3) Marks are indicating along with the questions.

1. Each part 2 marks:

(*a*) What are convolutional codes ?

(*b*) How the grouping of channels into subsets is done in frequency management ?

(*c*) Differentiate between dropped call and terminated call ?

(*d*) What are the advantages of GMSK over MSK modulation techniques ?

(*e*) What are glass mounted antennas ?

2. (*a*) Explain cell site antennas in detail. **(5)**

(*b*) Explain the formula of dropped call rate in detail. **(5)**

3. (*a*) What is the role of BSC and MSC in handoff ? **(3)**

(*b*) Distinguish between intersystem and intrasystem handoff ? **(2)**

(*c*) Explain GMSK modulation technique in detail. **(5)**

4. (*a*) What are cyclic codes ? **(3)**

(*b*) Differentiate between block codes and convolutional codes. **(2)**

(*c*) Design a block code with a minimum distance of three and a message block size of 8 bits. **(5)**

❑❑❑

REFERENCES

1. Wideband and Wireless Digital Communications, by **Andreas F. Molisch** *Pearson Education.*
2. Satellite Communications by **TIMOTHY PRATT IEREMY Allnutt** *John Willey & Sons.*
3. Digital Communications by **EDWARD A. Lee, David G. Messers Chmitt** *Allied Publications Limited.*
4. Introduction to Analog and Digital Communication by **SIMON Hakyin** *John Wiley and Sons.*
5. Electromagnetic Field Theory by **G.S.N. Raju,** *Pearson Education.*
6. Wireless Communications (Principle and Practice) by **Theodor S. Rappaport,** *Prentice Hall of India Pvt. Limited.*
7. Electronic Communications by **Roddy Coolen,** *Pearson Education.*
8. Bluetooth technology by **Jennifer Bray and Charls F. Sturman,** *Pearso Education.*
9. Wireless Communication and Networking by **William Staling,** *Pearso Education.*
10. Mobile Cellular Telecommunication by **William C.Y. Lee,** *Mc-Graw Hil International.*
11. Wireless Digital Communications by **Dr. Kamilo Feher** by *Pearso Education.*
12. Digital Communications (Fundamental and Analysis) by **Bernard Sktar** *Pearson Education.*
13. Radio Frequency and Microwave Electronics by **Matthew M. Radmanesh** *Pearson Education.*
14. Communication System by **R.P. Singh Sapre,** *Tata Mc-Graw Hills.*
15. Modern Communication System (Analog on Digital) by **B.P. Lathi,** *Oxfor Publications.*
16. Data Communication and Networking by **Behrouz A. Forouzan,** *Tata Mc Graw Hill.*
17. WWW.gsm.com.
18. Electronics for you.
19. WWW.IETF.ORG.
20. WWW.3GPP.ORG.
21. **Digital and Analog Communication by Brijesh Verma.**
22. **Wireless Communication System by Brijesh Verma.**
23. **Question Bank in Electronics and Communication by Brijesh Verma**

INDEX

A

Access Method: Time Division Multiple Access (TDMA) 44
Access Schemes 32
Additional Network Components 281
Adjacent Channel Interference (ACI) 61
Advantage of Using Digital Technology 44
Advantages of Asynchronous CDMA Over Other Techniques 328
Advantages of CDMA 331
Advantages of GSM 330
Advantages of Spread-Spectrum Wireless Systems 170
Advantages of the Digital Transmission 20
AGCH (Access Grant Channel) 14
Amplitude Modulation (AM) 144
Amplitude Shift Keying (ASK) 146
AMPS and ETACS Specifications 260
AMPS and ETACS 258
An Example on Comparison of CDMA, FDMA, TDMA Systems 237
Analog and Digital Transmission 43
Analog Cellular Phone Parameter 18
Analog Cellular System 16
Analog Information 43
Analog Modulation 143
Analog System in Europe 260
Antenna and Clusters 77
Antenna at Cell Site (Directional Antenna) 105
Antenna at Cell Site (Omnidirectional antenna) 105
Antenna Gain 104
Antenna Types 103
Antenna 82
Applications for GPRS 130
Applications of Spread Spectrum 178
Apply Area to Area Prediction 97
Area to Area Prediction Model 90
Asynchronous CDMA 327
Authentication in GPRS 128

B

Base Station Controller (BSC) 10, 278
Base Station Subsystem (BSS) 9
Base Station System (BSS) 278
Base Transceiver Station (BTS) 10, 279
Basic Cellular Concepts 50
Basic Cellular System 8
Basic Wireless Concepts 39
BCCH (Broadcast Control Channel) 14
Binary Code Space 206
Block Codes 205
Block Sum Check (BSC) 194
Bluetooth 249
Brewster Angle 86
Broadcast Channel (BCH) 299
Broadcast Control Channel (BCCH) 300
BSC 122
BTS 122
Burst Error Correction 200
Burst Formatting 306

C

Calculation of Near Far Field Propagation 96
Call Drop 16
Call Establishment 10
Call from an MS 288
Call From Mobile Phone to Fixed Land Line Number (PSTN Number) 16
Call to MS from PSTN 289
Calls to/from Fixed and Remote Mobile Subscriber 16
Capacity of a Cellular System 231
Capacity of Digital Cellular CDMA 236

Carrier Synchronization 161
CDMA (Code Division Multiple Access) 321
CDMA 2000 331
CDMA Background 245
CDMA Features 329
CDMA Information, Theory and Codes 246
CDMA2000 1x 332
CDMA2000 1xEV-DO 332
CDMA2000 1xEV-DV 333
CDMA2000 3x 332
Cell Capacity—How Many Channels Per Cell 61
Cell Coverage Area 99
Cell Coverage for Signal and Antenna Structure 88
Cell Re-Selection 287
Cell Selection 287
Cell Site Antenna Height 99
Cellular System Design Fundamentals 49
Cepstrum Vocoders 348
Changes with Respect to BSS 336
Changes with Respect to Core Network (NSS) 336
Changing Cells within an LA 284
Channel Assignment Schemes and Handoff 110
Channel Assignment Strategy 110
Channel Coding 305
Channel Vocoders 348
Character Suppression 201
Characteristics of Basic Antenna Structures 103
Chronology of First Generation Wireless Systems 24
Chronology of Second Generation Wireless Systems 26
Ciphering in GPRS 129
Ciphering 305
Classes of GPRS Terminals 124
Co-channel Interference (CCI) 57
Code Division Multiple Access (CDMA) 244
Code Division Multiplexing (Synchronous CDMA) 326
Coding Schemes 137
Combined Loss 47
Common Control Channel (CCCH) 301
Comparison between 2G and 3G Technologies 340
Comparison Between Analog and Digital Cellular System 19
Comparison Between Wireless Communication System 250
Comparison of Mobile Phone Standards 330
Comparison of spectrum efficiency for FDMA and TDMA Systems 234
Concept of Frequency Reuse Channels 52
Convolutional Codes 207
Corporate E-mail 133
Coverage and Applications 325
Critical Frequency 105
Cyclic Redundancy Check (CRC) or Frame Check Sequence (FCS) 189

D

Data Compression 200
Decoding 209
Dedicated Channels 14
Dedicated Control Channel (DCC) 301
Demodulation 174
Demodulation/Detection of ASK 148
Design Concept of CDMA 333
Design of a CDMA System 334
Detaching from the Network 287
Detection of DPSK 154
Detection of FSK Signal 150
Detection of PSK 153
Difference between GSM and CDMA 309
Differential Phase Shift Keying 154
Diffraction Loss 97
Digital Cellular System Parameter 19
Digital Cellular Systems 19

Digital Information 43
Digital Modulation and Demodulation Techniques 145
Digital Modulation 144
Digital Traffic Channel 19
Direct Sequence (DS) 174
Direct Sequence CDMA (DS-CDMA) 335
Disadvantages of CDMA 331
Disadvantages of GSM 330
Drop Call Rate for Handoff 139
Dropped Call Rates and Their Evaluations 139
Dynamic Channel Assignment Strategy 111

E

EDGE Benefits 138
Edge Technology 135
Effective Area 104
Efficiency of TDMA 229
Elements of Cellular Radio Systems Design and Interference 39
Encryption 20
Enhanced Full Rate (EFR) 15
Error Control 183
Error Correction 196
Error Detection and Correction 20
Error Detection and Error Correction 180
Error Detection 183
Evolution of Mobile Communication 3
Examples of Wireless Communication System 5

F

FACCH (Fast Associated Control Channel) 15
FCCH (Frequency Correction Channel) 14
FDMA 234
Features of GPRS 129
Features of WCDMA 340
Feedback Error Control 184
First Time Location Update (IMSI Attach) 284
Fixed Channel Assignment Strategy 110
Formant Vocoders 349
Forward Error Control 183
Fourth Generation Standards 32
Frame Structure of GSM 302
Frequency Correction Channel (FCCH) 300
Frequency Division Duplexing (FDD) 223
Frequency Hopping (FH) 177
Frequency Hopping Spread Spectrum 320
Frequency Hopping 49, 320
Frequency Management & Channel Assignment, Handoff & Dropped Calls 109
Frequency Modulation (FM) 144
Frequency Reuse and System Capacity 54
Frequency Reuse Distance (D) 55
Frequency Shift Keying (FSK) 149
Future Network 31

G

General Description of Transmission Problem 44
Generate a Standard Condition 90
Generation of ASK 147
Generation of Code for CDMA : Code Division Multiple Access 247
Generation of FSK 150
Generation of Phase Shift Keying 153
GGSN 122
GMSK Modulation 165
GMSK Modulation 293
GPRS (General Packet Radio Service) 120
GPRS Architecture 121
GPRS BSS 123
GPRS Call Flow 126
GPRS Multislot Classes 125
GPRS NSS 123
GPRS Services 127
GPRS Support Node 122

GPRS Terminals 123
GPRS vs EDGE/EGPRS 135
Grade of Services [GoS] 22
GSM (Global System for Mobile Communication) 117
GSM (Global System For Mobiles Communication) 264
GSM Geographical Network Structure 266
GSM Network Areas 294
GSM Spectrum Allocation in India 318
GSM Subscriber Services 297
GSM System Architecture 273
GSM Traffic Channels (TCHs) 299
GSM-850 and GSM-1900 268
GSM-900 and GSM-1800 267

H

Hamming Code 197
Handoff Strategies 111
Handoff 109
Handover Measurements and Procedures 344
Hard Handover (HHO) 342
Hata Model 85
History of Spread Spectrum 166
History of Wireless Communication 2
History of Wireless Technology 2
HLR 122
Home Automation 134
Horizontal Redundancy Check 188
Huffman Coding 201

I

IMPLICIT Detach 287
Important 4G Components 32
IMSI Detach 287
Incident Angle and Elevation Angle with Ground 89
Increasing the Capacity of Cellular Systems 69
Inter MSC Handover 117
Inter-cell–Inter BSC Handover 116
Inter-cell–Intra BSC Handover 116
Interference (Co-Channel Interference) 56
Interim Standard 95 (IS-95) 306
Interleaving 47, 305
International Traffic Cases 292
Internet E-mail 133
Inter-system HO 342
Intra-cell–Intra BSC Handover 116
Intra-system HO 342
IPv6 32
IS-136 262
IS-2000 262
IS-54 and IS-136 (D-Amps) 261
IS-95 262

L

Layered Network Architecture Advantages 30
Location of the Antenna Changed 99
Location Updating (Periodic Type Registration) 286
Location Updating (Same MSC/VLR) 284
Location Updating New MSC/VLR (Generic Location Update) 286
Long Distance and Near Distance Propagation of Mobile Signal 95
Long Distance Propagation 96

M

Maximum Usable Frequency 105
Microcell Zone 76
Microcells 77
Migrating From 2G to 3G 27
Minimum Shift Keying 161
Mobile Antenna 107
Mobile Assisted Handoff (MAHO) 113
Mobile Phone Initialization 15
Mobile Point to Point Model 89
Mobile Services Switching Centre (MSC) 11
Mobile Station (MS) 277

Mobile Technologies 30
Mobile to Mobile Propagation 101
Mobile Wireless Applications 4
Mobility Management (Location Update) 283
Mobility Management 10
Model of Transmission Medium 36
Modulation Method in Cellular Wireless System 155
Modulation Method 43
Modulation Methods and Coding for Error Detection and Correction 142
Modulation Technique 136
Modulation 173, 306
Moving of MS in the Network 284
MS in Idle State 283
MS Registration and Roaming 271
MSC/VLR 122
Multipath Fading 45
Multiple Access Techniques For Wireless Communication 224

N

Network Database 315
Network Initiated Handoff 112
Network Management Center (NMC) 280
Network Management System (NMS) 280
Network Switching Subsystem (NSS) 11
Network Switching Subsystem (NSS) 274
Network Technologies 5
New Applications, Better Applications 130
Non-Coherent Detection of FSK (Frequency Shift Keying) 151
Non-Coherent Detection 148
Non-linear Effects in FDMA 226
Nonobstructive Conditions 98
Number of Channels in a TDMA System 229
Number of Channels Supported by FDMA System 225

O

Objectives of Handover 343
Okumura Model 84
Operation and Maintenance Center (OMC) 280
Operation and Support System (OSS) 281
Operation of Cellular Systems 12
Operators and UMTS Networks 28
OQPSK (Offset QPSK) 157
OQPSK Waveform 158
Orthogonal Frequency Division Multiplexing (OFDM) 179
Overview of AMPS 17
Overview of Cellular Generations 22
Overview of Handover in Mobile Networks 342

P

Packet Handling 137
PACS Radio Interface 312
PACS System Architecture 311
Paging Protocol 7
Paging Systems 5
Paging 287
Parabolic Antenna 104
Parity Check (Vertical Redundancy Check) 186
PCH (Paging Channel) 14
PDP Context Activation 126
Personal Communication Access System (PACS) 311
Personal Digital Cellular/Pacific Digital Cellular (PDC) 312
Phase Modulation 144
Phase Shift Keying (PSK) 151
Planning a Cellular System 33
Point to Point Prediction Model 97
Practical Handoff Considerations 114
Prediction Model Considers Following Conditions 89
Principles of CDMA 245
Principles of Operation of an FDMA 224
Principles of Operation of TDMA 227

Principles of Soft Handover 345
Prioritizing Handoffs 113
Propagation Attenuation 35
Propagation in Near-in Distance 95
Propagation Over Water and in Flat Area or in Free Space 92
PTM (Point-To-Multipoint) Services 127
PTM-SC 122
PTP (Point-To-Point) Services 127
Pulse Modulation 144

Q

QPSK Detector 160
QPSK Generation 159
QPSK Signalling 156
Quadrature Phase Shift Keying (QPSK) 156
Queuing of Calls 113

R

RACH (Random Access Channel) 14
Radiation Patterns 103
Rayleigh Fading 46
Redundancy 185
Reflection Angle and Reflection Point 89
Relation Between Gain and Effective Area 105
Remote LAN Access 134
Room Mounted Antenna 108

S

SACCH (Slow Associated Control Channel) 15
Scattering 86
SCH (Synchronisation Channel) 14
SDCCH (Stand Alone Dedicated Control Channel) 14
Second Generation (2G) Cellular Network 24
Second Generation, Digital and Wireless Systems 258
Security in GPRS 128
Security in the GPRS Backbone 129
Setup Channel Antennas 106
Setup Channels 299
SGSN 122
Shadowing 45
Shannon-Fanno Encoding 204
Signal Processing in GSM 304
Single Bit Error Correction 196
Soft Handoff 329
Soft Handover (SHO) and Softer HO 342
Soft Handover (SHO) 345
Solutions to Transmission Problems 48
Space Diversity Antenna 106
Space Division Multiple Access (SDMA) 243
Spectrum Efficiency 317
Speech Coding 304
Spread Spectrum (SS) Communication Concepts 174
Spread Spectrum Block Diagram 176
Spread Spectrum 166, 323
Statistical Raw Data Collection 10
Summary of FDMA, TDMA and CDMA 239
Synchronization Channel (SCH) 300
System Capacity 55

T

TDMA 234
Text Messaging 118
Textual and Visual Information 131
The 1G (First Generation) 23
The 2G to 3G Network Standardization 29
The 3G (WCDMA) Architecture 336
The 3G Services 338
The 3G—Services and Required Data Rates 339
The 4G (Fourth Generation) 32
The 4G True Broadband Wireless System 33

The Future of AMPS 260
The Industrial, Scientific and Medical (ISM Band) 315
The SIM Card 118
Third Generation (3G) 28
Time Alignment 45
Time Dispersion 46
Time Division Duplexing (TDD) 224
Traffic Cases 282
Traffic Channels 15
Trans Coder (TC) 280
Transmission Errors 181
Transmission Rate 43
Transmitting from the Base Station 323
Tropospheric Reflection 96
Tropospheric Refraction 96
Trunking Efficiency and Grade of Service 65
Turbo Coding 211
Type of Handoff (Hand Over) 116
Types of Handover in 3G WCDMA Systems 342

U

UMTS, 3G Terminals 31
Uniqueness of Mobile Radio Environment 35
United States Digital Cellular (Is-54 and Is-136) 260

V

Various Types of Number for Network Identity 271
Vehicle Positioning 133
Vocoder Theory 347
Vocoders 347
Voice Excited Vocoder 349

W

Walsh Code 321
When MS in Active Mode 288
Wireless Personal Area Network (WPAN) 315
Wireless TDMA Design Consideration for mobile phone 240
Worst Case Co-Channel Interference 59

Win Prizes

ATTENTION READERS

Win Prizes

We request for your frank assessment regarding various aspects of the book as given below :

MOBILE COMMUNICATION

By : BRIJESH VERMA

GGSIPU

Please Fill up the Sheet in Neat and Clean Writing

(i) What topics of your syllabus, important for your examination point of view, are not covered in this book ?

....

....

....

....

(ii) What are the chapters wherein the treatment of the subject matter is not systematic or organised or updated ?

....

....

....

....

(iii) Have you come across misprints/mistakes/factual inaccuracies in the book ? Please specify the chapters and the page numbers.

....

....

....

....

(iv) What other books on the same subject have you found/heard better than the present book ? Please specify in terms of price and quality.

....

....

....

....

(v) Further suggestions and comments for the improvement of this book.

....

....

....

....

(PS : If need be please attach a separate sheet to write your views)

OTHER DETAILS

(i) Who recommended you this book ? (Please tick (✓) in the box near the cption relevant to you.

☐ Teacher ☐ Friends ☐ Bookseller

....

....

....

....

(ii) Name of the recommending teacher, his designation and address:

....

....

....

....

(iii) Name and address of the bookseller you purchased the book from :

....

....

....

....

(iv) Name and address of your institution (Please mention the University or Board as the case may be)

....

....

....

....

(v) Your name and complete address :

....

....

....

....

(vi) Course you are enrolled in :

....

....

....

....

The best assessment will be awarded each month : The award will be in the form of our publications as decided by the Editorial Board.

Please mail the filled up coupon at your earliest to :

S.K. KATARIA & SONS®

4885/109, Prakash Mahal,
Dr. Subhash Bhargav Lane,
Opposite Delhi Medical Association,
Daryaganj, New Delhi-110002
Phone: +91-11-23243489, 23269324
TeleFax : +91-11-23243489
email : katariabooks@yahoo.com